THE STORY OF LIFE IN 25 FOSSILS

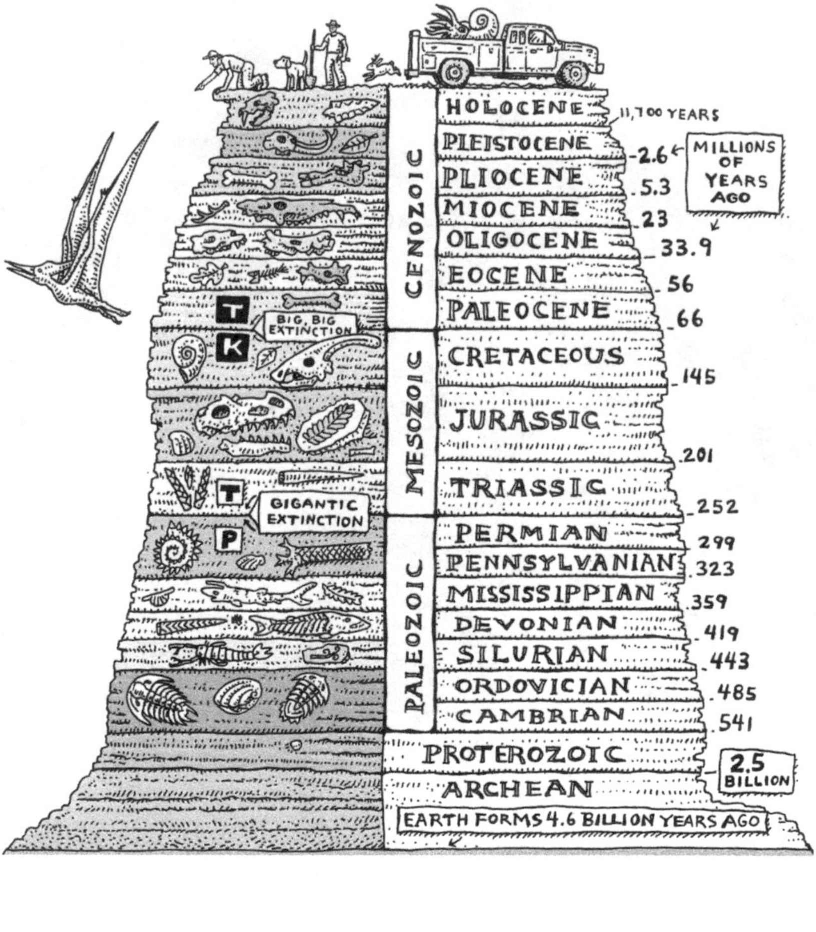
HOLOCENE
11,700 YEARS
PLEISTOCENE
2.6
MILLIONS OF YEARS AGO
PLIOCENE
5.3
MIOCENE
23
OLIGOCENE
33.9
EOCENE
56
PALEOCENE
66
CENOZOIC
T
BIG, BIG EXTINCTION
K
CRETACEOUS
145
MESOZOIC
JURASSIC
201
TRIASSIC
T
GIGANTIC EXTINCTION
252
P
PERMIAN
299
PENNSYLVANIAN
323
MISSISSIPPIAN
359
DEVONIAN
419
SILURIAN
443
ORDOVICIAN
485
CAMBRIAN
541
PALEOZOIC
PROTEROZOIC
2.5 BILLION
ARCHEAN
EARTH FORMS 4.6 BILLION YEARS AGO

THE STORY OF LIFE in 25 FOSSILS

TALES OF INTREPID FOSSIL HUNTERS AND THE WONDERS OF EVOLUTION

DONALD R. PROTHERO

COLUMBIA UNIVERSITY PRESS NEW YORK

COLUMBIA UNIVERSITY PRESS
Publishers Since 1893
New York Chichester, West Sussex

cup.columbia.edu

Paperback edition, 2018

Library of Congress Cataloging-in-Publication Data

Prothero, Donald R.
The story of life in 25 fossils : tales of intrepid fossil hunters and the wonders of evolution / Donald R. Prothero
pages cm
Includes bibliographical references and index.
ISBN 978-0-231-17190-8 (cloth)
ISBN 978-0-231-17191-5 (pbk.)
ISBN 978-0-231-53942-5 (e-book)
1. Fossils. 2. Paleontology. 3. Life—Origin. 4. Evolution (Biology) I. Title.

QE723.P76 2015
560—DC23

2015003667

FRONTISPIECE: THE GEOLOGICAL TIMESCALE. (COURTESY RAY TROLL)
COVER IMAGE: TRUDY NICHOLSON
COVER DESIGN: JULIA KUSHNIRSKY
BOOK DESIGN: VIN DANG

I DEDICATE THIS BOOK TO OUR GREAT POPULARIZERS AND ADVOCATES OF SCIENCE

NEIL SHUBIN

BILL NYE

NEIL DeGRASSE TYSON

AND

THE LATE CARL SAGAN AND STEPHEN JAY GOULD

CONTENTS

PREFACE

The history of life on Earth is an incredibly complex story. At the present moment, there are somewhere between 5 and 15 million species alive on our planet. Because more than 99 percent of all the species that ever lived are extinct, this suggests that hundreds of millions of species have lived on Earth, and probably a lot more, since the origin of life 3.5 billion years ago or even earlier.

Thus picking just 25 fossils to represent hundreds of millions of extinct species is not an easy task. I tried to focus on fossils that represent important landmarks in evolution. They show us the critical stages of how major groups first evolved or demonstrate the evolutionary transition from one group to another. In addition, life is more than just the origination of new groups. It is an amazing display of diversity in adaptations to size, ecological niches, and habitat. Thus I picked some of the most extreme examples of what life can achieve, from the largest land animal to the largest land predator, to several of the largest extinct creatures ever to swim in the oceans.

Naturally, such a hard choice leaves out many creatures, and I agonized over what to include and what to skip. I tried to focus on examples of fossils that are relatively complete and well known, which excludes many specimens that are too fragmentary to interpret reliably. Given the interests of nonscientist readers, I tended to favor dinosaurs and vertebrates in general. I apologize to all my paleobotanist and micropaleontologist friends for giving their disciplines short shrift with only one chapter apiece.

I hope you will forgive my sins of omission and commission, and embrace the creatures whose stories I have chosen to tell. May they illuminate your life!

ACKNOWLEDGMENTS

I thank Patrick Fitzgerald, Kathryn Schell, and Irene Pavitt at Columbia University Press for all their help with this project. Patrick deserves special thanks for coming up with the idea and making many valuable suggestions. Thanks also to Bruce Lieberman and David Archibald for reviewing the complete draft of the book, and to Mike Everhart and Tom Holtz for reviewing individual chapters. I especially thank Darren Naish for using his encyclopedic knowledge of tetrapods to carefully check the last 15 chapters. The many people who graciously provided the illustrations and photos for this book are acknowledged in the appropriate places. I thank Nobumichi Tamura, Carl Buell, and Mary Persis Williams for their incredible artwork that graces this book.

In addition, I thank my sons, Erik, Zachary, and Gabriel, for their love and support when I was writing it. I especially thank my wonderful wife, Dr. Teresa LeVelle, for her support and encouragement, and for helping me find quiet time to finish the book by deadline.

THE STORY OF LIFE IN 25 FOSSILS

PLANET OF THE SCUM

If the theory [of evolution] be true, it is indisputable that before the lowest Cambrian stratum was deposited, long periods elapsed . . . and the world swarmed with living creatures. [Yet] to the question why we do not find rich fossiliferous deposits belonging to these earliest periods . . . I can give no satisfactory answer.

CHARLES DARWIN, *ON THE ORIGIN OF SPECIES*

DARWIN'S DILEMMA

When Charles Darwin published *On the Origin of Species* in 1859, the fossil record was a weak spot in his argument. Almost no satisfactory transitional fossils were known, including none of the fossils discussed in this book. The first good one to be discovered was *Archaeopteryx* in 1861 (chapter 18). Even more troubling was the absence of any fossils that date to before the earliest period of the Paleozoic era, known as the Cambrian period (beginning about 550 million years ago [see frontispiece]). Of course, the fossil record was poorly known in the mid-nineteenth century, and it had been only 60 years since anyone had begun to note the sequence of fossils in detail. Still, Darwin was puzzled that in the few "Precambrian" beds below the earliest trilobites, there were no fossils that showed the transitions from simpler animals to trilobites and the other organisms of the Cambrian. Darwin said it all very clearly in the epigraph to this chapter.

Darwin attributed this puzzling lack of fossils to the "imperfection of the geological column" and the unlikely possibility that most organisms ever

fossilize. To a large extent, he was correct. He posed this question to his scientific peers, who for the next century tried desperately to find any kind of fossils older than the trilobites.

Many geologists already knew the problems with finding fossils that date to the Precambrian. Most Precambrian rocks are so old that they are deeply buried and long ago were heated and put under intense pressure that turned them into metamorphic rocks, so any fossils were likely to have been destroyed. Most rocks that are truly ancient are also likely to have been eroded away, another form of destruction. Even where they are relatively well preserved, the oldest rocks are usually buried under a thick layers of much younger rocks, so there are very limited exposures of them almost anywhere on Earth. All these factors conspired against the idea that we could just easily pick up fossils from Precambrian rocks, as we could from Cambrian rocks.

Still, there was more to the problem than this. It turns out that the conditions in the Precambrian (especially, little or no oxygen and no ozone layer) seem to have prevented early organisms from forming shells or other hard parts for a very long time. Instead, for 2 billion years, the world was dominated by mats of bacteria and (much later) algae, growing in the shallow waters of the shorelines and coating the rocks (figure 1.1). There *are* fossils in Precambrian rocks, only most of them are microscopic and cannot be seen without carefully grinding thin slices of rock on a microscope slide to see them at high magnification. To a field geologist, there are no visible fossils in most Precambrian rocks.

Nevertheless, there are many noticeable features in these rocks that people had been arguing about for a long time. For example, a structure that looks like a weird radiating pattern of grooves was described in 1848 by pioneering Canadian geologist Sir John William Dawson as *Oldhamia* (figure 1.2). He thought that it was the fossil of some kind of polyp. Yet Irish geologist John Joly was walking down a frozen muddy trail and found a similar pattern formed by ice crystals in the mud. In 1884, he argued that *Oldhamia* was just a feature produced by ice crystals, and not a fossil. More recently,

Figure 1.1 ◀

Reconstruction of the shallow tide pools on Earth as they looked for more than 80 percent of life's history, from 3.5 billion years ago to 550 million years ago. The only visible forms of life were mounds and domes of cyanobacterial mats, known as stromatolites. (Painting by Carl Buell; from Donald R. Prothero, *Evolution: What the Fossils Say and Why It Matters* [New York: Columbia University Press, 2007], fig. 7.1)

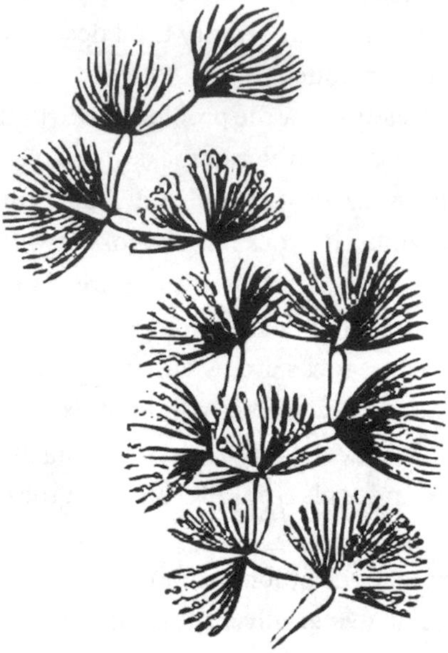

Figure 1.2 ▲
An original illustration of *Oldhamia*. (Redrawn by E. Prothero)

scientists have reevaluated *Oldhamia*, and now they conclude that it is the burrow of some kind of worm, so it is evidence of life after all—but this example shows how easily people can be fooled when they are so desperate to find signs of life in the Precambrian.

Another "creature" was discovered in 1868 by the legendary biologist (and Darwin's defender) Thomas Henry Huxley, who noticed a slimy "organism" in jars of mud recovered from the deep sea in 1857. He named this "creature" *Bathybius haeckeli* (the genus name from the Greek for "deep life," and the species name in honor of German biologist Ernst Haeckel). However, prominent British scientist Charles Wyville Thomson was not impressed, and he looked at the specimens and thought they were just fungal decay products. Another biologist, George Charles Wallich, proposed that the "organism" was the product of chemical disintegration of organic materials.

For this and many other reasons, Wyville Thomson and many other British scientists organized and funded the voyage of HMS *Challenger* from 1872 to 1876. The *Challenger*, a fully rigged sailing ship with steam power as well, was one of the first to actually conduct round-the-world oceanographic voyages. At that time, the British scientific community had no idea what the bottom of the ocean was like and thought that trilobites were still hiding in the deep oceans. They also sought answers to what *Bathybius* really was. The *Challenger* crew took more than 361 deep-ocean mud samples, without finding one *Bathybius*. Then the ship's chemist, John Young Buchanan, looked at some older samples and found something that resembled the mystery "slime." When he analyzed it, he realized that it was merely a reaction product of calcium sulfate with the alcohol used to preserve the sample. Wyville Thomson sent a polite letter to Huxley informing him about Buchanan's identification of the "organism." To his credit, Huxley published a letter in the journal *Nature* acknowledging his mistake. In 1879, at the 1879 meeting of the British Association for the Advancement of Science, Huxley took full responsibility for his error.

Yet another false alarm came in 1858, the year before Darwin's *On the Origin of Species* was published. Legendary Canadian geologist Sir William E. Logan (later director of the Geological Survey of Canada) found some unusual rocks from the banks of the Ottawa River near Montreal. Logan showed the specimens to scientists over many years, but most were unconvinced that they were proof of early life. The specimens then became the cause of Dawson, one of the most prominent scientists in Canada. In 1865, Dawson named Logan's layered structure *Eozoon canadense* (dawn animal of Canada) (figure 1.3). Dawson thought it was the fossilized remains of a

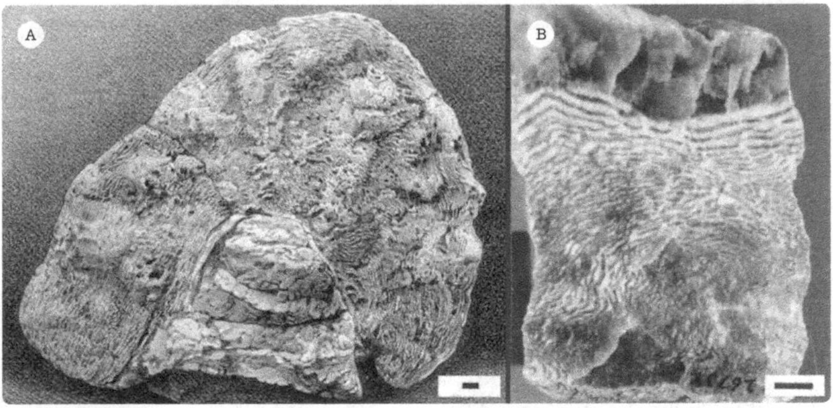

Figure 1.3 ▲

Eozoon canadense (dawn animal of Canada): (*A*) illustration in Dawson's *Dawn of Life*; (*B*) the holotype specimen at the Smithsonian Institution. Scale bars = 1 centimeter. ([*A*] from John W. Dawson, *The Dawn of Life* [London: Hodder and Stoughton, 1875]; courtesy J. W. Schopf)

huge foraminiferan (a group of amoeba-like, single-celled creatures that live in the oceans and make calcite shells). He called it "one of the brightest gems in the crown of the Geological Survey of Canada." Yet not long after that pronouncement, other geologists looked at the specimens more closely and at the geological setting. They found that *Eozoon* was just metamorphic layering of the minerals calcite and serpentine, not a fossil. The clincher was the discovery in 1894, near Mount Vesuvius in Italy, that the heat of volcanism can produce a similar structure in rocks.

CRYPTOZOON: YET ANOTHER FALSE ALARM?

Oldhamia, Bathybius, Eozoon. These and many other pseudofossils are among the discredited examples of Precambrian "life" that were once touted as the original ancestors of living things, and then debunked. Today, only historians of geology remember them.

In retrospect, it is easy to see why people were fooled. Most geologists learn early in their careers that the geologic landscape is full of pseudofossils, objects that appear to be possible fossils until you look closer (and know what to look for). Almost every amateur rock hound is fooled by the very plant-like patterns of pyrolusite dendrites, a mineral structure of manganese oxide that looks just like a branching fern. The most common pseudofossils are concretions, which are grains of sand or mud cemented together in a variety of shapes. Most are shaped like spheres or odd blobs, but many have bizarre forms that untrained amateurs visualize as a "fossil brain" or a "fossil phallus" or many other shapes that fool our tendency to see a "pattern" where there is none.

Like seeing "castles" in clouds or "animals" among the stars, humans are hardwired to infer meaning and pattern in nearly any collection of random images, a phenomenon known as *pareidolia*, or "patternicity." Thus experienced geologists learn to be very skeptical of interpreting just any odd-shaped rock as a fossil, and it takes years of experience to tell one from another. This was especially true in the early days of geology, when most sedimentary structures, and structures formed by burrowing, had not yet been defined and distinguished from true body fossils.

The next important figure in this story was Charles Doolittle Walcott, a self-trained geologist with the United States Geological Survey (figure 1.4). He had but ten years of schooling and never earned a degree, but received

Figure 1.4 ▲
Charles Doolittle Walcott, working in the Burgess Shale quarry in 1912. (Photograph courtesy Smithsonian Institution)

many honorary degrees later in life. Nevertheless, Walcott went on to become one of America's most important scientists in the early twentieth century. Almost single-handedly, he documented the entire Cambrian record of North America from New York State to the Grand Canyon, and became the founder of the study of Precambrian fossils as well. Later in life, he was legendary for multi-tasking on a scale scarcely imaginable today. He was director of the U.S. Geological Survey (1894–1907), and then was promoted to secretary (director) of the Smithsonian Institution (1907–1927), while also serving as president of the National Academy of Sciences (1917–1923). He also served as president of both the American Philosophical Society and (like Dawson) the American Association for the Advancement of Science. Despite this incredible administrative workload, he also managed to eke out a few weeks each summer to continue his grueling fieldwork in the Rocky Mountains and Colorado Plateau, describing huge mountains of Cambrian rock and amassing gigantic collections of fossils that he somehow found time to describe and publish. It was on one of those field trips that he accidentally stumbled on the Burgess Shale, a Middle Cambrian gold mine of soft-bodied fossils (chapter 5). He described these fossils superficially, but did not have time to really examine them, given his overcommitment to a crushing workload.

Walcott began his career working for the legendary James Hall, the first chief geologist and paleontologist of New York State. On a vacation in Saratoga, Walcott took a short field trip to Lester Park, only 5 kilometers (3 miles) west of Saratoga Springs. There, he was impressed by a layered structure in the very ancient Precambrian rocks he was studying (figure 1.5). In 1878, when he was only 28 years old, he began to describe in detail these layered, dome-like or cabbage-like structures, which Hall named *Cryptozoon* (hidden life) in 1883. They were common in nearly every Precambrian rock, so Walcott was convinced that they were the first evidence of life ever fossilized.

Most other scientists were very skeptical, however. Layered structures are very easily produced by natural means without organisms being involved, such as the layered structures in metamorphic rocks that fooled Dawson into identifying the "fossil" *Eozoon* or those formed during slow crystallization from a solution or by metamorphic foliation. The prominent botanist Sir Albert Charles Seward, the most influential man in paleobotany for many years, was a major critic of *Cryptozoon*. He correctly pointed out that there were no organic structures of plants or anything else preserved, making the case for *Cryptozoon* very shaky.

Nevertheless, many geologists were describing these layered, dome-like or cabbage-like structures, which were the only megascopic feature of most Precambrian rocks, and giving them names. In addition to *Cryptozoon*, there was another genus named *Collenia* for a differently shaped layered structure, and the name *Conophyton* was applied to layered structures with a conical rather than domed shape. Soviet geologists, who had huge areas of unmetamorphosed Precambrian rocks to study in Siberia, were especially fond of naming every shape of these layered structures. All these features were given the broader category name *stromatolite* (layered rock), even though most geologists were not certain that they were biologically produced.

Figure 1.5 ▶

The Lester Park stromatolites, called *Cryptozoon* by James Hall and Charles Doolittle Walcott. The top of these cabbage-like specimens were sliced off by a glacier, exposing their concentric internal layering. (Photograph by the author)

EverGrip

EUREKA!

For the first half of the twentieth century, the geological and paleontological community was deeply divided about what stromatolites were. Study after study had produced no signs of organic material or preserved cells in layers, so the case seemed weak. As long as no extant example of these structures was living and growing, there was no convincing evidence to silence the doubters.

In 1956, geologist Brian W. Logan of the University of Western Australia in Perth and some other geologists were exploring the northern coast of Western Australia. Logan and his colleagues came across a lagoon known as Shark Bay, about 800 kilometers (500 miles) north of Perth. When the tide went out in Hamelin Pool, on the southern shore of the bay, they saw a 500-million-year-old landscape that no scientists had seen on Earth (figure 1.6)! Lo and behold, the bottom of the bay was covered by 1- to 2-meter (3.3- to 6.6-foot) tall cylindrical towers with domed tops. They were dead ringers for many of the *Cryptozoon* and other Precambrian stromatolites—but they were still alive and growing! Closer inspection showed that these pillars and towers were made of millimeter-scale finely layered sediment, just like ancient stromatolites. On the top surface were the organisms that produced these mysterious structures. They were sticky mats of blue-green bacteria, or cyanobacteria (incorrectly called blue-green algae, even though they are *not* algae, which are true plants with nucleated eukaryotic cells). Blue-green

Figure 1.6 ▲
The domed stromatolites of Shark Bay, Australia. (Photograph courtesy R. N. Ginsburg)

bacteria not only are among the most primitive and simple forms of life on Earth, but probably were the first photosynthetic life on Earth. Most scientists think that cyanobacteria produced Earth's first atmospheric oxygen, so that one day more complex animals could evolve.

Further studies of the Shark Bay stromatolites revealed how they produce their finely layered structure. These slimy mats of blue-green bacteria grow very rapidly toward the sun when the tide comes in and immerses them during the day. The freshly growing mats have a sticky surface that traps sediment, especially at night or when the tide is going out and the cyanobacteria stop growing for a few hours. Then, when the tide comes in and the sun is up again, the bacteria grow new filaments reaching up to the sun, and they completely engulf the layer of sediment that accumulated the previous night. This goes on, day after day, year after year, so that in an area with favorable conditions, hundreds of individual growth layers of sediment are trapped by daily mat growth. Eventually, the organic material of the bacteria decays away, leaving just the layered sediments with no organic structures or chemical traces of their previous existence.

So if this process is so easy, why aren't stromatolites everywhere on Earth, as in the Precambrian? Shark Bay provided an answer to that question as well. The shallow water of Hamelin Pool is extremely salty because a bar of sand across the mouth of the bay restricts flow in and out. In addition, the subtropical desert-belt location of the bay is very hot and sunny. As the water evaporates, the sediments in the shallow bay just get saltier and saltier. They are so salty, in fact, that they have twice the salinity of the ocean (over 7 percent salt, rather than 3.5 percent), and only the cyanobacteria can tolerate these conditions. Grazing snails (like limpets and periwinkles and abalones in modern tide pools) that normally would eat such bacterial mats cannot live in such salty water, so the mats just keep growing, uncropped. This is very much like the world of the Precambrian, when more advanced marine grazers like snails had not yet evolved. For 3 billion years, the most complex forms of life were just microbial mats and eventually algal mats, with nothing to hinder their growth. As my friend J. William Schopf of UCLA says, early Earth was the "planet of the scum" (see figure 1.1).

Since the discovery in Shark Bay in 1956 (first published in 1961), living stromatolites have been found in many places on Earth. Most of them have one key feature in common: they grow in environments where the conditions are too hostile for more advanced forms of life (like grazing snails) to eat them. I've seen them close-up, growing in salty lagoons along the

Pacific coast of Baja California. They live in the salty water of the west coast of the Persian Gulf, and huge dome-topped pillars like those at Shark Bay also grow in the salty lagoons of Lagoa Salgada (Portuguese for "salty lagoon") in Brazil. Among the few that survive in water of normal salinity are those in Exuma Cays in the Bahamas, where the water currents are too strong for even limpets and periwinkles to hang on.

More and more fossil stromatolites also have been found, including some as old as life itself. These include probable stromatolites from the Warrawoona Group in western Australia (only a few hundred kilometers east of Shark Bay) that are 3.5 billion years old, along with the oldest microscopic evidence of cells of cyanobacteria. There are undoubted stromatolites from the 3.4-billion-year-old Fig Tree Group in South Africa. By 1.25 billion years ago, stromatolites were at the peak of their diversity in shape and size and abundance, and they are still the only visible evidence of life on the planet at that time. Then they began a slow decline through the next 500,000 years, and by the Cambrian they were only 20 percent of their original abundance—probably due to the huge number of new grazing creatures like snails that cropped them anywhere they grew in normal marine waters. (The Lester Park stromatolites, shown in figure 1.5, are in the Hoyt Limestone, which dates to the Middle Cambrian, so they are among the few exceptions of stromatolites that survived into the Cambrian.) By the time of the huge radiation of invertebrate life in the Ordovician (about 500 million years ago), they had nearly vanished from Earth.

However rare they have been for the past 500 million years, microbial mats are always ready to spring back and flourish any time their predators are suppressed. After three of Earth's great mass extinctions (the end-Ordovician, the Late Devonian, and the biggest mass extinction of all at the end of the Permian), stromatolites returned in abundance in the "aftermath" world when there were few survivors of the animals that had been clobbered by the extinctions. In each case, stromatolites grew like weeds, taking advantage of the wide-open landscape with the few opportunistic survivor species, and flourishing whenever the creatures that ate them were wiped out.

Finally, here's another thing to think about. For almost 85 percent of life's history (from 3.5 billion to about 630 million years ago), there were no creatures on this planet large enough to make visible fossils. Only stromatolites can be seen without the aid of a microscope. There are lots of different ideas about why life did not get going sooner, most of which are connected to the fact that the level of atmospheric oxygen was not high enough to support multicellular life until sometime in the Cambrian. Whatever the cause, for most of life's history the planet had microbial mats and domed stromatolites on its surface, and nothing else. If alien beings had actually landed on Earth, they would have seen them and gone away unimpressed.

Or consider the meteorite ALH84001, which was retrieved from the Allan Hills of Antarctica but had originally been blasted off Mars and eventually landed on Earth. In the 1990s, there was a big controversy over tiny rod-like and bead-like structures in the meteorite, and whether they were actually fossils of Martian life. The jury is still out on that question, but if there *had been* life on Mars, it is almost certainly now frozen, since Mars is too cold for liquid water. Earth would have looked much the same: until 600 million years ago, there were no organisms larger than single cells, so any piece of Earth rock or any sample of Earth's surface would have been just like Mars before it froze.

SEE IT FOR YOURSELF!

The original stromatolites that were the basis for James Hall and Charles Dolittle Walcott's *Cryptozoon* are visible in Lester Park, east of Saratoga Springs, New York. From downtown Saratoga Springs, take New York State Route 9N west. Turn left on Middle Grove Road, and then left again on Lester Park Road (also known as Petrified Gardens Road). Continue for about 500 feet. Once you enter the park, follow the signs to Petrified Gardens.

A number of museums have stromatolites on display or dioramas of stromatolites in the Precambrian. They include the Denver Museum of Nature and Science; Field Museum of Natural History, Chicago; Geology Museum, University of Wisconsin, Madison; National Museum of Natural History, Smithsonian Institution, Washington, D.C.; Natural History Museum of Utah, University of Utah, Salt Lake City; Raymond Alf Museum of Paleontology, Webb Schools, Claremont, California; Virginia Museum of Natural History, Martinsville; and Western Australian Museum, Perth.

FOR FURTHER READING

Grotzinger, John P., and Andrew H. Knoll. "Stromatolites in Precambrian Carbonates: Evolutionary Mileposts or Environmental Dipsticks?" *Annual Review of Earth and Planetary Sciences* 27 (1999): 313–358.

Knoll, Andrew H. *Life on a Young Planet: The First Three Billion Years of Evolution on Earth*. Princeton, N.J.: Princeton University Press, 2003.

Schopf, J. William. *Cradle of Life: The Discovery of Earth's Earliest Fossils*. Princeton, N.J.: Princeton University Press, 1999.

02 | THE FIRST MULTICELLULAR LIFE | *CHARNIA*

GARDEN OF EDIACARA

Aspiring paleontologists are typically attracted to the large, flashy specimens such as carnivorous dinosaurs and Pleistocene mammals. But to find the real monsters, the weird wonders of lost worlds, one must turn to invertebrate paleontology. Without question the strangest of all fossilized bodies are to be found among the Ediacarans.

MARK MCMENAMIN, *THE GARDEN OF EDIACARA*

FROM ONE CELL TO MANY

As we saw in chapter 1, the absence of any fossils from the Precambrian was long considered a problem for evolutionary biology. Charles Darwin agonized about it, as did many others until the discovery of undoubted microfossils in 1954 and the confirmation in the late 1950s that stromatolites were made by microbial mats. These discoveries showed that life had remained single-celled from about 3.5 billion to about 630 million years ago. There were still no fossils of multicellular life before the "Cambrian explosion." Many people thought that life would never be found in that puzzling and mysterious gap in the record before hard-shelled multicellular animals like trilobites.

But then some curious fossils began to show up in the rocks. Most of them were of fairly large (some almost 1 meter [3.3 feet] across) soft-bodied creatures that had not evolved hard parts. All had been fossilized as impressions in the sandstones or mudstones on the sea bottoms, so there were no actual complete body fossils (a problem when there are no shells or other

Figure 2.1 ▲
Reconstruction of *Charnia*. (Courtesy Nobumichi Tamura)

hard parts). Some were found in Namibia in the 1930s and in the Ediacara Hills of Australia in the 1940s, but they were not well dated at the time, so everyone assumed that these fossils were Early Cambrian.

Finally, in 1956, a 15-year-old schoolgirl named Tina Negus found a specimen in the Charnwood Forest, near Grantham in Lincolnshire, England (figure 2.1). As she describes it:

> During my teenage years, I came across a monograph on Charnwood Forest geology in my local library. We had often visited Charnwood, and many of the places mentioned were familiar to me. I copied out most of the maps from the book, and badgered my long-suffering parents for a visit as soon as possible. We parked and found our way to the quarry. I knew from my reading that the deposits here were of bedded volcanic ash, laid down underwater—a new concept to me. At that time the quarry was little visited, the footpath not much more than a sheep-trod. At the base I stood fingering the surface, and discovered just about head height . . . . a fossil! I had no doubts at all that it was indeed a fossil, but was very puzzled for all the books I had seen defined the

> Precambrian as the period before life began. I thought it was a fern, certainly some sort of frond, but did notice that the "leaflets" had no central rib, and that the cross-striped appearance of the "leaves" extended into the "stalk."
>
> At school the following day, I approached my Geography teacher, for I thought Geography the closest to Geology I could get. I told her I had found a fossil in Precambrian rocks at Charnwood Forest. She replied, "There are no fossils in Precambrian rocks!" I said I knew this, but it was because of this "fact" that I was interested and perplexed. She did not pause in her stride, nor look at me, but said "Then they are NOT Precambrian rocks." I assured her that they were, and she repeated the initial statement that Precambrian rocks contain no fossils—a truly circular argument, and a mind not open to anything new. I gave up, but asked my parents if we could go back there.

Negus did not have the tools or the experience to recover the specimen from such hard rock. But a year later, a local schoolboy named Roger Mason (who later became a geology professor) managed to extract the specimen from the rocks. He gave it to Trevor Ford, a local geologist, who officially published the specimen in 1958 in the *Yorkshire Geological Science Proceedings*. Ford named it *Charnia masoni* (the genus name for Charwood Forest, and the species name in honor of Roger Mason), and he thought that it was some kind of algal structure. Later geologists would argue that it was related to the coral relatives known as "sea pens," which look like a soft feather under the water. But the central "stem" of *Charnia* is not straight, as in a fern or "sea pen" or feather, but has a zigzag pattern. It is still not clear what kind of creature it really is, as we shall see. No matter what its identity, it was the first multicellular fossil (or, indeed, *any* kind of fossil) recovered from undoubted Precambrian rocks. As exemplified by Negus's geography teacher, most people before the late 1950s had a rather circular definition of what constituted a "Precambrian fossil." They were sure that there were no visible fossils from the Precambrian, so either the specimen was from Cambrian rocks or, alternatively, it was not really a fossil.

FOSSILS OF THE FLINDERS RANGES

Even before *Charnia* was formally described, geologists had been discovering fossils of large soft-bodied organisms in other places in the world. But since they were found in beds of uncertain age, they were routinely assigned to the Cambrian. As early as 1868, Scottish geologist Alexander

Murray had discovered frond-like fossils that resembled *Charnia* in the deep-marine sandstones of Mistaken Point in Newfoundland, but no one knew how to interpret them or how to date them, so they were forgotten. In 1933, German geologist Georg Gürich was mapping the geology of and prospecting for gold in Namibia (at that time, the South African colony of South-West Africa) when he found numerous fossils of curious soft-bodied creatures; but, again, no one knew their age, so they were assumed to be Cambrian.

The richest and best studied of these strange faunas came from the Ediacara Hills of the Flinders Ranges of South Australia, roughly 336 kilometers (227 miles) north of Adelaide. In 1946, Australian geologist Reginald Sprigg was working in the Ediacara Hills, mapping the geology and assessing the abandoned mines to decide whether new technology would justify their reopening. He sat down to eat lunch one day when he came across the first of these remarkable fossils. But he was not a paleontologist, nor had he been hired to collect fossils, so he passed the word about them to paleontologist Martin Glaessner of the University of Adelaide.

Glaessner was a remarkable man. Born on Christmas Day in 1906 in northwestern Bohemia (now in the Czech Republic), in the Austro-Hungarian Empire, he was educated at the University of Vienna, where he earned both a law degree and a doctorate in geology by the age of 25. During his early career, he was sent to Moscow to organize the study of micropaleontology for the State Petroleum Research Institute of the Soviet Academy of Sciences. Thus he was one of the pioneers of using microfossils for dating oil-bearing rocks and for determining ancient water depth. In Moscow, he met and married a Russian ballerina, Tina Tupikina, but this required that he either become a Soviet citizen or leave the Soviet Union. Returning to Austria in 1937, he had to flee almost immediately as Hitler's armies overtook the country (he was partially Jewish on his father's side). Glaessner and his wife ended up in Port Moresby, New Guinea, where he was asked to organize a micropaleontology department for the new Australian Petroleum Company. Then war came to New Guinea in 1942, so he and his wife fled to Australia, where he continued working in the oil industry until 1950. He spent the rest of his career as professor and department chair in geology and paleontology at the University of Adelaide.

There he took up the study of the mysterious fossils that Sprigg had sent him and organized large-scale collecting of many more specimens. After much hard work, he had described fossils that to him resembled sea jellies, sea pens, and a variety of weird "worms" (figure 2.2). Thanks to the discovery of *Charnia* in England and Australia, he was able to show that the Ediacaran fossils were latest Precambrian in age. This proved that there had been a worldwide diversification of these curious large soft-bodied organisms in many places (Africa, Australia, England, Newfoundland, and Russia near the White Sea, among many other places). In 1984, he published a summary of all his work in *The Dawn of Animal Life*, still regarded as a classic. Late in his career, Glaessner received numerous awards for his pioneering work on the earliest multicellular life.

Glaessner did his best to interpret these curious impressions and markings on the Flinders sandstones in terms of modern organisms (figure. 2.3). Round blobs looked like sea jellies, while the frond-like forms resembled sea pens. Some were extraordinarily large for the earliest multicellular life. For example, some of the broad leaf-shaped "worms" with a feather-like pattern of furrows known as *Dickinsonia* are nearly 1.5 meters (5 feet) in length (see figure 2.2)!

The extraordinary preservation of these normally easily decayed creatures suggests several things: few organisms served as scavengers in the Late Precambrian; the Ediacaran creatures may have been covered by mats of cyanobacteria that helped bury and preserve them; or many of them (especially at Mistaken Point, Newfoundland) were buried alive during submarine gravity slides of mud from shallow water.

WHITHER EDIACARA?

Subsequent scientists were not so sure that the Ediacaran fossils were so easily shoe-horned into living groups like worms and sea pens and sea jellies. They noted that the symmetry and construction of the "sea jellies" did not match those of any living sea jelly. Likewise, the "sea pens" did not have a straight shaft down the middle, but a zigzag shaft like *Charnia* (unlike any living sea pen). Most of the "worms" had no symmetry or construction like that of any modern group of worms, let alone the signs of a digestive tract or other organ systems that all worms have.

This peculiarity of their construction has led paleontologists to entertain other, less conventional explanations for the Ediacaran fossils. Some, like Adolf Seilacher of Yale University and the Universität Tübingen, have ar-

Figure 2.2 ▲

The Ediacara fossils consisted of large soft-bodied, quilted organisms, known from only their impressions on the seabed: (*A*) the large ribbed "worm" *Dickinsonia*; (*B*) the segmented "worm" *Spriggina*; (*C*) the shield-shaped possible trilobite relative *Parvancorina*. (Photographs courtesy Smithsonian Institution)

Figure 2.3 ▲

Diorama of the Ediacaran fauna reconstructed a sea pens, sea jellies, and "worms." (Courtesy Smithsonian Institution) gued that they are not related to modern animals at all. Instead, Seilacher suggested, they were an early experiment in multicellular creatures, with body plans unlike any of those today, that he called the Vendozoa or Vendobiota. (The Russians use the term "Vendian" for the entire latest Precambrian that produced these fossils, although the international geological organizations now call this time interval the Ediacaran.) Seilacher noted that they were built more like a water-filled air mattress, with a "quilted" construction, and no evidence of central nervous or digestive tracts, which even the simplest worms have. This suggests that the Ediacarans may have been some sort of creature that did not use organs like a digestive or respiratory or nervous system. Instead, these fluid-filled "mattresses" had the maximum surface area compared with their volume, thanks to the increase of surface due to their quilting. They absorbed all their food and oxygen directly through their highly folded "skin" while releasing waste products the same way.

Mark McMenamin of Mount Holyoke College suggested what he calls the "Garden of Ediacara" hypothesis. In his view, the huge surface area of these creatures, compared with their volume, may have allowed them to harbor large numbers of cyanobacteria or true algae as symbiotic creatures within their tissues. These photosynthetic symbionts would have provided lots of oxygen, while absorbing carbon dioxide waste, as do the algae that live in modern reef corals, giant clams, and many other marine organisms. Gregory Retallack of the University of Oregon, who specializes in fossil soils, argues that they were largely lichens or fungi, not plants or animals. More recently, he has suggested that many of these fossils are actually preserved soil structures.

Thus there is no shortage of opinions about the nature of these mysterious creatures from the dawn of animal life. Some still think of them as conventional sea jellies, sea pens, and worms, but most argue that they are like nothing living today. Whether they were truly a unique experimental assemblage of creatures called the Vendobiota, some sort of large endosymbiotic organism, or lichens or soils is still not easily resolved. After all, they are just impressions on the soft bedding surface of sands or muds on the sea bottom. We have a very limited idea of their three-dimensional structure with all its surfaces, let alone any internal structure or hard parts. That's just the problem: without hard parts, it was difficult to preserve these creatures in the fossil record. They were often folded or crushed or distorted precisely because they appear to have been blobs of water-filled tissue, much like sea jellies.

Whatever these creatures were, the important thing to remember is that they demonstrate beyond a doubt that the leap from single-celled life to large multicellular creatures had occurred by 630 million years ago. Their diversification was triggered as the planet warmed up following a "snowball Earth" glaciation that covered the planet with an ice sheet from the poles to the equator. For the next 90 million years, they were practically the only forms of life on Earth, until tiny shelled organisms began to appear during the end of their reign (chapter 3). Then their populations crashed as the simplest shelled organisms, and soon the trilobites, began to take over. By 500 million years ago, the Ediacaran creatures were gone completely, leaving the mystery of their biology behind.

SEE IT FOR YOURSELF!

The original specimen of *Charnia*—along with another fossil described by Trevor Ford in 1958, *Charniodiscus*, which is very similar to *Charnia*—are on display in a place of honor at the New Walk Museum and Art Gallery in Leicester, England.

Very few museums have displays of Ediacaran fossils, since they are not quite as eye-catching or glamorous as dinosaur skeletons. Among the few in the United States that do are the Denver Museum of Nature and Science; Field Museum of Natural History, Chicago; and National Museum of Natural History, Smithsonian Institution, Washington, D.C. Numerous museums in Australia have specimens from the Flinders Ranges on display, especially the South Australian Museum, Adelaide; and Western Australian Museum, Perth. In addition, there are the Mistaken Point Ecological Reserve, on the Avalon Peninsula, Newfoundland; and Senckenberg Naturmuseum, Frankfurt, Germany.

FOR FURTHER READING

Attenborough, David, with Matt Kaplan. *David Attenborough's First Life: A Journey Back in Time*. New York: HarperCollins, 2010.

Glaessner, Martin F. *The Dawn of Animal Life: A Biohistorical Study*. Cambridge: Cambridge University Press, 1984.

Knoll, Andrew H. *Life on a Young Planet: The First Three Billion Years of Evolution on Earth*. Princeton, N.J.: Princeton University Press, 2003.

McMenamin, Mark A. S. *The Garden of Ediacara.* New York: Columbia University Press, 1998.

Narbonne, Guy M. "The Ediacara Biota: A Terminal Neoproterozoic Experiment in the Evolution of Life." *GSA Today* 8 (1998): 1–6.

Schopf, J. William. *Cradle of Life: The Discovery of Earth's Earliest Fossils.* Princeton, N.J.: Princeton University Press, 1999.

Seilacher, Adolf. "Vendobionta and Psammocorallia: Lost Constructions of Precambrian Evolution." *Journal of the Geological Society, London* 149 (1992): 607–613.

——. "Vendozoa: Organismic Construction in the Proterozoic Biosphere." *Lethaia* 22 (1989): 229–239.

Valentine, James W. *On the Origin of Phyla.* Chicago: University of Chicago Press, 2004.

03 | THE FIRST SHELLS | *CLOUDINA*

"LITTLE SHELLIES"

The wave of discoveries that rewrote the story of the earliest Cambrian began when the former Soviet Union mustered sizable teams of scientists to explore geological resources in Siberia after the end of World War II. There, above thick sequences of Precambrian sedimentary rocks, lie thinner formations of early Cambrian sediments undisturbed by later mountain-building events (unlike the folded Cambrian of Wales). These rocks are beautifully exposed along the Lena and Aldan rivers, as well as in other parts of that vast and sparsely populated region. A team headed by Alexi Rozanov of the Paleontological Institute in Moscow discovered that the oldest limestones of Cambrian age contained a whole assortment of small and unfamiliar skeletons and skeletal components, few bigger than ½ in (1 cm) long. These fossils have been wrapped in strings of Latin syllables but have been more plainly baptized in English as the "small shelly fossils" (SSFs for short).

J. JOHN SEPKOSKI JR., "FOUNDATIONS: LIFE IN THE OCEANS"

THE SHELL BUILDERS

In chapter 1, we saw that the first answer to Charles Darwin's question about the "Cambrian explosion" was the discovery of the bacterial mats called stromatolites, which date to 3.5 billion years ago, and eventually of microfossils of cyanobacteria and other kinds of bacteria from beds of the same age. In chapter 2, we saw how single-celled life gave rise to multicellular soft-bodied creatures of the Ediacara fauna. But what about animals with shells? When did they arise?

The problem with growing a hard shell (*biomineralization*) is not as simple as you might suppose. For most animals, it is a daunting task to pull ions of calcium and carbonate, or silicon and oxygen, from the seawater and then to secrete them to construct calcite or silica shells. They need special biochemical pathways to make this kind of mineralization happen, and it is usually a very energetically expensive process.

The thick shell of a clam or a snail, for example, is built by a fleshy part of the body called the mantle, which lies just beneath the shell and surrounds the soft tissues of the mollusc. This organ has specialized structures and physiological mechanisms that allow it to pull calcium and carbonate ions from the ocean and turn them into calcium carbonate crystals. Molluscs can secrete this chemical in two kinds of minerals: calcite, the common mineral found in most limestones; and aragonite, or "mother of pearl," which most molluscs use to line the inner part of their shells. This is why there is an iridescent "pearly" luster on the inside of most mollusc shells, such as those of abalones. This is also the mechanism that grows pearls so valued by jewelry collectors. Pearls are simply layered structures of aragonite that are secreted around a central nucleus (like a grain of sand) trapped in the mantle of certain molluscs. The coating of aragonite is secreted so that the sand grain does not continue to irritate the mantle layer.

Based on the long duration of the Ediacaran fauna (more than 100 million years), we know that large soft-bodied organisms got along just fine without hard shells for a very long time. Judging from the data from the molecular clock of the divergence times of the major animal groups, most of the major phyla (sponges, sea jellies, and anemones; worms; segmented arthropods; brachiopods, or "lamp shells"; and molluscs) existed as soft-bodied forms well back into the Ediacaran, long before they added shells to allow the further diversification of body designs.

So if shells are such a burden, why evolve them at all? In most cases, the shell serves as protection against predators. Many paleontologists have argued that when shells started to appear, they were an adaptive response to new predators on the planet that were gobbling up all the vulnerable shell-less soft-bodied creatures. For some animals, the shells also serve as reservoirs of chemicals that the body needs. And some molluscs use their shells to secrete excess waste products of various metabolic processes.

Most important, mineralized shells also allow the diversification of body plans and thus greater ecological diversity and flexibility. The handful of living shell-less molluscs (such as solenogasters) are mostly shaped like worms, but with the addition of the shell, molluscs could evolve such diverse and distinct groups as chitons, clams, oysters, scallops, tusk shells, limpets, abalones, snails, cuttlefish, squid, and the chambered nautilus. These molluscs range from the slow and simple limpets and abalones, which creep along tide-pool rocks and graze on algae; to the headless filter-feeding clams; to the extremely intelligent and fast-moving octopi, squids, and cuttlefishes, which are predators.

THE "LITTLE SHELLIES" APPEAR

The late appearance of shells after the more than 100 million years of the evolution of large soft-bodied animals suggests that the development of shells was not an easy process. Nor would we expect large shells to have appeared all at once. Indeed, that is what we see in the fossil record.

For the longest time, there was no evidence of animals any simpler than trilobites from the Early Cambrian (chapter 4). To some, the "sudden appearance" of trilobites, with their complex segmented shells made of the protein chitin reinforced with calcite, suggested that they (and other groups of multicellular shelled animals) had arisen suddenly, without precursors, an event once called the "Cambrian explosion."

Shortly after World War II, the Soviets began to invest great effort in the geological exploration of remote regions like Siberia, mostly to find economic resources like coal, oil, uranium, and metals. In the process, they did a lot of basic geologic mapping and fossil collection in these areas. Along the Lena and Aldan rivers, which drain north out of Siberia into the Arctic Ocean, they found much more complete sequences of the Cambrian and Ediacaran rocks than were known anywhere else on Earth at the time. Soon, they began to describe an interval in the earliest Cambrian *before* the trilobites appeared in the third stage of the Cambrian (which the Soviet geologists called the Atdabanian). The two earliest stages of the Cambrian, which lay beneath the earliest trilobites, were called the Nemakit-Daldynian and the Tommotian.

Although these rocks yielded no trilobites, they did contain fossils of some of the other common large shelly Cambrian groups, such as the sponges, the sponge-like extinct archaeocyathans, and the "lamp shells," or brachiopods. But the most common finds were tiny (mostly smaller than 5 millimeters [0.2 inch] in diameter) fossils nicknamed the "little shellies" or the "small shelly fossils" (SSFs). These minute specimens were hard to find

Figure 3.1 ▲

Typical small shell fragments visible on the weathered surface of the dark band in the middle, from the Wood Canyon Formation in the White Mountains near Lida, Nevada. (Photograph by the author)

unless the fossil collector knew exactly what to look for, so it's no wonder that they were missed for decades by geologists accustomed to discovering large, flashy trilobites. Typically, dense concentrations of these tiny creatures populated the shelly layers (figure 3.1), and they were impossible to collect as complete specimens in the field. Instead, it was much easier to haul chunks of fossiliferous rock to the lab, and slowly dissolve the fossils out of the rock with acid. Or the chunks of fossiliferous limestone were sliced up and ground down into thin sections of rock only 30 microns thick and glued to a microscope slide. Observed through the microscope, these limestones were chock-full of a wide array of small but complex fossils (figure 3.2).

When these tiny fossils were discovered, it was not clear to what groups of familiar animals they belonged. Some were clearly shells of clam-like molluscs and snail-like molluscs. Others appeared to be pieces of "chain-link" armor for the bodies of much larger creatures. Many were the tiny needle-like or spiky elements known as spicules, which are woven together to form the only hard parts found in sponges.

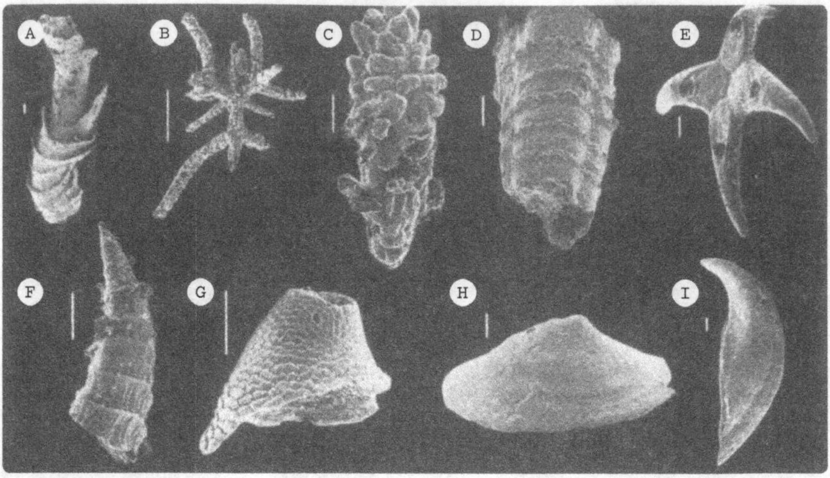

Figure 3.2 ▲

Rocks from the earliest stages of the Cambrian (Nemakit-Daldynian and Tommotian) do not produce trilobites, but are dominated by tiny phosphatic fossils nicknamed the "little shellies." Some may have been mollusc shells (*E*, *H*, and *I*), while others apparently were sponge spicules or pieces of the "chain-mail armor" of larger creatures, such as worms: (*A*) *Cloudina hartmannae*, one of the earliest known skeletal fossils, from the same beds that produce Ediacaran fossils in China; (*B*) spicule of a calcareous sponge; (*C*) spicule of a possible coral; (*D*) *Anabarites sexalox*, a tube-dwelling animal with triradial body symmetry; (*E*) spicule of a possible early mollusc; (*F*) *Lapworthella*, a cone-shaped organism of unknown relationships; (*G*) skeletal plate of *Stoibostromus crenulatus*, an organism of unknown relationships; (*H*) skeletal plate of *Mobergella*, a possible mollusc; (*I*) cap-shaped shell of *Cyrtochites*, a possible mollusc. Scale bars = 1 millimeter. (Photographs courtesy S. Bengston)

Significantly, many of them were made of calcium phosphate (the mineral apatite), not calcium carbonate, which most marine animals use to build their shells. Along with the earliest brachiopods (the lingulids) that used calcium phosphate to build their shells, this is suggestive of why it took so long for animals with large shells, such as trilobites, to evolve. It indicates that there were many hurdles and struggles to overcome before the process of mineralizing of shells got going in the Early Cambrian. First of all, none of these creatures secreted more than a few dozen tiny pieces of shell, so they were not yet ready to construct a shell as big as that of a trilobite. More important, a variety of lines of chemical evidence, along with the abundance of calcium phosphate (not calcium carbonate) shells, suggest that the atmosphere and oceans had not yet achieved the level of about 21 percent oxygen that is found on the planet today. Instead, it is estimated that the oxygen level was much lower still, which would have made it hard to run the geochemical and physiological mechanisms that allow molluscs to secrete minerals for shells.

PRESTON CLOUD'S PREDICTIONS

The field of Precambrian geology and paleontology was virtually nonexistent until 1954, when Stanley Tyler and Elso Barghoorn discovered and published the first evidence of Precambrian microscopic fossils. One man in particular became the pioneer and dominant figure of Precambrian biology and geology starting in the 1950s and 1960s, and remained so until his death: Preston H. Cloud. I met Pres several times in my career, and as both J. William Schopf, in *Cradle of Life*, and I recall, he was a towering figure in the field—even though he was only a slim 5 feet, 6 inches tall and had a shiny bald head and a bristly beard. But he was (in Schopf's words) "a giant, a wiry wonder, full of energy, ideas, opinions, and good hard work. And he was probably the greatest biogeosynthesist the United States ever produced. . . . Cloud was not given to idle chatter and struck some colleagues as a bit imperious (one of them referred to him as 'the little general,' though never to his face). Yet Cloud had an overriding saving grace. He was brilliant."

Cloud had a long career both in academia (especially at the University of California, Santa Barbara), and at the U.S. Geological Survey, where he built the paleontological branch into a powerhouse. Cloud's innovative and wide-ranging thinking made him an expert in many areas, from brachiopods to bauxite mining to oceanography to coral reefs to carbonate petrology. In 1974, he began writing books that warn about the future of the planet, about limited resources and peak oil, and about the ecological and environmental disasters that humans are creating on Earth. His two major books on this topic (*Cosmos, Earth, and Man: A Short History of the Universe* [1978] and *Oasis in Space: Earth History from the Beginning* [1988]) were the first to connect his broad understanding of 4.5 billion years of Earth history with predictions about how humans are likely to destroy the planet.

Long before anyone else was working on the evidence for early life, Cloud pushed for more and more studies of Precambrian microfossils and stromatolites, as well as for the search for more Ediacaran fossils. Even more important, he created the framework of our understanding of Precambrian Earth—the period of 3 billion years of low oxygen levels, the slow evolution of single-celled life, and the explosion of eukaryotic cells during the "oxygen holocaust" between 2 and 1.8 billion years ago—and he came up with many innovative ideas for how Precambrian geochemistry, atmospheres, and oceans had worked. His famous paper "A Working Model of the Primitive Earth" (1972) has been the foundation of nearly every study on the Precambrian in the past forty-plus years.

CLOUDINA

Like many other geologists, Cloud was frustrated with the big difference between the large but unshelled Ediacaran creatures and the shelled trilobites. Late in his life, he was overjoyed with the discovery and description of the Early Cambrian "little shellies," closing most of that gap. Still, why were there no shelled fossils before the Cambrian? Why did there appear to have been this evolutionary break between Edicarans and SSFs?

Then, in 1972, Gerard J. B. Germs described fossils from the Nama Group in Namibia (at that time, the South African colony of South-West Africa), which dates to the Late Precambrian. He reported a strange calcareous fossil about 6 millimeters (0.2 inch) across and about 150 millimeters (6 inches) long. It was constructed of a set of nesting conical shells, with a hollow tubular cavity inside (figure 3.3). There is still no agreement as to which modern group of animals it belongs to (such as a worm group that secretes a tubular skeleton), or even if it belongs to a modern group at all. The organisms are usually found associated with stromatolites, so they preferred shallow-water microbial-mat habitats. And there is some evidence of other creatures nibbling on them, so true predation had begun.

Whatever these mysterious creatures were, they were the first shelled animals on the planet (along with a Chinese tubular fossil called *Sinotubulites*), and they occurred around the world in the latest Precambrian: not only in Namibia, but also in Antarctica, Argentina, Brazil, California, Canada, China, Mexico, Nevada, Oman, Spain, Uruguay, and especially Russia. Appropriately, in 1972, Germs named it *Cloudina*, in honor of Preston Cloud and his huge number of contributions to Precambrian biogeology. Although subsequent years brought waves of argumentation about and reinterpretation of these frustratingly simple and incomplete fossil, it seems very appropriate that the oldest shelled animal on Earth was named after Preston Cloud.

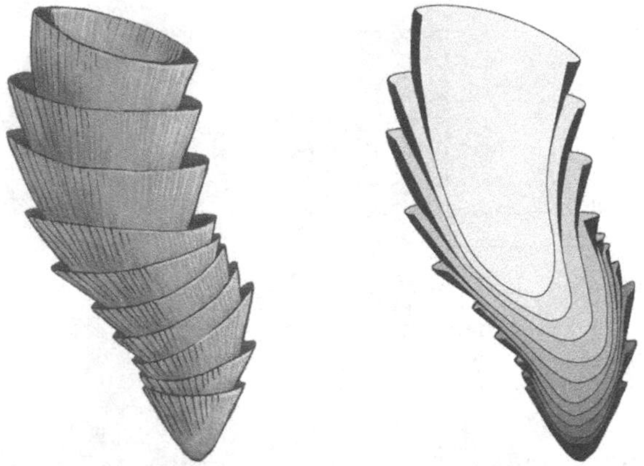

Figure 3.3 ▲

Reconstruction of *Cloudina*, showing the cone-in-cone outer structure and the cylindrical internal chamber, which was occupied by the soft-bodied shell maker. (Drawing by Mary P. Williams, based on several sources)

THE "SLOW FUSE"

The "Cambrian explosion" was not an explosion at all, but a "slow fuse" (figure 3.4). From about 600 to about 545 million years ago, the only multicellular life on the planet was the large soft-bodied, shell-less Ediacarans. Apparently, the geochemical conditions (especially low oxygen level) did not allow for the evolution of large shelled animals. Along with the mysterious Ediacarans, the precursors of the "little shellies," especially *Cloudina* and *Sinotubulites*, lived among the stromatolitic mats.

Then, between 545 and 520 million years ago (Nemakit-Daldynian and Tommotian stages), the largest creatures on the planet were soft-bodied animals with tiny bits of mineralized armor in their skins, or sponges woven of small spicules, as well as little shelled molluscs and brachiopods. At 520 million years ago, at least 80 million years after larger multicellular animals first appeared, we *finally* get animals with large calcified shells: the trilobites. Thus there was no "Cambrian explosion," unless you count 80 million years (beginning of the Ediacaran to the Atdabanian) or 25 million years (duration of the first two stages of the Early Cambrian) as an "explosion."

Creationists and others are determined to ignore this evidence and distort the fossil record for their own purposes by promoting a false version of the "Cambrian explosion." As Harvard paleontologist Andrew Knoll put it:

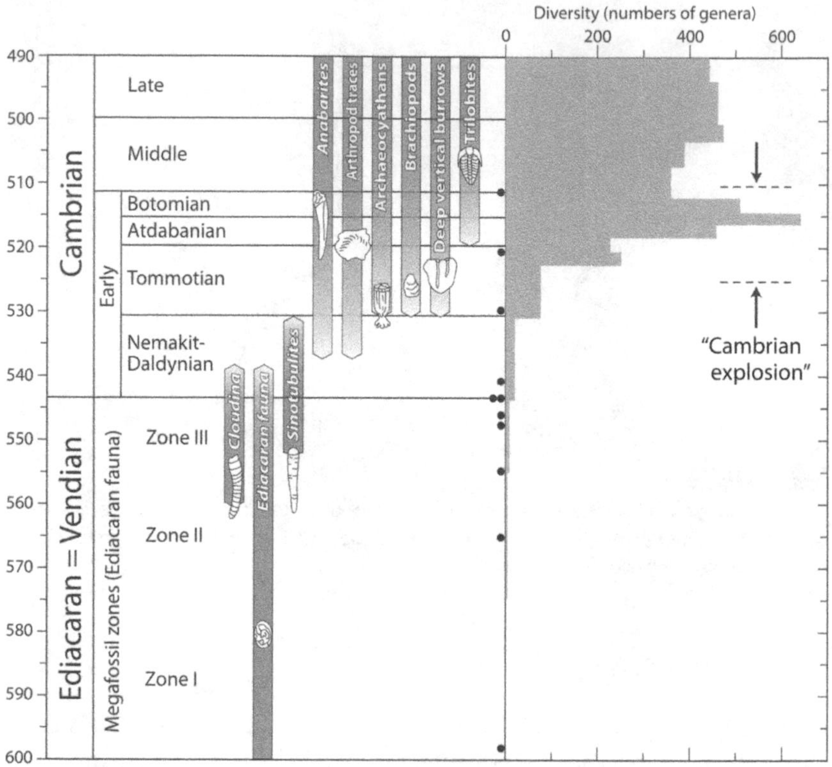

Figure 3.4 ▲

A detailed examination of the stratigraphic record of fossils through the late Precambrian and the Cambrian shows that life did not "explode" in the Cambrian, but appeared in a number of steps spanning about 100 million years. The large soft-bodied Ediacaran fossils first appeared 600 million years ago, in the Vendian stage of the Late Precambrian (see figure 2.2). Toward the end of their reign, we see the first tiny shelly fossils, including the simple conical *Cloudina* and *Sinotubulites*. The Nemakit-Daldynian and Tommotian stages of the Cambrian are dominated by the "little shellies" (see figure 3.2), plus the earliest brachiopods, the conical sponge-like archaeocyathans, and many burrows showing that worm-like animals without hard skeletons were also common. Finally, in the Atdabanian stage, around 520 million years ago, we see the radiation of trilobites and a big diversification in the total number of genera, thanks to the mineralized shells of trilobites, which preserve particularly well (histograms on the right side of the diagram). Thus the "Cambrian explosion" took place over more than 80 million years and thus was not a "sudden" event, even by geological standards. (Redrawn from Donald R. Prothero and Robert H. Dott Jr., *Evolution of the Earth*, 7th ed. [Dubuque, Iowa: McGraw-Hill, 2004], fig., 9.14)

Was there really a Cambrian Explosion? Some have treated the issue as semantic—anything that plays out over tens of millions of years cannot be "explosive," and if the Cambrian animals didn't "explode," perhaps they did nothing at all out of the ordinary. Cambrian evolution was certainly not cartoonishly fast. . . . Do we need to posit some unique but poorly understood evolutionary process to explain the emergence of modern animals? I don't think so. The Cambrian Period contains plenty of time to accomplish what the Proterozoic didn't without invoking processes unknown to population geneticists—20 million years is a long time for organisms that produce a new generation every year or two.

FOR FURTHER READING

Attenborough, David, with Matt Kaplan. *David Attenborough's First Life: A Journey Back in Time*. New York: HarperCollins, 2010.

Conway Morris, Simon. "The Cambrian 'Explosion': Slow-fuse or Megatonnage?" *Proceedings of the National Academy of Sciences* 97 (2000): 4426–4429.

——. *The Crucible of Creation: The Burgess Shale and the Rise of Animals*. Oxford: Oxford University Press, 1998.

Erwin, Douglas H., and James W. Valentine. *The Cambrian Explosion: The Construction of Animal Biodiversity*. Greenwood Village, Colo.: Roberts, 2013.

Foster, John H. *Cambrian Ocean World: Ancient Sea Life of North America*. Bloomington: Indiana University Press, 2014.

Grotzinger, John P., Samuel A. Bowring, Beverly Z. Saylor, and Alan J. Kaufman. "Biostratigraphic and Geochronologic Constraints on Early Animal Evolution." *Science*, October 27, 1995, 598–604.

Knoll, Andrew H. *Life on a Young Planet: The First Three Billion Years of Evolution on Earth*. Princeton, N.J.: Princeton University Press, 2003.

Knoll, Andrew H., and Sean B. Carroll. "Early Animal Evolution: Emerging Views from Comparative Biology and Geology." *Science*, June 25, 1999, 2129–2137.

Runnegar, Bruce. "Evolution of the Earliest Animals." In *Major Events in the History of Life*, edited by J. William Schopf, 65–93. Boston: Jones and Bartlett, 1992.

Schopf, J. William. *Cradle of Life: The Discovery of Earth's Earliest Fossils*. Princeton, N.J.: Princeton University Press, 1999.

Schopf, J. William, and Cornelis Klein, eds. *The Proterozoic Biosphere; A Multidisciplinary Study*. Cambridge: Cambridge University Press, 1992.

Valentine, James W. *On the Origin of Phyla*. Chicago: University of Chicago Press, 2004.

OH, GIVE ME A HOME, WHEN THE TRILOBITES ROAMED

Trilobites tell me of ancient marine shores teeming with budding life, when silence was only broken by the wind, the breaking of the waves, or by the thunder of storms and volcanoes. The struggle of survival already had its toll in the seas, but only natural laws and events determined the fate of evolving life forms. No footprints were to be found on those shores, as life had not yet conquered land. Genocide had not been invented as yet, and the threat to life on Earth resided only with the comets and asteroids. All fossils are, in a way, time capsules that can transport our imagination to unseen shores, lost in the sea of eons that preceded us. The time of trilobites is unimaginably far away, and yet, with relatively little effort, we can dig out these messengers of our past and hold them in our hand. And if we can learn the language, we can read the message.

RICCARDO LEVI-SETTI, *TRILOBITES*

AMBASSADORS OF DEEP TIME

One of the most popular of all fossils for amateur collectors and professional paleontologists alike are the trilobites. These creatures lasted from 550 to 250 million years ago and, over those 300 million years, evolved more than 5000 genera and 15,000 species, all of which are now extinct (figure 4.1). They range from the tiny *Acanthopleurella* (barely 1 millimeter [0.04 inch] in length) to the giant *Isotelus rex* (more than 70 centimeters [2.3 feet] long). Since they are relatively easy to collect in many places, and extraordinarily abundant almost everywhere in beds of the Early Paleozoic (especially the Cambrian), they often become the core of many amateur fossil collections. Their wonderfully complex shapes, elaborate ornamentation, bizarre struc-

Figure 4.1 ▲
Reconstruction of two trilobites as they may have appeared in life. (Courtesy Nobumichi Tamura)

tures of the eyes and many other parts of their anatomy, and surprising features make them irresistible to most fossil collectors.

This fascination is not confined to modern times. A trilobite from the Silurian carved into an amulet was found in a rock shelter more than 15,000 years old. A trilobite from the Cambrian preserved in chert was carried a long way by Australian Aborigines and carved into an implement. The Ute peoples used to carve the common trilobite *Elrathia kingi*, from the House Range of Utah, into amulets. They called them *timpe khanitza pachavee* (little water bug in stone house). *Elrathia kingi* are so abundant in this locality that they are commercially mined with backhoes and are sold in huge quantities to nearly every rock shop and fossil dealer around the world.

Even more important, trilobites were the first large shelled organisms on Earth. There is abundant evidence from the genetic divergence times of their close relatives that soft-shelled trilobites were around in the earliest Cambrian and developed mineralized shells in the Atdabanian, the third stage of the Early Cambrian (see figure 3.4). This may be because atmospheric oxygen levels were finally high enough that trilobites could crystallize calcite in their shells. Most of the creatures that preceded them either were soft-bodied, with no hard parts or shells, or had tiny and inconspicuous shells (chapters 2 and 3); therefore, they were fossilized only in environments with conditions that favored preservation, not decomposition (chapter 5). Not only did trilobites have a large complex shell made of the protein chitin (as do crabs, lobsters, shrimps, insects, spiders, scorpions, and all other arthropods), but this relatively soft and easily decayed shell was fortified by layers of the mineral calcite. Thus trilobites were much more likely to be fossilized than any other Cambrian creature, since they were one of the few groups with mineralized shells. The appearance of hard-shelled trilobites in the Atdabanian makes them overrepresented in the fossil record, and they give the false impression that there was a "Cambrian explosion" of life between the Tommotian and the Atdabanian (see figure 3.4). Instead, it was an "explosion" of animals with mineralized skeletons.

The abundance of easily fossilized trilobites in deposits of Late Cambrian age meant that more than 300 genera in 65 families were recognized, completely overwhelming all other fossil groups known from that time. In just about any Cambrian deposit, the majority of fossils are trilobites, so paleontologists use the stages of trilobite evolution to tell time in the Cambrian.

WHAT IS A TRILOBITE?

Trilobites are the earliest known fossilized arthropods, the phylum that includes insects, spiders, scorpions, crustaceans, and many other creatures (chapter 5), and they clearly display all the features of that phylum. Like all other arthropods, they had a jointed exoskeleton that fell apart when they molted, so the fossils are often incomplete pieces of molted shell, and not the complete animal, which likely lived on to molt again. Unlike that of most other arthropods, however, the chitinous exoskeleton of trilobites was reinforced with mineralized calcite, which made them much more fossilizable than insects or spiders or scorpions or most crustaceans.

The "head" of most arthropods is called the *cephalon* (Greek for "head) in trilobites (figure 4.2*A*). It is usually a broad structure with two "cheek regions" on each side of a central lobe ("nose") called the *glabella*. On each side of the glabella are typically two eyes. Some trilobites had tiny eyes or none at all, so they had limited vision or were blind; others had huge eyes that wrapped around and gave a 360-degree range of view to spot any predators. Many advanced trilobites had lenses made of two crystals of calcite,

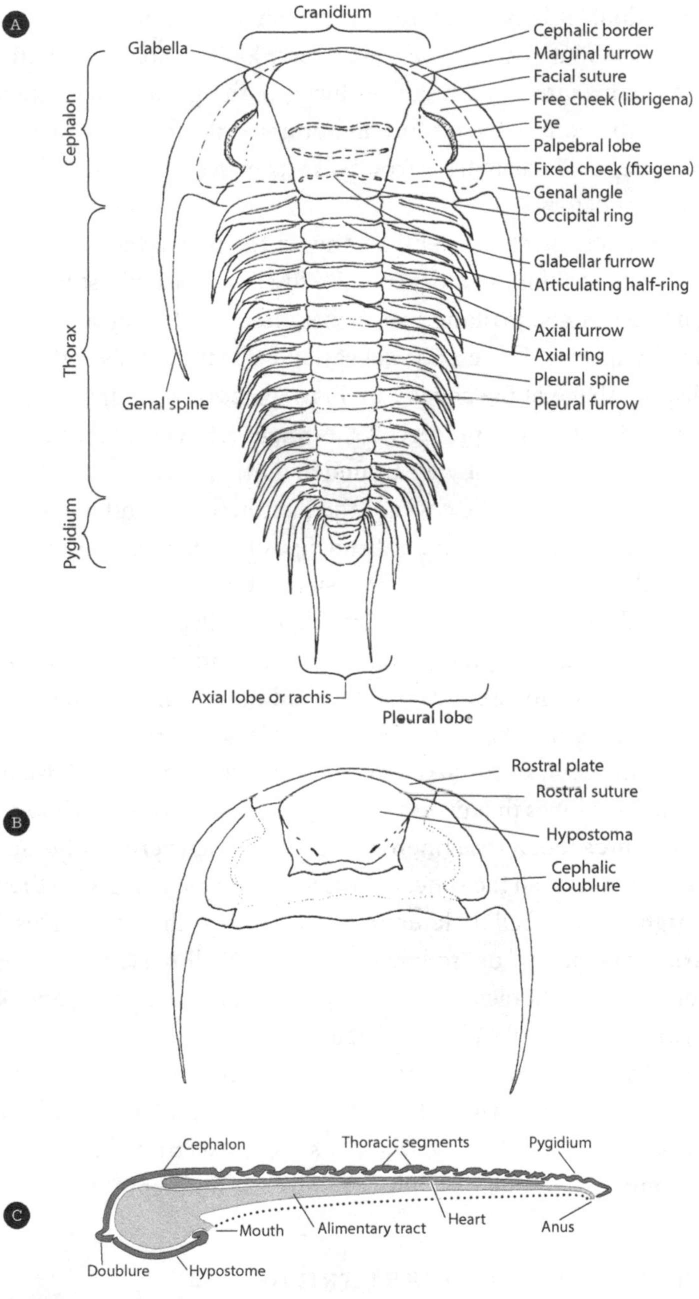

Figure 4.2 ▲

Basic anatomy of a trilobite: (*A*) top view of the complete exoskeleton; (*B*) bottom view of the cephalon; (*C*) cross section through the axis, with skeletal parts indicated in black. (Modified from several sources) forming a doublet lens structure that corrected for spherical aberration in thick lenses. About 400 million years after trilobites evolved these features, they were reinvented by Christian Huygens, the great Dutch scientist, in the sixteenth century. Even more important, trilobites were probably the first creatures on Earth to have true eyes and to use visual clues to find food and avoid predators.

The cheek regions broke off from the central part of the cephalon (*cranidium*) during molting, so most trilobite fossils consist of just the center of the cephalon. A good trilobite specialist often needs only a cranidium to identify the species. There are great variations in the details of the eyes, the glabella, the shape of the cheeks, and the spines on the edge of the cephalon. On the front of well-preserved specimens are two antennae that trilobites used for feeling their way around in dark muddy waters. On the bottom is a plate that partially covered the mouth, with mouthparts that were used for sucking up food-rich mud and digesting the nutrients out of it (see figure 4.2*B*). Most trilobites were deposit feeders or mud grubbers.

The middle part of the body of a trilobite is called the *thorax*, as in most arthropods. In trilobites, the thorax is divided into segments, allowing the middle of the body to flex and curl as the animal moved or to roll up for protection. Each segment has two lobes on the sides (*pleural lobes*) and one that runs down the central axis (*axial lobe*). It is this side-by-side division of the body into three lobes that gives the group their name. Some trilobites have just two or three thoracic segments, so they were not very flexible and must have lain flat nearly all the time. Others have many segments, so they could roll up tightly into a ball to deter predators, such as the isopod crustaceans known as "roly-polies" or "sowbugs" do today. Well-preserved fossils show that beneath each thoracic segment was a pair of walking legs, and on each side a pair of feather-like gills attached to the base of the legs.

The tail end of the trilobite is not called the abdomen, as in many arthropods. Instead, it is known as the *pygidium* (Greek for "little tail"). In most trilobites, the last few thoracic segments are fused into a single large plate-like pygidium. The olenellids, however, are very different.

OLENELLUS AND THE FIRST TRILOBITES

Once you acquire a discerning eye for telling trilobites apart, *Olenellus* is one of the easiest trilobites to recognize (figure 4.3). It was named and studied in detail by none other than the pioneering Cambrian expert Charles

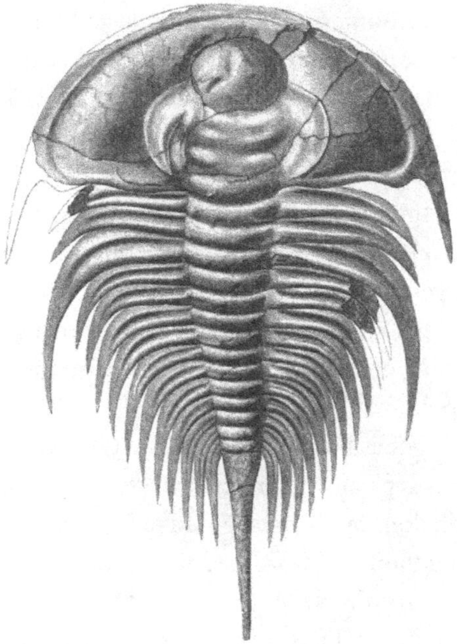

Figure 4.3 ▲
A specimen of *Olenellus*, showing the characteristic *D*-shaped cephalon, bulbous tip of the glabella, large crescent-shaped compound eyes, spines on the tips of the cephalon and on certain thoracic segments, and absence of a large fused pygidium, or tail segment. (Photograph courtesy Wikimedia Commons)

Doolittle Walcott, in 1910 (chapter 1). Its most distinctive feature is the lack of fusion of the last few segments of the thorax into a single pygidium. Instead, *Olenellus* has a long spike on the tail. This is a very primitive feature, which is no surprise, since the olenellids are the oldest trilobites known.

In addition to the absence of a pygidium, there are many other distinctive or primitive features in *Olenellus*. The cephalon is large and shaped roughly like a capital *D*. There is no line of rupture (*cranial suture*) on the top of the cephalon, which separated the cheeks from the cranidium when *Olenellus* molted. Two big crescent-shaped eyes wrap around each side of a furrowed glabella, which has a bulbous knob at the tip in front. The eyes are simple, with many tiny lenses made of calcite rods packed closely together, like the typical compound eyes of most insects and many other arthropods. The eyes could not have formed a clear photographic image, but would have alerted the trilobites to large areas of light and darkness and movement near them. Studies of the extraordinary Cambrian faunas such as those preserved in the Burgess Shale in Canada and the Maotianshan Shale in China show that there were almost no large predators at that time (chapter 5). The largest may have been the 1-meter (3.3-foot) long *Anomalocaris*. Fossils show that it clearly took bites out of the trilobites found in the Middle Cambrian Burgess Shale, where it was discovered. But compared with the later Paleozoic, there was not a lot of predation pressure, and trilobites were relatively simple and unspecialized in the Cambrian. Not until the Ordovician do we get super-predators, such as nautiloids with shells over 6 meters (20 feet) long. Only then did trilobites evolve distinctive shells specialized for burrowing or swimming or rolling up into to a ball as a defense against tougher predators.

One other striking feature of *Olenellus* is that it is very spiny on the edge of its shell. There is typically a spine (*genal spine*) sticking out of the back corners of the cephalon. Many have broad spines sticking out of their thoracic segments, and then backward, usually segment number 3. Some have additional spines protruding from the front of the cephalon as well. These spines often help paleontologists recognize different genera and species within the olenelloids.

Once olenelloids appeared in the Atdabanian stage of the Early Cambrian (about 520 million years ago), they flourished into multiple genera and species found almost everywhere around the world in that stage and in the succeeding Botomian stage of the Early Cambrian. They then vanished at the end of the Toyonian stage (about 509 million years ago). Bruce Lieberman of the University of Kansas has analyzed thousands of specimens of olenellids and concluded that their ancestors originated in what is now western Russia or Siberia at the beginning of the Cambrian, but, like other trilobites, were not calcified or fossilized until the Atdabanian stage.

WHAT HAPPENED TO THE TRILOBITES?

Through the rest of the Paleozoic, the trilobites were hammered by a series of mass extinction events (figure 4.4). They include the multiple minor extinctions in the Late Cambrian, when several pulses of disasters wiped out the

Figure 4.4 ▶

Diversification and extinction of the trilobites. (From several sources)

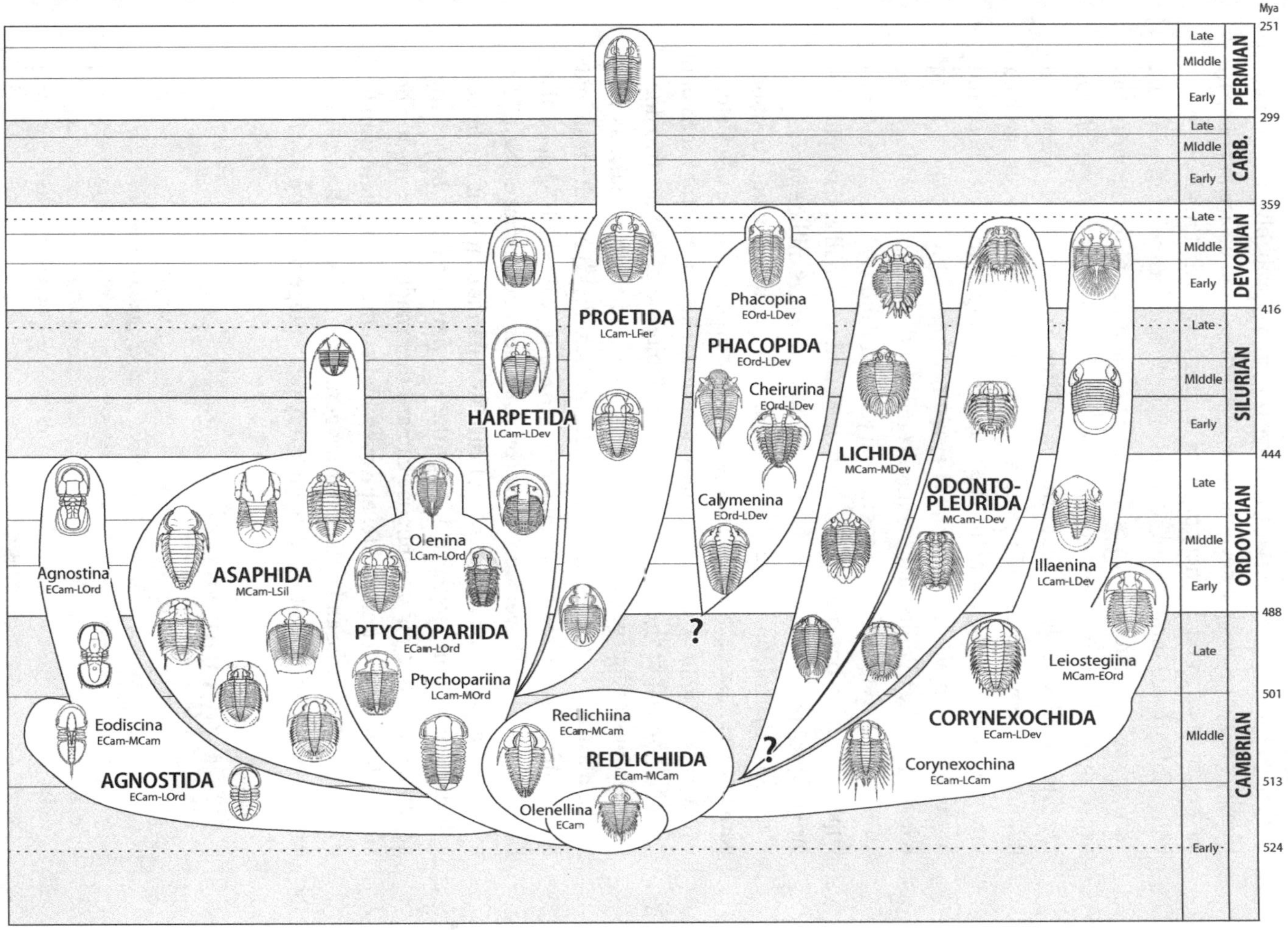
Mya
251
299
359
416
444
488
501
513
524
PERMIAN
CARB.
DEVONIAN
SILURIAN
ORDOVICIAN
CAMBRIAN
Late
Middle
Early
PROETIDA
LCam-LPer
HARPETIDA
LCam-LDev
Phacopina
EOrd-LDev
PHACOPIDA
EOrd-LDev
Cheirurina
EOrd-LDev
Calymenina
EOrd-LDev
LICHIDA
MCam-MDev
ODONTO-
PLEURIDA
MCam-LDev
Illaenina
LCam-LDev
Leiostegiina
MCam-EOrd
CORYNEXOCHIDA
ECam-LDev
Corynexochina
ECam-LCam
ASAPHIDA
MCam-LSil
Olenina
LCam-LOrd
PTYCHOPARIIDA
ECam-LOrd
Ptychopariina
LCam-MOrd
Agnostina
ECam-LOrd
Eodiscina
ECam-MCam
AGNOSTIDA
ECam-LOrd
Redlichiina
ECam-MCam
REDLICHIIDA
ECam-MCam
Olenellina
ECam
?
?

diversity of trilobites, wave after wave. During the Ordovician, trilobites experienced their first encounter with large predators, probably gigantic nautiloids. Trilobites quickly became more specialized and easier to tell apart as they soon adopted a variety of shapes and lifestyles that made them less vulnerable to predation. These adaptations included burrowing (the smooth "snowplow" trilobites known as Asaphida or Illaenida), rolling up into a ball (the Calymenida), or becoming tiny (the Trinucleida, such as *Cryptolithus*, the thumbnail-size "lace collar" trilobite). Then came the Late Ordovician extinction (about 450 million years ago), and only a few lineages survived into the Silurian and Devonian. The final flourishing of trilobites occurred in the Devonian, when the complex-eyed Phacopida were common, along with large spiky trilobites like *Terataspis* (about 0.5 meter [1.5 feet] long). The Late Devonian extinctions at 375 million years ago and 357 million years ago wiped out all but one order of trilobites, the relatively small and simple Proetida, which persisted in the background for another 125 million years.

Finally, the trilobites disappeared during the great Permian extinction, some 250 million years ago, the largest mass extinction in Earth's history, when 95 percent of all marine species vanished. This huge event wiped out not only the last of the stragglers among trilobites, but also the two dominant groups of Paleozoic corals (the tabulates and the rugosids), as well as the blastoids (relatives of the crinoids, or "sea lilies") and the fusulinid foraminiferans (incredibly abundant protozoans with shells shaped like rice grains). There have been many controversies about what caused "the mother of all mass extinctions," but current data indicate that the largest volcanic eruption in geological history, the Siberian lava flows, helped trigger an extremely rapid greenhouse climate that made the oceans too hot and acidic to support much life, and overcharged the atmosphere with too much carbon dioxide and not enough oxygen. These, along with some other catastrophic events, destroyed all but a tiny percentage of life on Earth.

SEE IT FOR YOURSELF!

A number of museums have trilobite fossils on display, although most do not show very good complete specimens of olenellids. Some that do include the Denver Museum of Nature and Science; Field Museum of Natural History, Chicago; Geology Museum, University of Wisconsin, Madison; and National Museum of Natural History, Smithsonian Institution, Washington, D.C.

Olenellids are so abundant at some localities around the world that they are easy to collect for yourself. Here are three famous and easy-to-reach localities in the United States. Consult fossil-collecting guidebooks and Web sites for more such areas.

- **MARBLE MOUNTAINS, CALIFORNIA.** Take Interstate 40 (either eastbound or westbound) to exit 78, Kelbaker Road; leave the interstate; and drive 1 mile south to the "T" junction. Turn left (east) and drive to the ghost town of Chambless along the National Trails Highway (former U.S. Route 66). Turn southeast on the road to Cadiz. After the paved road curves due east and just as the paved road goes due south to cross the railroad tracks, turn onto a dirt road on the left that goes due north. Drive about 1 mile north on the dirt road until you reach a junction with a well-traveled east–west dirt road, and then turn east. About half a mile along this road, you will see the dirt road heading northeast toward an old quarry. Follow this road as far as it is passable, and then hike up to the Latham Shale (the brown shale unit) below the gray cliff-forming Chambless Limestone. Look for old "glory holes" of serious collectors, and turn over the larger shale pieces. You will see many good cephala of every size, although complete trilobites are extremely rare. Two good Web sites are "Trilobites in the Marble Mountains, Mojave Desert, California" (http://inyo.coffeecup.com/site/latham/latham.html) and "Trilobites of the Latham Shale, California" (http://www.trilobites.info/CA.htm).
- **EMIGRANT PASS, NOPAH RANGE, CALIFORNIA.** Take Interstate 15 to Baker, California; leave the interstate; and drive north for 48 miles on California State Route 127 (Death Valley Road) to Old Spanish Trail. Turn right on Old Spanish Trail and proceed through Tecopa to Emigrant Pass. The exposure is just to the west of the summit of the pass on the south side of the road (GPS coordinates = 35.8856N, -116.0603W). Two good Web sites are "Trilobites in the Nopah Range, Inyo County, California" (http://inyo.coffeecup.com/site/cf/carfieldtrip.html) and "Ollenelid Trilobites at Emigrant Pass, Nopah Range, CA" (http://donaldkenney.x10.mx/SITES/CANOPAH/CANOPAH.HTM).
- **OAK SPRING SUMMIT, LINCOLN COUNTY, NEVADA.** From Caliente, Nevada, take U.S. Route 93 west for 10 miles, or east for 33 miles from the junction with Nevada State Route 375 and 318 (between Hiko and Ash Springs). Look for a turnoff on the north side of the highway, signaled by a prominent Bureau of Land Management sign that reads "Oak Springs Trilobite Site." Turn north, drive along the dirt road, park in the gravel lot, and then hike up the Trilobite Trail from the trailhead across the sagebrush until you arrive at a flat area covered with pieces of the Pioche Shale. Turn over shale, and you will find many fine cephala, occasionally better specimens, of every age and size. A good Web site is "Oak Spring Summit" (http://tyra-rex.com/collecting/oaksprings.html).

FOR FURTHER READING

Erwin, Douglas H., and James W. Valentine. *The Cambrian Explosion: The Construction of Animal Biodiversity*. Greenwood Village, Colo.: Roberts, 2013.

Fortey, Richard. *Trilobite: Eyewitness to Evolution*. New York: Vintage, 2001.

Foster, John H. *Cambrian Ocean World: Ancient Sea Life of North America*. Bloomington: Indiana University Press, 2014.

Lawrance, Pete, and Sinclair Stammers. *Trilobites of the World: An Atlas of 1000 Photographs*. New York: Siri Scientific Press, 2014.

Levi-Setti, Ricardo. *The Trilobite Book: A Visual Journey*. Chicago: University of Chicago Press, 2014.

IS IT A WORM OR AN ARTHROPOD?

There are vastly more kinds of invertebrates than vertebrates. Recent estimates have placed the number of invertebrate species on Earth as high as 10 million and possibly more. . . . Invertebrates also rule the earth by virtue of their sheer body mass. For example, in tropical rain forest near Manaus, in the Brazilian Amazon, each hectare (or 2.5 acres) contains a few dozen birds and mammals but well over a billion species of invertebrates, of which the vast majority are mites and springtails. There are about 200 kilograms [440 pounds] by dry weight of animal tissue in a hectare, of which 93% consists of invertebrates. The ants and termites alone comprise one-third of this biomass. So when you walk through a tropical forest, or most other terrestrial habitats for that matter, vertebrates may catch your eye most of the time but you are visiting a primarily invertebrate world.

EDWARD O. WILSON, "THE LITTLE THINGS THAT RUN THE WORLD"

WONDERS OF THE BURGESS SHALE

One of the most amazing fossil localities in the world is the legendary Burgess Shale, in the Rocky Mountains near Field, British Columbia. It was accidentally discovered by pioneering Cambrian paleontologist Charles Doolittle Walcott (chapter 1), who was working on rocks of the Middle Cambrian (about 505 million years old) in the area in the summer of 1909. On August 30, his horse stumbled on a large rock on the trail. Walcott dismounted, pushed away the slab, and found that the underside was covered with fossils preserved as delicate films. He traced the slab to where it had fallen from up the slope, and soon began a large quarrying operation (see figure 1.4). Each summer until 1924, he returned to the Burgess Shale, eventually amassing a collection of more than 65,000 specimens in the Smithsonian Institution. Almost all the fossils from the Burgess Shale are those of from soft-bodied organisms that had been buried in a submarine landslide during the Middle Cambrian. Not only were they buried, but the bottom waters were apparently low in oxygen, preventing the usual scavengers and decomposers from doing their work. Consequently, the Burgess Shale preserves the delicate soft tissues that are seldom seen in the fossil record.

But Walcott was far too busy running the Smithsonian and fulfilling his many other commitments, so he managed only a superficial description of the fossils before he died in 1927. Many of the fossils remained unstudied, filed away in cabinet drawers. The fossils that Walcott did study and publish were described and then assigned to familiar groups like arthropods and worms, without time to adequately prepare the specimens or examine their fine detail.

So the fossils remained for decades. In 1949, legendary British trilobite expert Harry Whittington accepted a position at Harvard University. He soon realized that the enormous Burgess Shale collection in the cabinets in his office that had never been examined. When Whittington returned to England and became the Woodwardian Professor of Palaeontology at Cambridge University in 1966, he launched a large-scale project on the Burgess Shale fossils. He and his students returned to Walcott's quarry and unearthed hundreds of new specimens. They also took much more care than had Walcott in preparing out the details of the fossils, often digging below their surfaces to see the three-dimensional structures underneath that Walcott had missed. Over the next few years, Whittington and his students (especially Derek Briggs, who focused on the arthropod-like animals, and Simon Conway Morris, who was assigned the weird things lumped in the wastebasket category "worms") made revolutionary discoveries that Walcott had never noticed.

Once you look closer at the Burgess Shale fauna, and excavate the fossils in three dimensions, it turns out that many of them had body plans unlike those of any animal on Earth. *Opabinia*, for example, had five eyes in the middle of its forehead, a long segmented body, and a vacuum-like nozzle in the front for feeding (figure 5.1). The largest predator, *Anomalocaris*, reached over 1 meter (3.3 feet) in length, with long branched feeding appendages, a segmented body with swim flaps on the sides, and a mouth that looked like a pineapple slice but worked like the iris in a camera lens (it had

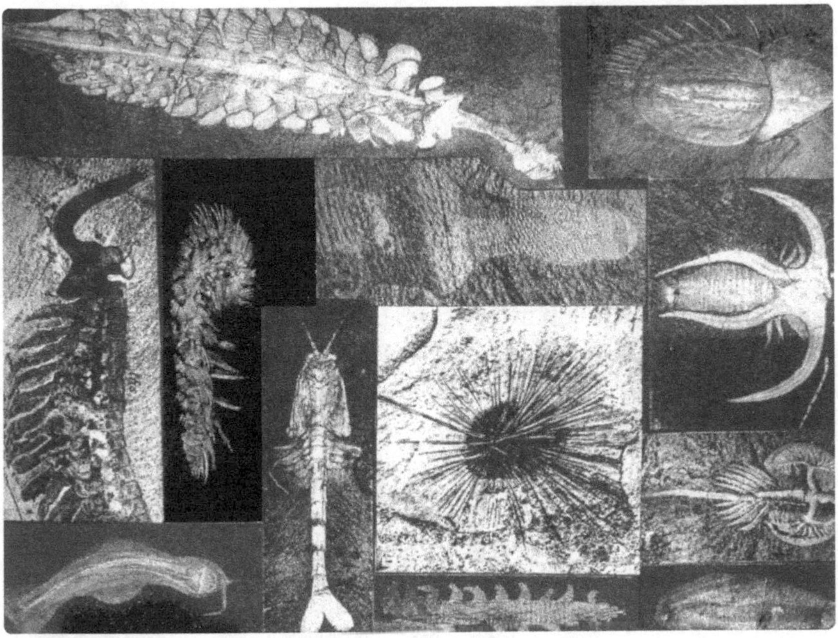

Figure 5.1 ▲
Fossils of the Burgess Shale, including the nozzle-nosed *Opabinia* (*top left and far left*). (Photographs courtesy Smithsonian Institution)

been mistaken for a sea jelly by Walcott). *Wiwaxia* was a little domed creature with a row of spines protruding from its body. *Dinomischus* looked like a soft-bodied version of the hard-shelled crinoids, or "sea lilies." As Whittington, Briggs, and Conway Morris pointed out, many of these creatures seemed to belong to brand-new phyla and could not be shoehorned into such existing groups as arthropods and worms.

In addition to these oddities, there were, of course, many soft-bodied creatures that resembled perfectly good shrimp and other arthropods. And as at any other Cambrian locality, there were plenty of Middle Cambrian trilobites, the only fossils of hard-shelled animals in the Burgess Shale. But their presence demonstrates how most fossil localities are biased for these hard-shelled organisms, leaving only trilobites with an abundant Cambrian fossil record. Without the Burgess Shale and other sites of extraordinary preservation, such as the Maotianshan Shale in China and the Sirius Passet locality in Greenland, we would never know that the seafloor had once been inhabited by a full range of bizarre and unexpected animals with unknown body plans, since they were soft-bodied and seldom fossilized.

In 1989, Stephen Jay Gould published his best-selling book *Wonderful Life: The Burgess Shale and the Nature of History*. Most of it describes the amazing fossils of this locality (the first time ever for a general audience) and details how much the work of Whittington, Briggs, and Conway Morris had changed what we thought about the nature of these creatures. Gould also pointed out how mistaken Walcott had been to try to squeeze each of these extinct animals into living phyla. Instead of a gradual unfolding and diversification and expansion of life since the Cambrian, the Burgess Shale taught us that life had diversified into its maximum range of shapes and number of body plans by the Middle Cambrian, and then extinction in the Devonian had pruned away all but a few survivors (arthropods, molluscs, and some others).

But Gould made a larger point as well. To him, the Burgess Shale underscored the importance of *contingency*, lucky accidents of life that determine how all the events that follow will play out. If we look at the broad panoply of strange animals that swam in the seas of the Middle Cambrian, who would guess that most of these incredible creatures were experimental animals that would not even survive the end of the Cambrian? And who would guess that the tiny insignificant fossil known as *Pikaia* (chapter 8) was a representative of our lineage, the vertebrates, which would eventually come to rule the planet (along with the arthropods)?

If by some accident, vertebrates had vanished in the Cambrian along with most of the experimental forms, how would the history of life have unfolded? There certainly would have been no dinosaurs, nor would there be mammals—or humans. Each time you replay the tape of life's history, it comes out differently. If the random, unpredictable effects of an asteroid impact in Mexico and huge eruptions of lava in India 65 million years ago had not wiped out the dinosaurs, the mammals would never have grown any bigger than they had during the 120 million years of the Age of Dinosaurs, and humans would not be here, either. The modern world is an improbable, lucky accident, one of millions of possible ways in which the scenario of life could have progressed. All living organisms are not the inevitable outcome of long-term evolution, but the descendants of ancestors that happened to survive many mass extinctions and other random events.

In his book, Gould makes an analogy with the famous Frank Capra Christmas classic, *It's a Wonderful Life*, with Jimmy Stewart and Donna

Reed. In the movie, Stewart's character, George Bailey, is given a chance to see how the world would have been if he had never lived—and discovers that every human life and every little event has unpredictable consequences.

HALLUCINATION IN STONE

Among the strangest and most difficult to interpret of the Burgess Shale fossils was a "worm" that Walcott had assigned to the polychaete worm genus *Canadia*. When Conway Morris began to work on the miscellaneous "worms" that Walcott had neglected, a few specimens stood out (figure 5.2*A*). They did look somewhat worm-like, with a long trunk of some kind, but they had pairs of straight pointed protrusions on one side of the body and what appeared to be a single row of "legs" or "tentacles" on the other side. There was a discolored "blob" at one end of the body that may have been the head—and there was not much more to go on. Clearly, this creature was unlike any worm on the planet (extinct or living). Conway Morris's original reconstruction had the body supported by the pairs of straight spiky appendages, with a row of the "tentacles" atop the body (see figure 5.2*B*). In 1977, he renamed this fossil *Hallucigenia*, because it seemed like a creature that would be seen only in a nightmare or hallucination.

Other scientists were not so sure. Some thought that the fossil was actually that of the appendage of a larger animal. This had already happened with *Anomalocaris*, whose appendages in front of the mouth had been mistaken for shrimp-like creatures. But the prevailing opinion was that *Hallucigenia* was a member of the Lobopodia, a wastebasket group for a number of marine "worms with legs" that had turned up in Early Paleozoic rocks around the world.

In 1991, Lars Ramskold and Hou Xianguang described and published another hallucigenid, *Microdictyon*, from the Lower Cambrian Maotianshan Shale of China (figure 5.3*B*). This specimen was much better preserved than any fossils of *Hallucigenia*. The better preservation showed that Conway Morris had reconstructed *Hallucigenia* upside down (see figure 5.2*C*)! *Microdictyon* sported a row of paired spines along the top, and the "legs" of Conway Morris's *Hallucigenia* actually were spines along its back. The floppy little "appendages" down the back of Conway Morris's reconstruction of *Hallucigenia* were the real legs, which were paired, as would be expected. Even more surprising, *Microdictyon* had a series of small armored plates along its body, which had been known for a long time from the Early

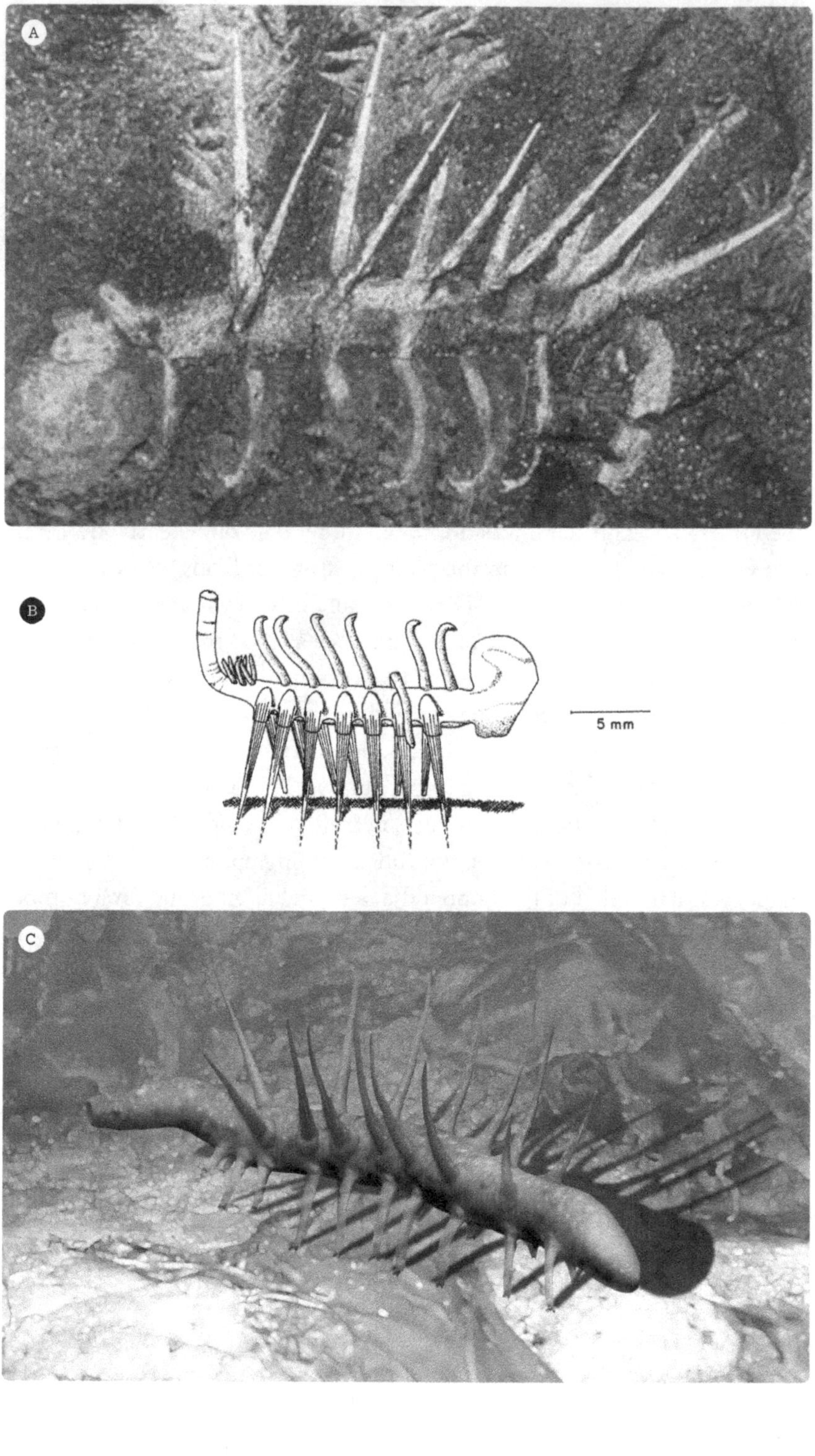
A
B
5 mm
C

Cambrian "little shellies" (chapter 3), but nobody knew what creature they had belonged to!

In addition to flipping *Hallucigenia* upside down, *Microdictyon* solved another puzzle: the mystery of their origins. With both animals right side up, it was clear that they were lobopods and that such creatures had been common on the seafloor in the Cambrian. In fact, an even better preserved, unquestioned lobopod already was known from the Burgess Shale: *Aysheaia* (see figure 5.3*A*). And with the better-preserved specimens, scientists could finally figure out what lobopods were as well. It turns out that they were ancient relatives of a living phylum that creeps in the jungles, a group known as the velvet worms.

WHAT IS AN ARTHROPOD?

Insects, spiders, scorpions, crustaceans, barnacles, horseshoe crabs, and trilobites are members of the largest phylum of animals on Earth: the Arthropoda, or "joint-legged" animals. (In Greek, *arthros* means "joint" [as in "arthritis"], and *podos* is "foot" or "appendage.") By any measure, arthropods have been and always will be the dominant animals on Earth—even though we like to think of ourselves as rulers of the planet. With over 1 million species (and probably a lot more still uncounted), arthropods make up more than 85 percent of the roughly 1.4 million (and counting) animal species (figure 5.4). There are almost 900,000 species of insects and over 340,000 species of beetles alone. When asked what his knowledge of biology taught him about the Creator, the great biologist J. B. S. Haldane said, "God must have had an inordinate fondness for beetles." By contrast, our phylum, Chordata, contains fewer than 45,000 species, over half of which are fish. There are barely more than 4000 species of mammals.

If the total species diversity does not impress you, what about abundance? Arthropods are legendary for reproducing quickly when the conditions are right and multiplying to astonishing numbers. Think of the plagues of locusts or the speed with which aphids can overwhelm a plant or the immense number of individuals in an ant colony or a termite nest. If

Figure 5.2 ◀

The "worm" *Hallucigenia*: (*A*) Burgess Shale fossil; (*B*) original reconstruction, showing the spines as "legs"; (*C*) current reconstruction, showing the spines on the back. ([*A–B*] courtesy S. Conway Morris, Cambridge University; [*C*] courtesy Nobumichi Tamura)

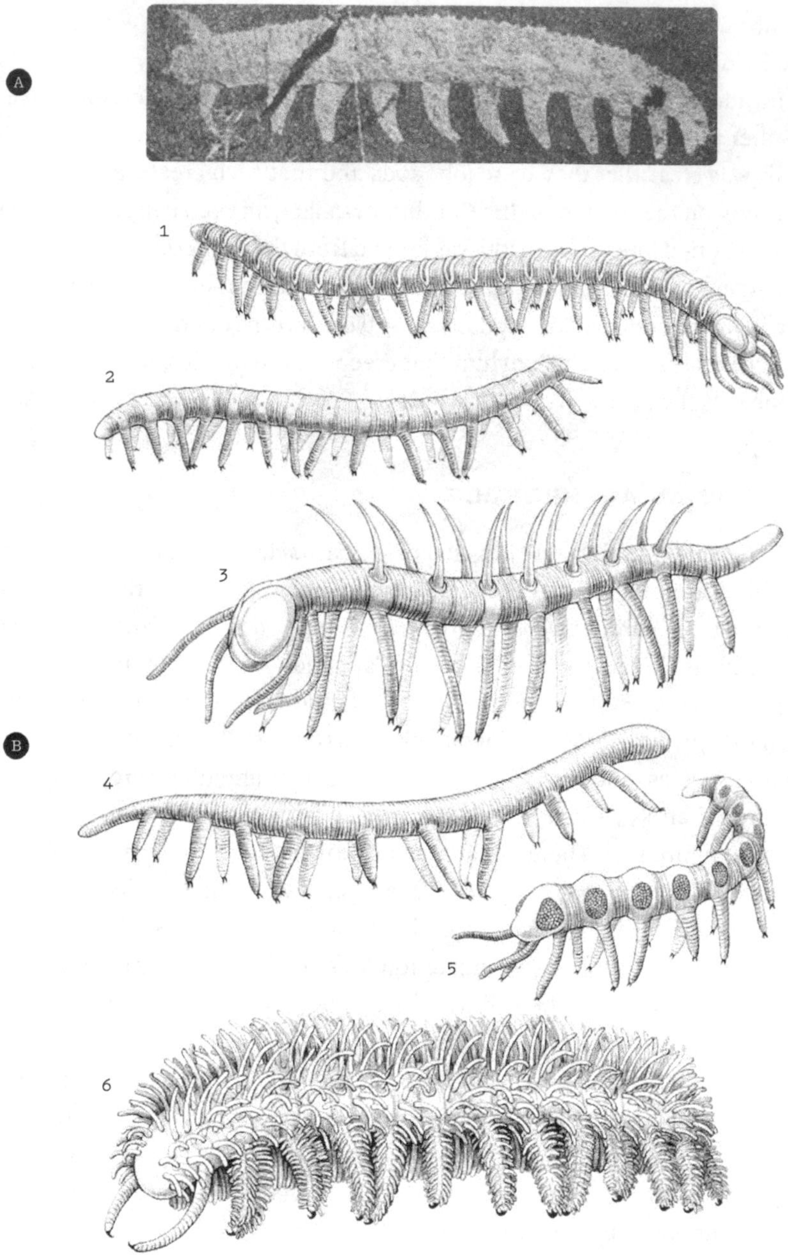

Figure 5.3 ▲

Onychophorans and lobopods: (*A*) the Burgess Shale fossil *Aysheaia*, a primitive onychophoran; (*B*) reconstruction of a number of lobopod fossils: (*1*) *Cardiodictyon*; (*2*) *Luolishania*; (*3*) *Hallucigenia*; (*4*) *Paucipodia*; (*5*) *Microdictyon*; (*6*) *Onychodictyon*. (Courtesy S. Conway Morris, Cambridge University)

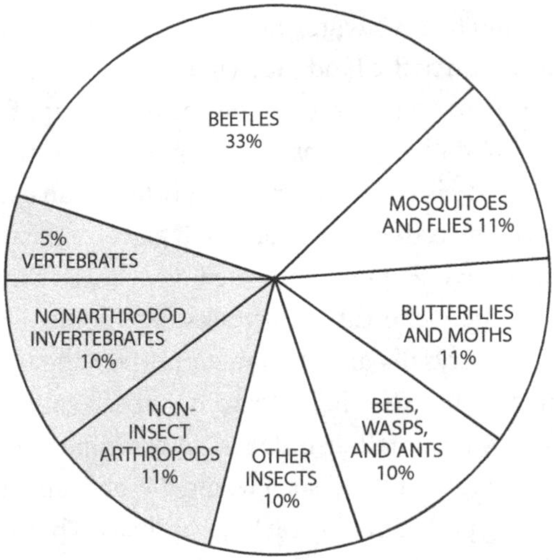

Figure 5.4 ▲
The diversity of animal species and the dominance of arthropods, especially insects. (Drawing by Pat Linse, based on several sources)

not held in check, a single pair of cockroaches can have 164 billion offspring in just seven months! In the tropics, a few acres might support a few dozen birds or mammals, but over 1 billion arthropods, including mites, beetles, wasps, moths, and flies. A single ant colony may contain 1 million individuals. In the richest parts of the ocean, there can be millions of tiny planktonic arthropods (shrimps, copepods, krill, and ostracodes) in 1 cubic meter (35 cubic feet) of water.

Arthropods are also extremely adaptable and can occupy nearly every niche on the planet—except for those that allow large body size. There are arthropods that fly; arthropods that live in fresh- and salt water; arthropods that tolerate extremes in temperature, from subfreezing to almost boiling; and arthropods that are internal and external parasites on other organisms. The key to this adaptability is their construction. They are modular creatures, with many segments that can easily be added to or subtracted from or can be modified in shape. On each segment is a pair of jointed appendages, which can be refashioned into mouthparts, legs, antennae, pincers, paddles, wings, and many other structures. Most characteristic of all, they have a hard external shell, or exoskeleton, with muscles and soft tissues inside, rather than the internal skeleton surrounded by muscles that vertebrates have. The exoskeleton confers many advantages: it provides protection against predators and forms a waterproof covering that allowed arthropods to move from the ocean to the land. But a hard shell does not grow, so every once in a while, the arthropod must molt, or break out of its exoskeleton and form a newer, slightly larger one. During the short time after it an arthropod molts, its body is soft and is not supported by an exoskeleton.

Molting is a key constraint in arthropods. It can be a great advantage or a significant disadvantage. For example, the body shape of many insects can change completely between different molts, exemplified by a caterpillar transforming into a chrysalis and then metamorphosing into a butterfly or moth, with a completely different body from that of a caterpillar.

Molting also dictates small size. Once an arthropod reaches a certain size, it can grow no larger. If it did so, it would dissolve into jelly as it molted due to the increasing pull of gravity on large animals. That's why there have never been land arthropods bigger than the huge dragonfly *Meganeura*, with a wingspan of 1 meter (3.3 feet), or the millipede *Arthropleura*, which was 3 meters (10 feet) long. Even though the bodies of marine arthropods can get slightly larger than those of land arthropods because they are supported by water, there are none bigger than the king crab or some of the huge marine "sea scorpions" of the Silurian, which reached 3 meters in length. The next time you see a low-budget horror film with a gigantic ant or praying mantis, you can laugh because such creatures are biologically impossible. Sadly, few screenwriters know enough science to realize this.

"VELVET WORMS" AND ARTHROPODS

Most people have never seen a "velvet worm" in life, unless they live in the tropical jungles of the Southern Hemisphere and have a habit of combing through decaying leaf litter at night (figure 5.5). Nevertheless, there is an entire phylum of these tiny creatures known as the Onychophora (on-ee-KOFF-o-ra). About 180 species live in the forests of Africa, South America, and Southeast Asia. Most are tiny (0.5 centimeter [0.2 inch]), but some reach lengths of 20 centimeters (8 inches). They look vaguely like caterpillars: two rows of multiple short stumpy legs run along the bottom of their long worm-like body, and at the end of each leg is a hard hooked claw, much as in insects and other arthropods. Their head has mouthparts and, like many arthropods (but no worms), a pair of antennae. Their simple eyes are like

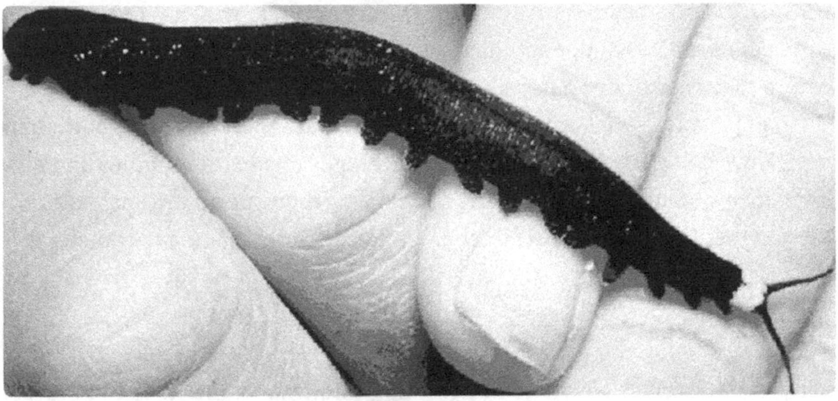

Figure 5.5 ▲
An onychophoran, or "velvet worm." (IMSI Master Photo Collection)

the medial ocelli in arthropods, with some image-forming capability, but they do not need excellent vision in the dark moist world they inhabit.

"Velvet worms" are ambush predators that feed on small insects, millipedes, snails, and worms in the leaf litter. Most of the time, they detect their prey by tiny changes in air currents. They creep up on their quarry with their smooth, graceful fluid motion, and then touch it gently several times to determine if it is small enough to be prey or large and a probable predator. If it is a potential meal, they produce a nasty slime from glands along their body to capture and subdue it; the mucus also makes them distasteful to predators. Once they attack their prey, they will stop at nothing to find it again if it has escaped. As soon as the prey is ensnared, they kill it with a bite from their strong jaws, and then wait for the enzymes in the slime to liquefy its innards so they can digest it.

"Velvet worms" have no hard chitinous shell, as do arthropods, but a thin skin of dermis and epidermis, and their body (like those of most worms) is supported by the hydrostatic pressure of the fluid in their internal cavities. Their flexible skin allows them to squeeze into tiny cracks for protection against predators. This burrowing strategy also protects them against desiccation, or they burrow into the soil if that is available. Their skin is covered with hundreds of tiny soft fiber-like bristles, which give them the appearance and feel of velvet. They are so small that they conduct much of their gas exchange by diffusion through their skin. Simple trachaea in their skin serve as respiratory organs, but (unlike arthropods, which can close their tracheae) are always open, which restricts them to moist tropical habitats where they cannot dry out.

In most ways, "velvet worms" seem unremarkable until you get to their reproduction. Many species incubate their eggs inside their bodies and give birth to live offspring. In a few, the males carry their sperm in a special structure on their head and insert the head into a female's vagina to transfer it.

MACROEVOLUTION BETWEEN PHYLA

Why is the "velvet worm" so interesting and important? It is the perfect transitional form between worms and arthropods. It has the long soft body of many worms, as well as many advanced features of arthropods: partial segmentation of its body, arthropod-like eyes and antennae, and hook-like "feet" on its stumpy caterpillar-like legs, among several other anatomical similarities.

More important, it must molt its skin in order to grow. It shares this feature with only arthropods and a few other groups of invertebrates: the tardigrades, or "water bears"; roundworms (nematodes); and several other kinds of worms. This characteristic is so fundamental to the embryology and body plans of many animals that it is strong evidence that they are all closely related. In fact, a large group of phyla, including all the molting animals (arthropods, onychophorans, roundworms, tardigrades, and the rest), has been named the Ecdysozoa (shedding animals). (*Ecdysis* is the Greek word for "shedding the outer layer," and an "ecdysiast" is a fancy name for a strip-tease artist.) If that were not enough, in recent years the DNA and other molecular systems of the animals have been closely studied. Sure enough, the Ecdysozoa share unique sequences of DNA and other molecular similarities that confirm their close relationship.

Both primitive arthropods and the tiny plates of lobopods are known from the two earliest stages of the Cambrian—the Nemakit-Daldynian and the Tommotian—long before the "Cambrian explosion" in the Atdabanian. But whereas the arthropods blasted off—first during the Cambrian, with trilobites, and then by the Silurian, with the first millipedes, scorpions, and insects on land—the lobopods had vanished by the Devonian. Some time before they did, though, their descendants, the "velvet worms," crawled onto land. With their soft bodies, they had little chance of fossilization, but a "velvet worm" known as *Ilyodes*, which dates to the Carboniferous, establishes that they were on land by 360 million years ago. "Velvet worms" have been living on this planet inconspicuously in the jungles ever since.

SEE IT FOR YOURSELF!

The Burgess Shale locality is in Yoho National Park in British Columbia, and is a hard hike from any road, so it is open to only qualified researchers. However, a handful of museums have displays of the Burgess Shale fossils. The Field Museum of Natural History, in Chicago, has a computer animation, projected onto three screens, depicting a Cambrian underwater scene of Burgess Shale fauna, including a *Pikaia* swimming, *Hallucigenia* and *Wiwaxia* walking, an *Opabinia* trying to catch the priapulid worm *Ottoia*, a swarm of *Marella*, and an *Anomalocaris* catching a trilobite. Below this animation are interpretive panels and 24 fossils from the Burgess Shale.

Other museums in the United States include the Denver Museum of Nature and Science; Geology Museum, University of Wisconsin, Madison; Sam Noble Oklahoma Museum of Natural History, University of Oklahoma, Norman; and National Museum of Natural History, Smithsonian Institution, Washington, D.C. In Canada, Burgess Shale fossils are in the collections of the Canadian Museum of Nature, Ottawa, Ontario; Royal Ontario Museum, Toronto; and Royal Tyrrell Museum, Drumheller, Alberta. In Europe, the fossils can be seen at the Sedgwick Museum of Earth Sciences, Cambridge University; and Natürhistorisch Museum, Vienna, Austria.

FOR FURTHER READING

Conway Morris, Simon. *The Crucible of Creation: The Burgess Shale and the Rise of Animals*. Oxford: Oxford University Press, 1998.

Erwin, Douglas H., and James W. Valentine. *The Cambrian Explosion: The Construction of Animal Biodiversity*. Greenwood Village, Colo.: Roberts, 2013.

Foster, John H. *Cambrian Ocean World: Ancient Sea Life of North America*. Bloomington: Indiana University Press, 2014.

Gould, Stephen Jay. 1989. *Wonderful Life: The Burgess Shale and the Nature of History*. New York: Norton: 1989.

06 | THE ORIGIN OF MOLLUSCS | *PILINA*

IS IT A WORM OR A MOLLUSC?

If there were competitions among invertebrates for size, speed, and intelligence, most of the gold and silver medals would go to the squids and octopuses. But it is not these flashy prizewinners that make the phylum Mollusca the second largest of the animal kingdom, with more than 100,000 described species. That honor has been won for the phylum mostly by the slow and steady snails, with some help from the even slower clams and oysters. The name Mollusca means "soft-bodied," and the tender succulent flesh of molluscs, more than any other invertebrates, is widely enjoyed by humans. But many molluscs are better known for the hard shells that these slow-moving, vulnerable animals secrete as protection against potential predators. Ironically, it is for the beauty and value of these shells that many molluscs are most ardently hunted by humans, in some cases nearly to extinction.

RALPH BUCHSBAUM AND MILDRED BUCHSBAUM, *ANIMALS WITHOUT BACKBONES*

MISSING LINKS FOUND

The fossil record is full of amazing transformational sequences that show, for example, the evolution of horses from small four-toed ancestors and that of mammals from non-mammals (chapters 19 and 22). But many people are not satisfied with this huge mountain of evidence and ask another question: How did all the discrete phyla of animals (molluscs, worms, arthropods, echinoderms, and so on) evolve from a common ancestor? Where is the evidence for such a large-scale change in body plan, or macroevolution?

For the longest time, there was no fossil evidence to indicate how this happened, other than the clear-cut anatomical features in these creatures that show they evolved from a common ancestor. For example, the connection between the arthropods and the "velvet worms" was established by the similarity of the living animals, long before we had a fossil record to confirm this change, and the recent molecular evidence that finally proved their close relationship (chapter 5).

Or let's take another example: the molluscs. Today, the phylum Mollusca includes more than 100,000 described species, more than any other phylum except the arthropods. Molluscs range from such slow and simple creatures as chitons and limpets, which cling to rocks in tide pools and creep along, grazing on algae; through headless clams and oysters, which stay in one place, filter-feeding with their gills; to squids and octopi, which are extremely fast-moving and intelligent, communicate through flashing patterns on their skin, and can solve quite difficult problems. Like arthropods, molluscs have conquered most niches on Earth, including floating in the plankton and living on the seafloor bottom as well as on land (for example, land snails and slugs). Although most molluscs are small, some can be huge—such as the giant squid, which reaches about 18 meters (60 feet) in length; the giant clam, with a shell over 1 meter (3.3 feet) across; and the giant marine snail *Campanile giganteum*, with a huge spiraled shell over 1 meter long.

But what did the common ancestor of all this huge diversity of snails, clams, and squids look like? What kind of animal has the basic building blocks of all these body plans? And where did the molluscs come from among all the rest of the phyla of animals on Earth?

Most mollusc specialists speculated that the common ancestor of molluscs would have had a body plan based on the elements found in all the members of the phylum (figure 6.1). They often called such a creature the "hypothetical ancestral mollusc," based on its simple construction at the nexus of the different molluscan body plans. Such a creature would have had a fleshy layer around its body, the mantle, which secreted a simple cap-shaped shell like that of the limpets, among the most primitive of living molluscs. This creature would have had a broad fleshy "foot" along its bottom that allowed it to cling tightly to rocks for protection and to creep slowly along, feeding in safer conditions.

All living molluscs have a digestive tract that runs from the mouth to the anus and a respiratory system with feather-like gills for extracting oxygen from seawater and releasing carbon dioxide, found in a pocket in the

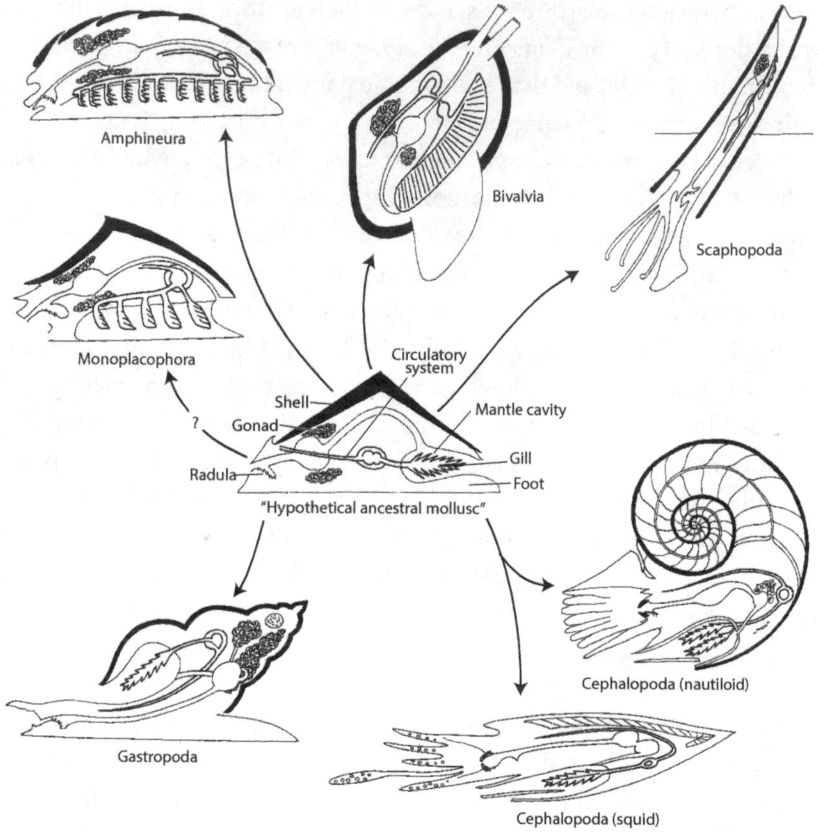

Figure 6.1 ▲

Radiation of the molluscs from the "hypothetical ancestral mollusc." (Modified from Euan N. K. Clarkson, *Invertebrate Palaeontology and Evolution*, 4th ed. [Oxford: Blackwell, 1993]; from Donald R. Prothero, *Bringing Fossils to Life: An Introduction to Paleobiology*, 3rd ed. [New York: Columbia University Press, 2013], fig. 16.3)

mantle called the mantle cavity. The ancestral mollusc must have had all these features, as well as some sort of excretory and reproductive systems. So the earliest molluscs would have been very limpet-like: a simple cap-shaped shell secreted by the mantle, a broad foot for clinging to rocks and creeping, a one-way digestive tract from mouth to anus, a respiratory system, and most of the other systems found in the major molluscan groups (excretory, reproductive, and so on).

THE FIRST MOLLUSCS

Marine biologists have all the benefits of studying living molluscs. They can watch them in action, both in marine aquariums and in nature. They have all the soft tissues to dissect and study in detail. Molecular geneticists can obtain the DNA sequence of molluscs from tiny tissue samples and learn what organisms are most closely related to them. All these things give us a clear answer: the closest living relatives of molluscs are the segmented worms, such as the earthworms that live in the soil and the polychaete worms that are extremely common in almost every marine habitat. But there is still a huge gap: How does an earthworm-like creature evolve into a limpet, with its hard shell and unsegmented body?

The problem is compounded by the fact that most worms never leave fossils, except as burrows, which do not say much about the burrow maker. And the only hard parts of most molluscs are their shells, which provide only a fraction of the information offered by soft tissues. Yet paleontologists have become remarkably adept at working with the simple shells of early molluscs and finding all sorts of clues that the soft tissues leave behind.

As early as the 1880s, paleontologists began to describe simple cap-shaped molluscs from the Early Paleozoic (figure 6.2). The fossils were not well preserved, so it was difficult to say much about them other than they had shells much like that of modern limpets, so must have lived much like a limpet as well. In 1880, the Swedish paleontologist Gustaf Lindström described a fossil shell from the Silurian of Gotland that he called *Triblidium unguis* (the species name from the Latin for "hoof" or "nail," since the shell looked like a fingernail). By 1925, this fossil had been renamed *Pilina unguis*. None of the early paleontologists could say very much about this fossil except that it was very limpet-like, and thus it was thought to be a very primitive limpet. However, on the inside of well-preserved shells were two rows of scars, suggesting that the mollusc had had paired muscles. Without soft tissues, however, they could go no further with this fossil.

Over time, a number of fossils of these simple cap-shaped creatures accumulated in beds that date from the Cambrian to the Devonian. Some paleontologists thought that these fossils might those of be the earliest, most primitive molluscs, but the specimens were still too incomplete to tell. More recently, the simple cap-shaped, clam-shaped, and coiled shells found in the "little shellies" (chapter 3) suggest that there were mollusc

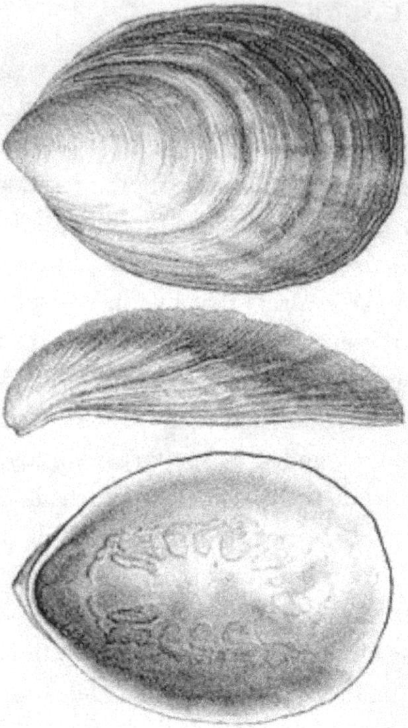

Figure 6.2 ▲

Fossil of the simple cap-shaped, limpet-like *Pilina*, showing the two rows of muscle scars on the inside of the shell. (Courtesy Wikimedia Commons)

predecessors in the Early Cambrian (see figure 3.2). Yet paleontologists have only the shape of the shell and some of its detailed structures on which to base this argument.

GALATHEA TRANSFORMS BIOLOGY

In the late 1940s, oceanography and marine geology were enjoying a huge phase of growth. The battles with submarines during World War II had taught the nations of the world that we knew almost nothing about the 70 percent of Earth's surface covered by oceans. Soon after the war ended, many governments (especially those of the United States, Great Britain, and Denmark) began to fund large-scale scientific expeditions to map the ocean floor, determine what lay at the bottom of the sea, and recover samples of rocks and marine life from all over the world. War-surplus destroyers were refitted and re-commissioned to the task of mapping the ocean. They carried proton-precession magnetometers originally designed to find submarines; these instruments would eventually produce the key evidence for seafloor spreading and plate tectonics. They routinely took sediment cores from nearly every part of the seafloor, bounced sound waves off the bottom to record the depth, and tossed sticks of dynamite off the fantail to bounce sound waves through the upper layers of the sea-bottom sediments and determine their structure.

Among these pioneering postwar efforts was the Second *Galathea* Expedition, mounted by the Danes from 1950 to 1952. The ship was named after the Greek myth of Pygmalion and Galatea. According to the story, the sculptor Pygmalion carved a perfect woman out of marble, named her Galatea, and fell in love with her. He was so enamored of his creation that the gods transformed her into a living woman, in answer to Pygmalion's prayers. Some might recognize this plot device in the Broadway musical *My Fair Lady*, in which Professor Henry Higgins (Pygmalion) transforms the poor slum girl Liza Doolittle (Galatea) into an elegant, aristocratic woman. The musical, in turn, was derived from George Bernard Shaw's famous play *Pygmalion*, which was based on the Greek myth.

The First *Galathea* Expedition had been undertaken between 1845 and 1847, using a three-masted sailing ship to explore the waters off the major Danish colonies around the world. In 1941, journalist Hakon Mielche and oceanographer-ichthyologist Anton Frederik Brunn were pushing to fund a second expedition in order to further Danish scientific and commercial interests. However, World War II and the Nazi invasion of Denmark put their planning on hold.

In June 1945, just after the war ended, the Danish scientific community resumed serious fund-raising and planning. They purchased the retired British sloop HMS *Leith*, a vessel with a long and distinguished record of escorting ships back and forth across the Atlantic during the war and sinking U-boats. The Danes refitted it for oceanographic purposes and renamed it HMDS *Galathea* 2. Unlike the first *Galathea*, this ship was designed to do extreme deep-sea surveys, dredging sediments from and measuring depths of the deepest parts of the ocean. It visited some of the places the mid-nineteenth-century expedition had visited, but the highlights of the mid-twentieth-century voyage around the world was dredging in waters more than

10,190 meters (33,430 feet) deep in the Philippine Trench (the deepest samples ever obtained back then), as well as in many other deep parts of the ocean, yielding creatures never before seen by scientists.

Along with many spectacular and bizarre deep-sea fishes and other marine creatures was a curious-looking mollusc, brought up in 1952 from waters over 6000 meters (19,700 feet) deep in the Costa Rica Trench (figure 6.3). When expedition zoologist Hennig Lemche got a chance to publish the specimen in 1957, he realized that it was truly revolutionary. He named it *Neopilina galatheae*, in honor of the fossil *Pilina* and the ship that had found it. It was indeed a relative of the mysterious cap-shaped fossils from the Early Paleozoic, and its soft tissues allowed paleontologists to interpret the mysterious marks and scars on the fossils. The prominent zoologist Enrico Schwabe called it "one of the greatest sensations of the twentieth century."

Lemche pointed out that *Neopilina* is a true "living fossil," a late-surviving genus in a class of molluscs called the Monoplacophora (from the Greek for "carrying a single shell"), which vanished from the fossil record in the Devonian. And what amazing information was revealed when the specimen was studied! As indicated by the two rows of muscle scars on the fossils, *Neopilina* has paired muscles that produce those scars, suggesting that it had segmented muscles just like segmented worms. Not only are the muscles segmented, but so are the gills, the kidneys, the multiple hearts, the paired nerve cords, and the gonads. In short, *Neopilina* shows that the mysterious monoplacophoran fossils were half mollusc, half worm: they had the segmentation of all their organ systems, like their worm-like ancestors, but they also had a mantle, a shell, a broad foot, and other features found in primitive shelled molluscs like limpets and chitons.

Since the description of *Neopilina* in 1957, many more living and fossil monoplacophorans have been found. There are now 23 extant species. These "living fossils" are live mostly in waters between 1800 meters (6000 feet) and 6500 meters (21,000 feet) in depth, but a few occur in waters only 175 meters (575 feet) deep. Little is known about their life habitats, because

Figure 6.3 ▶

The "living fossil" *Neopilina*, a relict of the Early Cambrian and a transitional form between segmented worms and molluscs: (*A*) the segmented paired gills on either side of the foot in the center of the body; there are also paired segmented retractor muscles and other organ systems; (*right*) a modern chiton; (*B–C*) living *Neopilina*. (Courtesy J. B. Burch, University of Michigan)

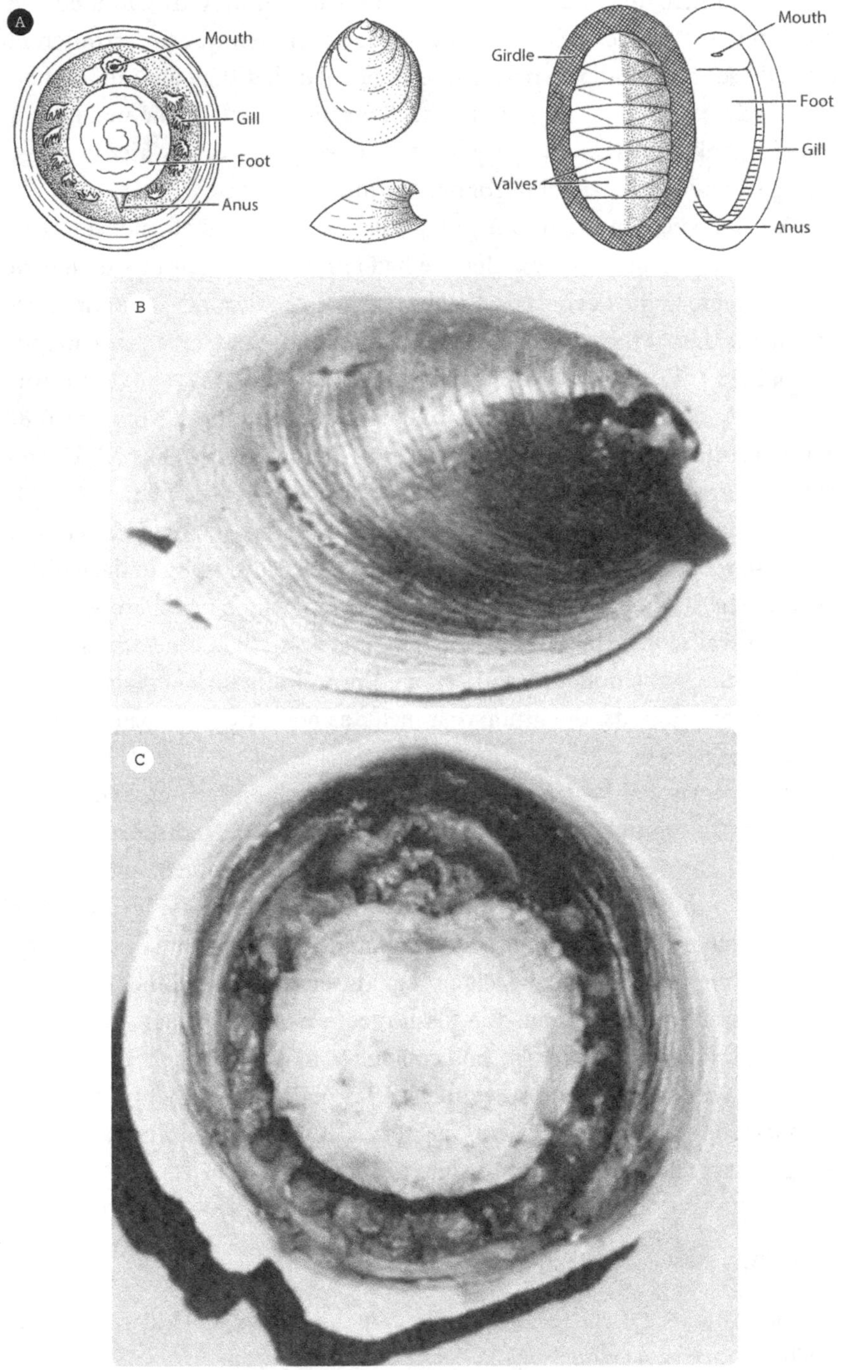
A
Mouth
Gill
Foot
Anus
Girdle
Valves
Mouth
Foot
Gill
Anus
B
C

they live in such deep water and cannot survive after they are captured and brought to the surface, where the pressure and temperature are so different from those in the deepest ocean. It is presumed that they are muddy-bottom feeders, grubbing through the seafloor muds for organic material or trapping sinking plankton, as are most creatures that live in water too deep for light to penetrate and thus for photosynthesis to occur.

How did such an important group escape the notice of science for so long? The biggest reason was that we had almost no means of studying or collecting life in the deepest part of the oceans. The Second *Galathea* Expedition was one of the earliest to undertake that task. In fact, a living monoplacophoran, *Veleropilina zografi*, had been discovered in 1896, but it was mistakenly described as an ordinary limpet and forgotten. Not until 1983 was it restudied, and scientists realized that their predecessors had seen an extant monoplacophoran long before the discovery of *Neopilina*.

Not only have 23 living species of monoplacophorans been found, but the fossil record of the class has improved as well. In addition to the earliest fossils to be studied are fossils like *Knightoconus*, which has chambers with dividing walls, like the chambered nautilus. Some paleontologists argue that it is the transitional fossil between the primitive monoplacophorans and the cephalopods, the group that includes not only nautilus but squids and octopi as well.

The discovery of *Neopilina* ranks as one of the classic examples of a mysterious fossil group long thought to be extinct that was rediscovered alive and well in the deep ocean. More important, the description of many extant and extinct monoplacophorans has shown how molluscs evolved from an ancestor shared with segmented worms, and then lost that segmentation as they diversified into snails, clams, squids, and so many other groups in this important phylum. Thus the fossil record has confirmed what anatomists and molecular biologists had concluded as a result of their research: molluscs are descended from segmented worms, and members of the class Monoplacophora are the "transitional forms" that demonstrate the macroevolutionary change from one phylum to another.

FOR FURTHER READING

Ghiselin, Michael T. "The Origin of Molluscs in the Light of Molecular Evidence." *Oxford Surveys in Evolutionary Biology* 5 (1988): 66–95.

Giribet, Gonzalo, Akiko Okusu, Annie R. Lindgren, Stephanie W. Huff, Michael Schrödl, and Michele K. Nishiguchi. "Evidence for a Clade Composed of Molluscs with Serially Repeated Structures: Monoplacophorans Are Related to Chitons." *Proceedings of the National Academy of Sciences* 103 (2006): 7723–7728.

Morton, John Edward. *Molluscs*. London: Hutchinson, 1965.

Passamaneck, Yale J., Christoffer Schander, and Kenneth M. Halanych. "Investigation of Molluscan Phylogeny Using Large-subunit and Small-subunit Nuclear rRNA Sequences." *Molecular Phylogenetics and Evolution* 32 (2004): 25–38.

Pojeta, John, Jr. "Molluscan Phylogeny." *Tulane Studies in Geology and Paleontology* 16 (1980): 55–80.

Runnegar, Bruce. "Early Evolution of the Mollusca: The Fossil Record." In *Origin and Evolutionary Radiation of the Mollusca*, edited by John D. Taylor, 77–87. Oxford: Oxford University Press, 1996.

Runnegar, Bruce, and Peter A. Jell. "Australian Middle Cambrian Molluscs and Their Bearing on Early Molluscan Evolution." *Alcheringa* 1 (1976): 109–138.

Runnegar, Bruce, and John Pojeta Jr. "Molluscan Phylogeny: The Paleontological Viewpoint." *Science*, October 25, 1974, 311–317.

Salvini-Plawen, Luitfried V. "Origin, Phylogeny, and Classification of the Phylum Mollusca." *Iberus* 9 (1991): 1–33.

Sigwart, Julia D., and Mark D. Sutton. "Deep Molluscan Phylogeny: Synthesis of Palaeontological and Neontological Data." *Proceedings of the Royal Society* B 247 (2007): 2413–2419.

Yonge, C. M., and T. E. Thompson. *Living Marine Molluscs*. London: Collins, 1976.

07 | THE ORIGIN OF LAND PLANTS | *COOKSONIA*

GROWING FROM THE SEA

The most convincing evidence of plant evolution is the record of fossil plants. Documented deep in the earth's crust are the progressive changes and modifications undergone by various groups of the plant kingdom through millions of years. Every year, students of fossil plants unearth new specimens that help piece together what paleobotanists hope some day will be a continuous story of the development of the plant kingdom from an age of more than one billion years ago to the present time. During that long period of time profound changes have occurred in the plant world. Groups have arisen, flourished, and become extinct; without the fossil record present-day botanists would be unaware that such groups of plants ever existed.

THEODORE DELEVORYAS, *MORPHOLOGY AND EVOLUTION OF FOSSIL PLANTS*

A STERILE EARTH

We look at the amazing forests and grasslands of Earth and glorify in the "green planet" that grows so much plant material that can sustain so many different kinds of animal life. But it has not always been this way. Earth was a hostile, barren place for most of its 4.5-billion-year history. There were no land plants that could live on its harsh surface, so bare rock was exposed to intense chemical weathering, releasing all its nutrients into the ocean without any marine organisms to absorb them. The only photosynthesizing organisms for the first 1.5 billion years of life's history were blue-green bacteria (cyanobacteria), which lived in the shallow waters of the oceans and formed stromatolites (chapter 1). Then, about 1.8 billion years ago, we see the first evidence of algae, which are true plants with eukaryotic cells (having a discrete nucleus for their DNA, plus organelles such as chloroplasts for their photosynthesis). Both cyanobacteria and algae continued to grow huge mats of slime on the shallow seafloor.

The extremes of heat and cold, the intensity of rainstorms and runoff without the protection of plant cover, plus the absence of an ozone layer (because of the lack of free oxygen in the atmosphere) meant that few plants could venture out of the water and onto land. As long as there was no ozone layer, both plant and animal cells would be bombarded with high levels of ultraviolet radiation, which causes mutations in genes and eventually kills cells. Only the protection of being immersed in water screens most life from ultraviolet light without the protection of the ozone layer.

Based on chemical evidence, it appears that about 1.2 billion years ago the first organisms began to colonize land. They were probably very simple associations of algae and fungi called cryptogamic soils, which are very similar to the crusts of organic material found on the desert surface when it is not disturbed. The lichens that break down bare bedrock are an example of this because lichens are not an organism, but a symbiotic association of algae and fungi. The cryptogamic soils would have been the only life on Earth's surface and would have served to help bind and stabilize the land against erosion by wind and rain, even as they helped marine algae and cyanobacteria pump more and more oxygen into the atmosphere.

Naturally, with no significant plant resources to consume on land, there was no animal life on land, either. Animal life needs not only food to eat, but also enough free oxygen in the air to breathe—which apparently did not accumulate in the atmosphere until about 530 million years ago. The combination of extreme heat and cold, lack of shelter and food, and unchecked erosion made the land a dangerous habitat that most creatures could not yet exploit.

THE FIRST LAND PLANTS

Thus the verdant planet we take for granted has not been this way for very long. For plants to begin to conquer the land, they had to be more than mats of low-growing algae, immersed in water. Algae grow well as long as they are submerged, but once they are on land, they must be kept moist or they die.

Algae must also be immersed in water to reproduce. The sperm of aquatic algae simply swim directly to the egg through the water. Green algae and

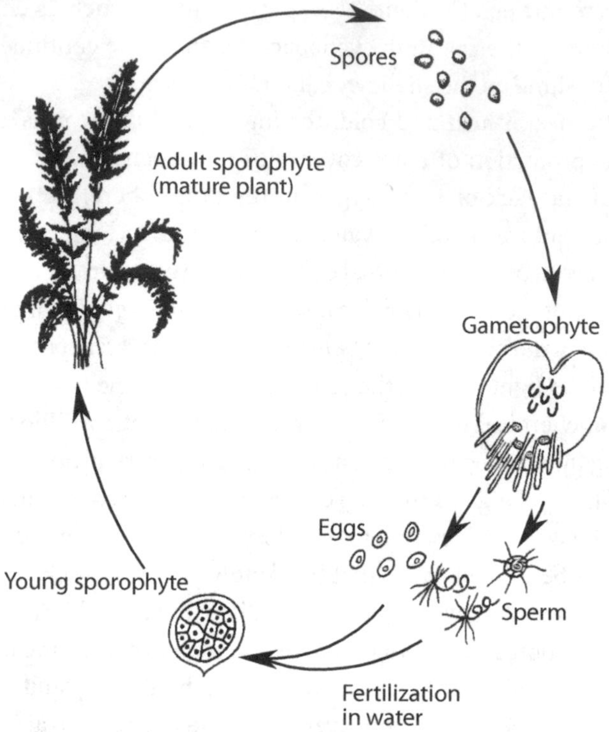

Figure 7.1 ▲
Generalized life history of a seedless vascular plant: the adult sporophyte produces spores, which grow into a gametophyte; it, in turn produces eggs and sperm, which combine to produce another sporophyte. (From Donald R. Prothero and Robert H. Dott Jr., *Evolution of the Earth*, 6th ed. [New York: McGraw-Hill, 2001])

many other primitive plants, for example, alternate between sexual generations (when haploid sperm and eggs are released) and asexual generations (when they clone themselves without using sex) (figure 7.1). The diploid (with two sets of chromosomes) plant is called a *sporophyte*, on which meiosis takes place to create spores within a *sporangium* and results in sexual reproduction. The haploid plant (with one set of chromosomes, after having gone through meiosis) is called a *gametophyte*. It generates separate sperm, eggs, or both within separate specialized structures. Alternation of generations is a common reproductive mechanism in many groups of primitive plants and animals, including most corals and anemones and sea jellies, as well as in a group of tiny shelled marine amoebas called foraminiferans.

The sporophyte in primitive land plants (such as ferns) is the visible part of the plant. It releases airborne haploid spores produced by meiosis that may land in a moist spot and germinate to form a tiny (less than 1 centimeter [0.4 inch] tall) gametophyte plant. The gametophyte bears separate sperm and eggs, and the sperm can swim to the eggs only where it is moist, which restricts the options of the most primitive land plants. This "weak link" in their reproduction prevented them from exploiting drier habitats.

The possibility of desiccation, or drying out, is another challenge faced by land plants. If it is not bathed in water, the surface of a plant dries up like a stranded alga unless it is protected by some sort of waxy covering, or *cuticle*, to conserve the water. But the cuticle also reduces water exchange on the surface, so the plant now has more difficulty taking in carbon dioxide and releasing oxygen, as well as regulating the transpiration of water vapor. Tiny pores called *stomata* provide openings through the cuticle. They can be opened or closed to regulate water and gas exchange through the cuticle. However, in the process of opening their stomata, water is lost as well.

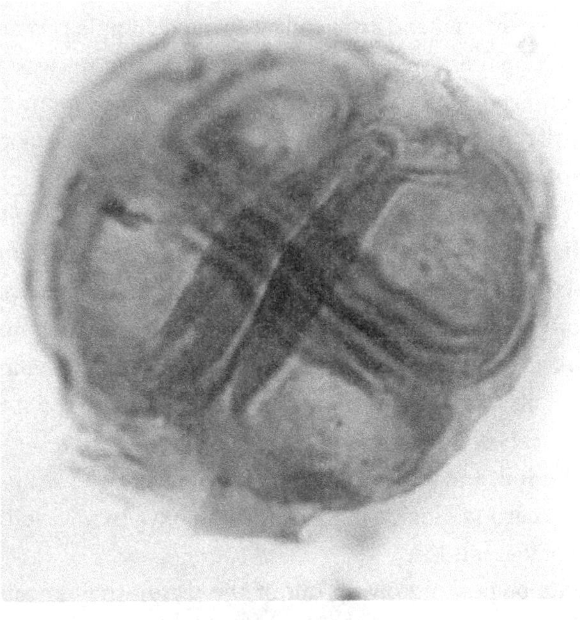

Figure 7.2 ▲

Four-part spores from the Late Ordovician of Libya, the earliest evidence of land plants. Magnification × 1500. (Photograph courtesy Jane Gray)

So what does the fossil record show about how plants invaded the land? The first fossil evidence comes from spores that came from mosses and liverworts, two low-growing plants still found in most habitats (figure 7.2). The fossil spores are Ordovician in age (about 450 million years old), although there may also be some possible spores of Middle Cambrian age (about 520 million years old). There are some 900 genera and 25,000 living species of these most primitive land plants. They have invaded nearly every land niche, even the cool moist shorelines of Antarctica. However, they cannot live in salt water. They have many key adaptations that help them survive on land, including the ability to shut down their metabolism in adverse conditions, such as drought or extreme temperatures; the tendency to grow in clumps; the capacity to propagate vegetatively through fragments that become new plants; and the ability to colonize barren areas of exposed rock where there is little soil or to grow on the surfaces of other organisms, such as trees.

UPRIGHT PIONEERS: VASCULAR PLANTS

For plants to live on land and grow tall, they need complex organ systems to transport fluids against gravity, aid in respiration, remove wastes, and support them. A marine alga such as kelp can have strands many meters long, but because all of it is constantly bathed in seawater, it does not need a system to transport water from one end to the other.

The plants that do have such systems are known as *vascular plants* because they have a network of tubes to carry fluids and nutrients from one part of the plant to another—just like our own cardiovascular system carries a fluid (blood) to all parts of our body to supply them with nutrients and take away waste products. Vascular plants, however, are being "stretched on the rack." The water and nutrients are down in the soil, but the sunlight for photosynthesis comes from above. The root end picks up nutrients and water from the soil, and moves them up to the leaves, where photosynthesis takes place (so carbon dioxide is absorbed and oxygen released), and a certain amount of water is lost.

Once plants began to grow up out of the water, they encountered two problems. First, moisture and nutrients had be transported to the higher part of the plant. Second, the plant was attempting to stand up against the force of gravity, which kept tugging it down. The solution lay in the evolution of elongate conducting cells, or *tracheids*, lined with a metabolic water by-product, *lignin*. Lignin is very rigid, thus lending support. It is also hydrophobic, with a surface that repels water rather than absorbing it (like waxed paper), thus speeding water through the tracheids. This conducting tissue occurs as a single central strand within the stem. In more advanced plants, tracheids can become massed to form larger woody trunks. Such vascular plants are formally known as tracheophytes, because they have tracheids inside them.

ISABEL COOKSON'S DISCOVERY

The earliest fossils of tracheophytes are tiny and not easily preserved, since the plants were made of soft organic material with no woody tissues that enhance the chances of preservation. None are known yet from the Ordovician, but by the Silurian (about 433 to 393 million years ago) there were simple plants known as *Cooksonia* (figure 7.3). Paleobotanist William Henry Lang named them in 1937 to honor the avid collector Isabel Cookson, who found the first specimens in Perton Quarry in Wales.

Cooksonia was about as simple as a vascular plant can be. Most of the specimens are crushed flat and show just a simple stem (usually less than 3 millimeter [0.12 inch] in diameter) that branches into two smaller stems. Most were no longer than 10 centimeters (4 inches) long. Many of the branched stems are topped by what, on the original compressed fossils, looked like small spheres, where the spores would form, so they are sporangia. However, better specimens and more detailed work has recently shown that the sporangia were not shaped like little round blobs, but more like a funnel or trumpet, with a conical opening in the center and a "lid" on top of the opening that disintegrated to release the spores (see figure 7.3*C*).

Cooksonia had no leaves. It must have performed photosynthesis through its entire surface. It certainly had no more advanced structures like seeds and flowers. Instead of individual roots, it appears that *Cooksonia* sprouted out of short horizontal connecting stems, or *rhizomes*, as do many living plants that have underground runners, creating numerous clones and reproducing vegetatively. There are dark areas along these flattened, poorly preserved specimens that may be the traces of vascular tissue, although it is not well preserved enough to be certain. In addition, at least some specimens seem to have had stomata as well, further confirming that *Cooksonia*

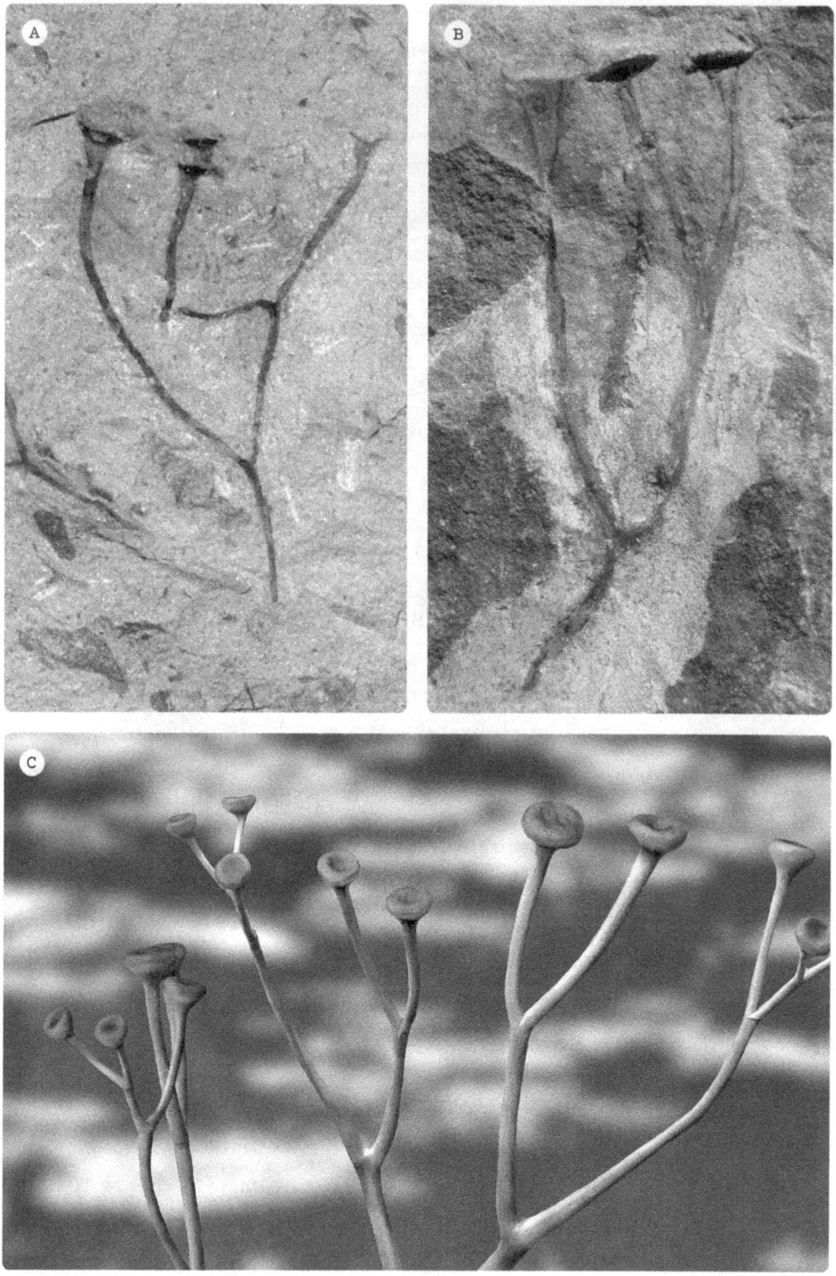

Figure 7.3 ▲

Cooksonia: (*A–B*) fossils; (*C*) reconstruction of its appearance in life, showing the funnel-shaped sporangia. ([*A–B*] courtesy Hans Steuer; [*C*] courtesy Nobumici Tamura) photosynthesized over its entire surface, while more advanced plants focus their photosynthesis in organs like leaves or needles.

At least four spore types are now associated with plants called *Cooksonia*, so most paleobotanists regard that genus as a "taxonomic wastebasket" for multiple lineages of very primitive plants. However, the preservation and details of the specimens are not good enough to confidently split *Cooksonia* into a number of genera, as taxonomy requires. Someday, however, it will be classified into multiple genera, as most other taxonomic wastebaskets are eventually.

THE GREENING OF THE PLANET

Other than paleobotanists, most people might not find such a simple tiny plant very exciting. But *Cooksonia* and the origin of vascular plants represent a monumental ecological and evolutionary breakthrough. The existence of vascular land plants and green habitats on land opened the landscape for many more opportunities, especially for animals. In the Late Ordovician, we see soils with burrows that were probably made by millipedes, most likely the first land animals of all. Then from the Silurian, there are fossils of many other land arthropods, including scorpions, spiders, centipedes, and the first wingless insects. The land was no longer barren, but was beginning to develop a complex food web of plant eaters and a diversity of predatory arthropods that ate the herbivores and one another. Finally, about 100 million years after arthropods colonized the land, the first amphibians crawled out of the water as well (chapter 10). The land was never again completely barren, but always had a green mantle of plants.

As we go through the later Silurian, there was an even greater variety of simple vascular plants. Then in the Devonian, the plants exploded in diversity, with the first forests appearing by the Late Devonian. In addition to mosses and liverworts, much more advanced plants, such as ferns, evolved. Two other important groups of living plants also appeared in the Late Silurian or the Devonian. One was the lycophytes, or "club mosses," which creep along the ground. These living fossils are low and unimpressive, but their ancestors in the Late Paleozoic grew in gigantic forests made up mostly of "club mosses" more than 36 meters (118 feet) tall, the largest land plants the world had ever seen up to that point (figure 7.4).

Figure 7.4 ▲

Lycophytes: (*A*) living *Lycopodium*, or "club mosses," which today are mostly small, low-growing plants; (*B*) reconstruction of the 50-meter (164-foot) *Lepidodendron*, a lycophyte tree that grew in the swamps of the Carboniferous; the details of the trunk, bark, leaves, cones, spores, and seeds are reconstructed from isolated finds. (Courtesy Bruce Tiffney)

The other important new group was the "horsetails," "scouring rushes," or sphenopsids (figure 7.5). Today, these primitive plants (one genus, a living fossil called *Equisetum*) grow in great abundance in sandy and gravelly soils close to water. Their fibrous stems contain tiny particles of abrasive silica, so they are hard for animals to eat. Early pioneers called them "scouring rushes" because a crushed handful of them made a good scouring pad for pots and pans. Horsetails are very distinctive because each long hollow stem segment is covered by a series of flutings or ridges along its length and is separated from adjacent segments by a distinct joint, from which all the leaves sprout. Each horsetail stem branches from a rhizome, which sprouts many clones through vegetative reproduction. *Equisetum* is a notoriously tough plant and grows rapidly in the right habitat. It quickly invades the wet parts of an entire garden if not kept in its own pot, and its underground stem is almost impossible to eliminate, so a horsetail comes back no matter what happens to it. The extinct sphenopsids of the Carbon-

Figure 7.5 ▲

Sphenophytes: (*A*) the giant Carboniferous horsetail *Calamites*, which reached 20 meters (65 feet) in height; (*B*) living *Equisetum*, showing the leaves radiating from the joints in the stems. (Courtesy Bruce Tiffney)

iferous included horsetails that were over 20 meters (66 feet) in height (see figure 7.5*A*).

In addition to all these primitive spore-bearing plants, the Late Devonian yields the first plants that reproduced with seeds, which have a hard coating that helps them germinate without being immersed in water. Some of these extinct "seed ferns" (not true ferns, but a more advanced fern-like plant that bore seeds) formed the first large trees, up to 12 meters (40 feet) in height.

The Devonian forests of "seed ferns" were succeeded in the Carboniferous (360 to 303 million years ago) by a gigantic explosion in the diversity of ferns, mosses, club mosses, horsetails, and "seed ferns." The Carboniferous coal swamps in which they grew produced huge volumes of vegetation across large areas of the tropical areas of North America, Europe, and Asia. When these plants died and sank into the muck, they were not quickly reduced to nothing, as happens in swamps today. There were almost no animals (like termites) that had evolved to digest the hard woody tissues of lignin that made up the trees, so they just accumulated without decomposing, were compressed and subjected to high temperatures, and turned into coal.

This enormous volume of organic matter locked into the coal in the crust then transformed Earth's atmosphere and climate. As the coal accumulated, it pulled carbon dioxide out of the atmosphere and sealed it inside the planet's crust. Soon, the "greenhouse" climate of the Early Carboniferous (with ice-free poles, high carbon dioxide, and high sea levels that drowned most of the continents) was transformed into an "icehouse" climate by the Late Carboniferous (with ice caps on the South Pole, lower carbon dioxide, and much lower sea levels as all the polar ice pulled water out of the ocean basins). Earth remained in the grip of these "icehouse" conditions for almost 200 million years longer, until the Middle Jurassic (middle part of the Age of Dinosaurs), when it flipped back from "icehouse" to "greenhouse" due to huge changes in the mantle and in the ocean basins (chapter 14).

The cycle from "greenhouse" to "icehouse" climate and back again has happened several times over the past billion years of the planet's history. In fact, the existence of plants and animals is why Earth is habitable, and not a runaway "greenhouse" like Venus or a frozen "icehouse" like Mars. Earth's living systems produce carbon reservoirs in the form of limestones (mostly by animals) and coals (by plants) that lock up carbon dioxide in the crust. This acts as a thermostat, preventing the planet from becoming a runaway "greenhouse" or a runaway "icehouse."

Sadly, we have been unintentionally changing the planet by undoing this natural cycle. Since the beginning of the Industrial Revolution, we have burned many millions of tons of coal and released the carbon dioxide once locked in it. Now that carbon dioxide is out of control, driving our human-induced "super-greenhouse" at rates never seen in the geological past. Without knowing it, we have upset the delicate balance of carbon in Earth's atmosphere, oceans, and crust. Our planet is already showing the extreme weather events that come from climate change, and our children and grandchildren will be paying the price for the dangerous experiment we performed when we broke the planetary thermostat.

SEE IT FOR YOURSELF!

Very few museums have displays about the earliest plants. The Field Museum of Natural History, in Chicago, has specimens of *Rhynia*, a close relative of *Cooksonia*, on display, as well as excellent fossils and dioramas of the coal-swamp forests. The Denver Museum of Nature and Science has an exhibition on primitive plants and a diorama of a coal swamp, as does the National Museum of Natural History, Smithsonian Institution, in Washington, D.C.

The oldest forest on Earth grew near present-day Gilboa, New York, during the Devonian (380 million years ago), and fossils of various parts of the trees are displayed at the Gilboa Museum (http://www.gilboafossils.org), Gilboa Town Hall, and New York State Museum, Albany.

FOR FURTHER READING

Gensel, Patricia G., and Henry N. Andrews. "The Evolution of Early Land Plants." *American Scientist* 75 (1987): 468–477.

Gray, Jane, and Arthur J. Boucot. "Early Vascular Land Plants: Proof and Conjecture." *Lethaia* 10 (1977): 145–174.

Niklas, Karl J. *The Evolutionary Biology of Plants.* Chicago: University of Chicago Press, 1997.

Stewart, Wilson N., and Gar W. Rothwell. *Paleobotany and the Evolution of Plants.* 2nd ed. Cambridge: Cambridge University Press, 1993.

Taylor, Thomas N., and Edith L. Taylor. *The Biology and Evolution of Fossil Plants.* Englewood Cliffs, N.J.: Prentice-Hall, 1993.

A FISHY TALE

Gill-slits, tongue bars, synapticulae
Endostyle and notochord: all these you will agree
Mark the protochordate from the fishes in the sea,
And tell alike for them and us their lowly pedigree.
Thyroid, thymus, subnotochordal rod;
These we share with lampreys, the dogfish and the cod-
Relics of the food-trap that served our early meals,
And of tongue-bars that multiplied the primal water-wheels.

WALTER GARSTANG, *LARVAL FORMS WITH OTHER ZOOLOGICAL VERSES*

HUGH MILLER AND THE OLD RED SANDSTONE

We—along with all other mammals, as well as birds, reptiles, amphibians, and fish—are *vertebrates*, animals with backbones. Where do the vertebrates come from? What do the oldest fossil fish show us about the origin of our phylum? To find that answer, we must go back to Scotland at the end of the eighteenth century.

In the late eighteenth century, the young science of geology began to emerge, primarily in Great Britain. The pioneering Scottish naturalist James Hutton first laid the foundations of modern geology with his trips around Scotland. Eventually, he published *Theory of the Earth* (1788), and the scientific approach to understanding the Earth was born.

One of the British rock units that Hutton studied extensively is a thick sequence of gritty rocks known as the Old Red Sandstone. It is widely exposed in Scotland and is found in many places in nearly all of eastern and central England as well. The more Hutton looked at it, the more he could see evidence of a huge mountain range that had been eroded away and deposited in streams and rivers to form the gravels and sandstones of the Old Red. In many places, it lies almost horizontally across an erosional surface cut into older rocks that were first tilted on their side and then eroded off after being turned from horizontal to vertical. This example of an angular unconformity convinced Hutton that the world was unimaginably old, "no vestige of a beginning," in his words. It was not, as conventional people thought back then, only 6000 years old, as suggested by the Bible.

Hutton's insights were not far off. Today, we can date the Old Red Sandstone to the Devonian (about 400 to 360 million years ago). The tilted rocks beneath the unconformity are Silurian in age (about 425 million years old). The collision that produced the tilting of the Silurian rocks occurred during the Caledonian Orogeny (after Caledonia, the Roman name for Scotland), which was caused when the core of Europe (known as the Baltic platform) collided with what is now northeastern Canada and Greenland. This huge mountain-building event crumpled all the Silurian rocks formed just before it occurred; the resulting Caledonian Mountains then eroded, producing river sands that eventually became the Old Red Sandstone. (The Catskill Sandstone in New York State was similarly formed by the erosion of the Acadian Mountains, which formed a belt with the Caledonian Mountains.)

A generation after Hutton, the Old Red Sandstone became famous thanks to the attention of a humble Scottish stonemason named Hugh Miller. He was the son of a sea captain, but attended school only until age 17, so he never had the formal education required to study fossils seriously. Pictures of him show a burly man with broad, strong shoulders (probably from working stone for many years), a thick bushy curls, and curly sideburns as well (figure 8.1) . Miller spent his younger years working the rock quarries, especially in the Old Red Sandstone. During the slack months in the quarries, he combed the seashore exposures of the Old Red Sandstone, where he found beautiful fossilized fish, one after another. Others working in the Old Red Sandstone soon had collected many specimens as well, so Miller set out to study them. By 1834, the silica dust from the quarries was beginning to destroy his lungs, so he quit the stonemason's life and moved to Edinburgh to be a banker and writer.

Even though he had a limited education, he became one of the first popular writers in the history of paleontology. In 1834, he published *Scenes and*

Figure 8.1 ▲
A contemporary portrait of Hugh Miller. (Courtesy Wikimedia Commons)

Legends of North Scotland, which was a best-selling popularization of the geology and natural history of Scotland, written for the ever-expanding audience for natural history books at that time. He followed that work in 1841 with *The Old Red Sandstone, or, New Walks in an Old Field*, which describes the rock unit and its amazing fossil fishes and "sea scorpions," fully illustrated by Miller as well (figure 8.2). This passage captures his style perfectly:

> Half my closet walls are covered with the peculiar fossils of the Lower Old Red Sandstone; and certainly a stranger assemblage of forms have rarely been grouped together; creatures whose very type is lost, fantastic and uncouth, and which puzzle the naturalist to assign them even their class;—boat-like animals, furnished with oars and a rudder;—fish plated over, like the tortoise,

> above and below, with a strong armor of bone, and furnished with but one solitary rudder-like fin; other fish less equivocal in their form, but with the membranes of their fins thickly covered with scales;— creatures bristling over with thorns; others glistening in an enamelled coat, as if beautifully japanned—the tail, in every instance among the less equivocal shapes, formed not equally, as in existing fish, on each side of the vertebral column, but chiefly on the lower side—the column sending out its diminished vertebrae to the extreme termination of the fin. All the forms testify of a remote antiquity—of a period whose "fashions have passed away."

Miller's books soon made him a celebrity among natural historians, but he was not a trained paleontologist. Luckily, he met the legendary Swiss fish paleontologist Louis Agassiz at a meeting of the British Association for the Advancement of Science. Then he gave his specimens to someone who could analyze them, and Agassiz soon named and described all of Miller's remarkable fossils

Miller used his books to assert his religious views and to fight the creeping tendency for French evolutionary thinking to blossom in Britain. His

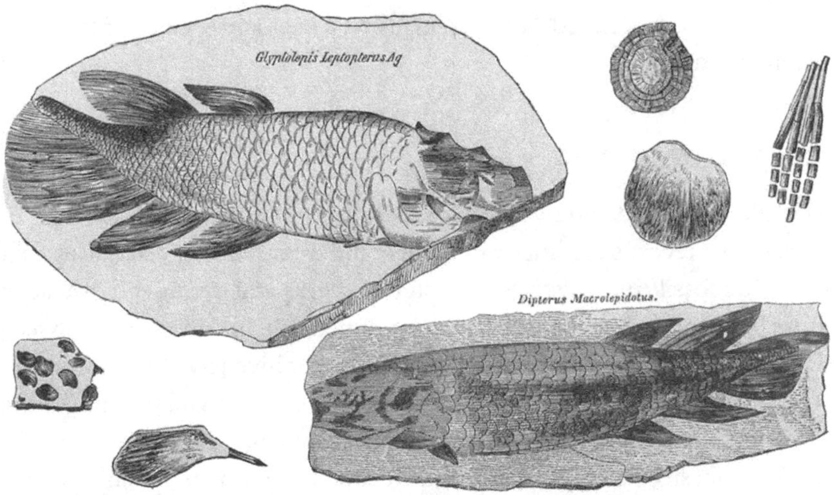

Figure 8.2 ▲

The lobe-finned fish *Glyptolepis* (distantly related to amphibians) and the fossil lungfish *Dipterus*. (Plate from Hugh Miller, *The Old Red Sandstone, or, New Walks in an Old Field* [Edinburgh: Johnstone, 1841])

book *The Foot-prints of the Creator: or, The* Asterolepis *of Stromness*, published in 1849, was an attack on the sensational evolutionary ideas propounded by Scottish publisher Robert Chambers in his book *Vestiges of the Natural History of Creation*, published in 1844.

But Miller was no biblical literalist. Like most British geologists at the time, he viewed Noah's flood as a local event in Mesopotamia and the fossil record as showing a series of creations and extinctions that are not mentioned in the Bible. Although he admitted that the fossil record shows changes through time, he denied that later species were descended from earlier ones.

Unfortunately, in 1856, at the age of 54, he began to suffer from mysterious severe headaches and mental illness, and he shot himself in the chest just after sending the proofs of his last book, *The Testimony of the Rocks*, to the publisher. The scientific world mourned him, and he had one of the largest funeral processions in the history of Edinburgh. Sir David Brewster wrote this about him: "Mr. Miller is one of the few individuals in the history of Scottish science who have raised themselves above the labors of an humble profession, by the force of their genius and the excellence of their character, to a comparatively high place in the social scale." Numerous fossils are named after him, including the "sea scorpion" *Hughmilleria* and the primitive fish now called *Millerosteus*, as well as many species of fish with the name *milleri*.

THE AGE OF FISHES

The Old Red Sandstone was deposited during the Devonian, the Age of Fishes, so it records the huge radiation of different types of fish during that time. The fossils include those not only of sharks and ray-finned fish, such as we have today, but also of many lobe-finned fish, including lungfish (see figure 8.2). There was an entire radiation of primitive jawed fish known as placoderms, which had plates of armor over their head and thorax. All the placoderms were extinct by the end of the Devonian.

These fossils also included the first evidence of a huge radiation of armored jawless fish. In the 1830s and 1840s, Agassiz described several of them, including *Pteraspis* and *Cephalaspis* (figure 8.3). Miller claimed that his fish fossils showed no evidence of evolution, but he was not enough of an anatomist to know what he was talking about. Nevertheless, the striking presence of these jawless vertebrates in the Devonian was an indication

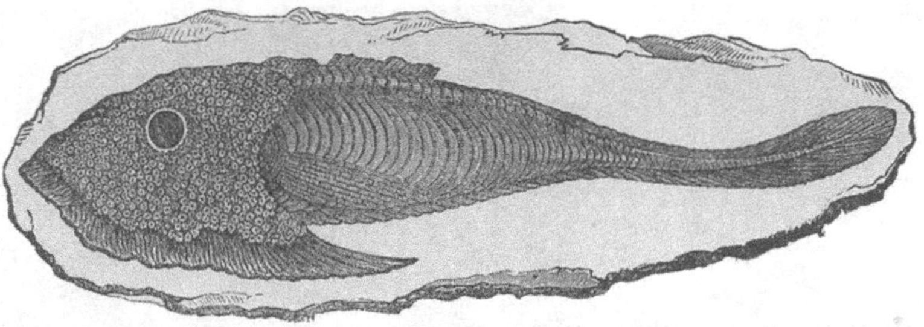

Figure 8.3 ▲

The armored jawless fish *Cephalaspis*. (Plate from Hugh Miller, *The Old Red Sandstone, or, New Walks in an Old Field* [Edinburgh: Johnstone, 1841])

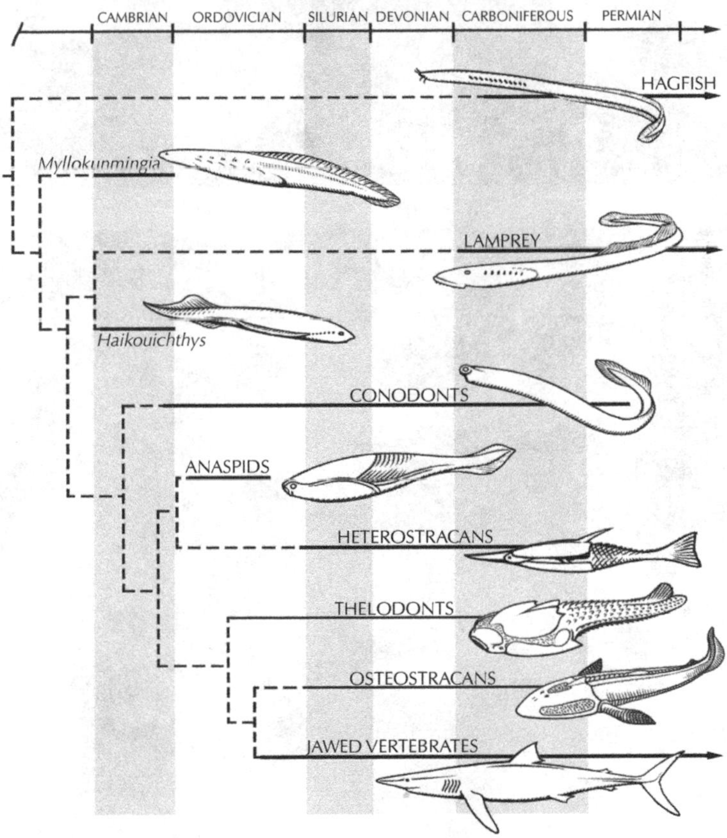

Figure 8.4 ▲

Family tree of jawless fish, showing the different groups. (Drawing by Carl Buell; from Donald R. Prothero, *Evolution: What the Fossils Say and Why It Matters* [New York: Columbia University Press, 2007], fig. 9.8)

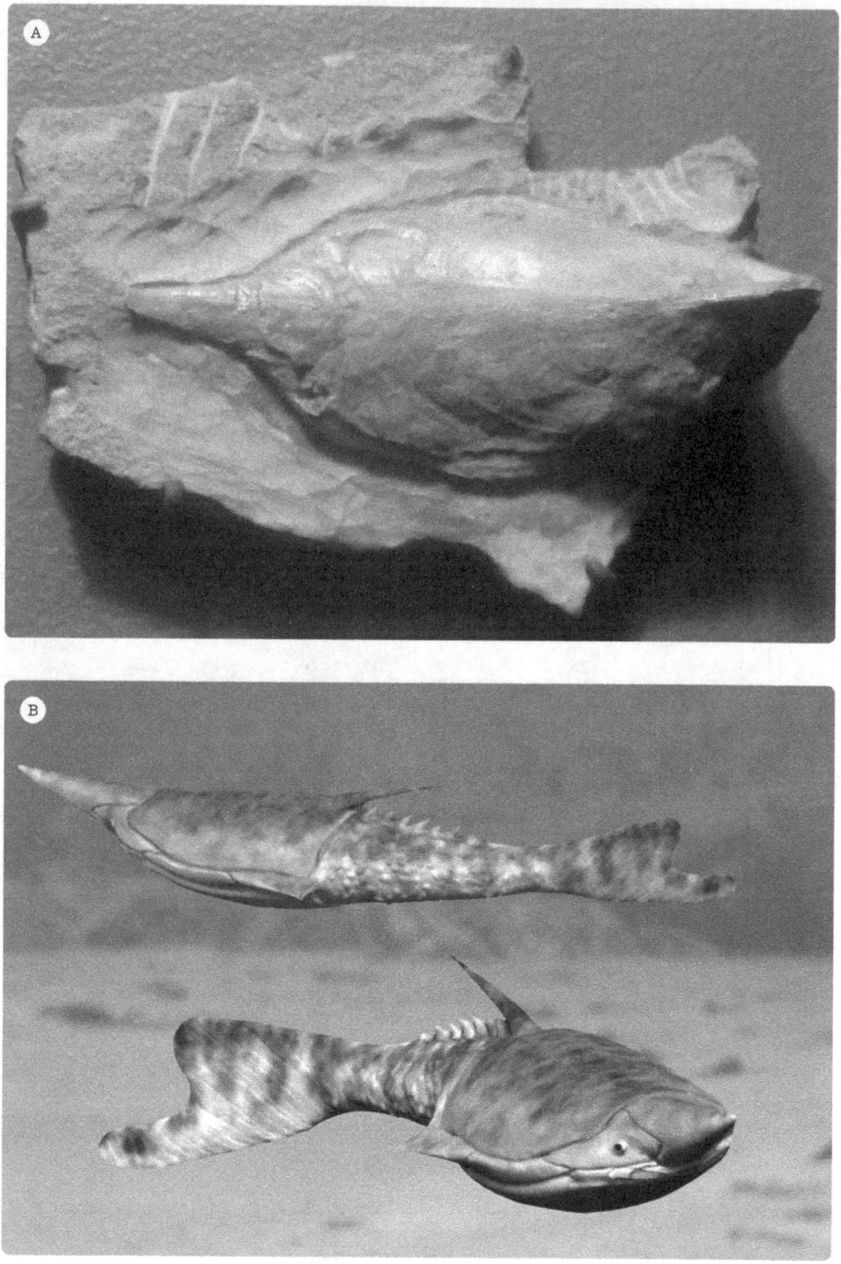

Figure 8.5 ▲

The jawless armored fish *Pteraspis*, a heterostracan: (*A*) head shield; (B) reconstruction of its appearance in life. ([*A*] courtesy Wikimedia Commons; [*B*] courtesy Nobumichi Tamura) that there were several steps in the evolution of modern jawed fish from jawless invertebrates.

Fossils of these armored jawless fish were soon discovered in many other localities, and they provided further evidence of how jawed vertebrates had evolved from jawless ancestors (figure 8.4). *Pteraspis* and its relatives (the heterostracans) tended to have streamlined, torpedo-shaped bodies covered in armor, often with long spines protruding from the sides or back, and a tail with the main lobe pointed downward (figure 8.5). Heterostracans had just a tiny slit-like mouth and no jaws, nor did they have strong muscular fins for steering, so they are thought to have swum like tadpoles, sucking in water and filter feeding on the particles in the water as it passed into their mouth and over the gills. By contrast, *Cephalaspis* (see figures 8.3 and 8.4) and its relatives (the osteostracans, or "ostracoderms") had a domed head with a flat bottom and a tail with the main lobe pointed upward (like in modern sharks). They are thought to have cruised along the bottom, grubbing for food in the mud as it was sucked through their jawless mouths.

FISHING BACK IN TIME

Over the years, more and more fossils of these armored jawless fish were found in Devonian and, eventually, Silurian beds around the world. But the only part of them that was easily fossilized was their external bony armor. Like sharks and most primitive fish, they did not have a bony skeleton. Instead, they had a skeleton made of cartilage, which does not fossilize well. If it were not for their armor, almost none of these fish would appear in the fossil record at all.

For the longest time, there was no evidence of jawless fish (or any other kind of fish) before the Silurian. The Ordovician seas were dominated by large predators, such as the 5.5-meter (18-foot) long nautiloids, but despite the abundant record of Ordovician marine fossils, not a trace of bone could be found. About the only clues were rare occurrences, such as in the Harding Sandstone near Canyon City, Colorado, which dates to the Middle Ordovician and is full of tiny pieces of the bony armor of a jawless fish called *Astraspis*. By the 1970s and 1980s, however, complete specimens of these earliest vertebrates had been found, such as *Arandaspsis* from Australia and *Sacambaspis* from South America (and now Australia as well).

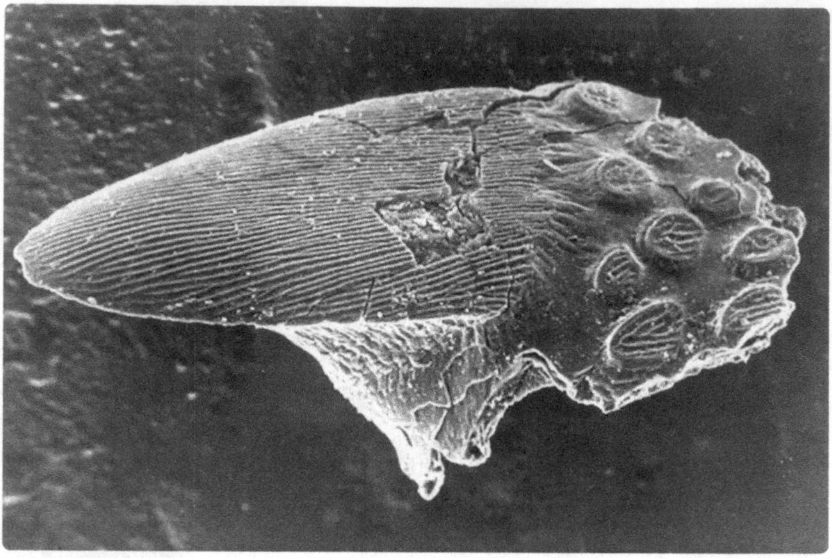

Figure 8.6 ▲

Isolated small plate fragment (about 1 millimeter in diameter) from the dermal armor of the Cambrian jawless fish *Anatolepis*, one of the earliest vertebrates to produce bone. (Courtesy U.S. Geological Survey)

All these Ordovician jawless fish can be described as little more than simple suction tubes of a filter-feeding fish, covered with tiny plates of bony armor. They had broad flat bodies with almost no fin protrusions or spikes of any kind, a broad slit-like mouth for sucking in food-rich water, and a simple asymmetrical tail. Instead of the plate-like armor found in *Pteraspis*, these fish were covered with hundreds of tiny pieces of bone, somewhat like chain-mail armor. They had tiny eyes and a series of canals on the outside of the body (lateral lines) that fish use for sensing motion in the water around them. All these Ordovician fish are extremely rare compared with fossils of most other animals of the time. Even more frustrating, none of them were known from the Cambrian.

Finally, in the 1970s, Jack Repetski, a paleontologist with the U.S. Geological Survey, was working on tiny microfossils known as conodonts from the Deadwood Sandstone of Wyoming, which dates to the Late Cambrian. While dissolving out the calcareous fossils to find the conodonts (which are made of calcium phosphate, just like vertebrate bone), he found some funny-shaped pieces that he realized were dermal armor from a jawless fish called *Anatolepis* (figure 8.6). Although there was a long argument as to whether the specimens were really from a vertebrate, this has been resolved and *Anatolepis* is currently the oldest known vertebrate for which we have bony-tissue fossils.

CONNECTING THE LINKS

Thus the trail of finding fossils of vertebrates in older and older rocks goes cold once we are in rocks formed before bone evolved. To date, the pieces of dermal armor from *Anatolepis* are still the oldest fossils known from bony specimens. Any older animals were soft-bodied, made of cartilage and softer tissues, and would have been very unlikely to fossilize, except in the best of conditions.

Since there was no further evidence to come from bony fossils, biologists and paleontologists trying to connect the dots between the vertebrates and their ancestors decided to work from the bottom up instead.

Here, we have an abundant record because many of the transitional animals that link vertebrates to the rest of the animal kingdom are still alive—and many have left behind abundant fossils as well. Mammals, birds, reptiles, amphibians, and fish are members of the phylum Chordata. Chordates are so named because as embryos (and sometimes as adults), they have a long flexible rod of cartilage (*notochord*) along their back to support their body; the notochord is the predecessor of the backbone.

The nearest relatives of the Chordata come from a different group, the phylum Hemichordata (half chordates) (figure 8.7). Today, they are represented by the acorn worms and the pterobranchs. Acorn worms (enteropneusts) look vaguely like any other worm to the casual onlooker, but they have the embryonic precursor of the notochord and the true throat region (*pharynx*) shared by all chordates. In addition, their nerve cord runs along their back, while their digestive tract runs along their belly, the configuration found in chordates and the opposite of what is found in most invertebrates (nerve cord along the belly, digestive tract along the back). These anatomical similarities are supported by an embryology like that of chordates. Finally, molecular analyses of their DNA shows that they are very close to the common ancestor of vertebrates plus their nearest invertebrate relatives, the echinoderms (sea stars, sea cucumbers, sea urchins, and their relatives).

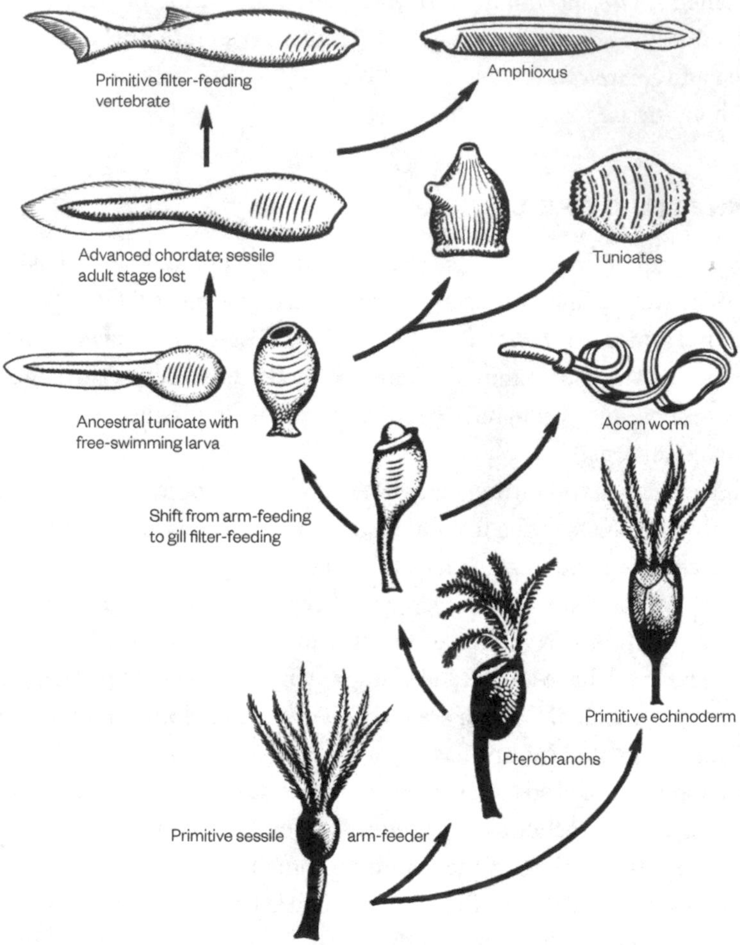

Figure 8.7 ▲

The evolution of chordates from invertebrates, as originally conceived by Walter Garstang and Alfred S. Romer more than a century ago. Many of the adult body forms were evolutionary dead ends (such as adult tunicates), but the larval tunicate retains the long tail and other features that led to more advanced chordates. (Drawing by Carl Buell; from Donald R. Prothero, *Evolution: What the Fossils Say and Why It Matters* [New York: Columbia University Press, 2007], fig. 9.4)

The next step toward vertebrates is a group represented by more than 2000 species all over the world's oceans: the tunicates, or "sea squirts" (see figure 8.7). Like acorn worms, sea squirts do not look much like a fish to the casual viewer, but surface appearances are deceiving. The adults are unimpressive, just a little sac of jelly that filters seawater through a basket that makes up their body. But the larvae of sea squirts look very much like fish or tadpoles, with a well-developed notochord, a long muscular tail with paired muscles, and a head end with a large pharynx, among many other key features. Once again, it is the embryological evidence that shows us the pathway. This is confirmed by the molecular evidence, which clearly demonstrates that tunicates are more closely related to vertebrates than are any other invertebrate in the sea.

The final stage linking invertebrates to vertebrates is another inconspicuous creature in the oceans: the lancelet, or amphioxus (*Branchiostoma*) (see figure 8.7). This insignificant sliver of flesh is only a few centimeters long, but a close examination shows that it is extremely fish-like without being a true fish. Lancelets have a long flexible notochord that supports their entire body, with numerous *V*-shaped muscle bands down the length of the body, which makes them good swimmers. The nerves run along the back and the digestive tract along the belly, as in all chordates. They do not have jaws or teeth, but their mouth leads to a pharynx and a "gill basket," which traps food particles. They do not have true eyes, but a light-sensitive pigment spot on the front helps them detect light and shadows. These creatures live with their tail end burrowed into the seafloor, leaving only their head sticking out in order to catch floating food particles.

Finally, several good fossils of lancelets show that they were around in the Early Cambrian, just as fish evolution was getting started. These include *Pikaia* from the Burgess Shale of Canada (chapter 6) and a similar fossil, *Yunnanozoon*, from the Chengjiang fauna of China, which dates to the Early Cambrian (518 million years ago).

THE FISHY LINK

We have traced the ancestry of vertebrates back to jawless fish from the Ordovician through the Devonian, with the oldest evidence of bone coming from the Late Cambrian. But the oldest fish was soft-bodied, so there is no further evidence to be obtained from fossils of bone. We have climbed up from the base of the soft-bodied chordate tree—from hemichordates like acorn worms; through tunicates; to lancelets, which are almost completely fish-like, but lack crucial anatomical traits (such as a distinct "head," a two-chambered heart, and a key embryological feature called neural crest

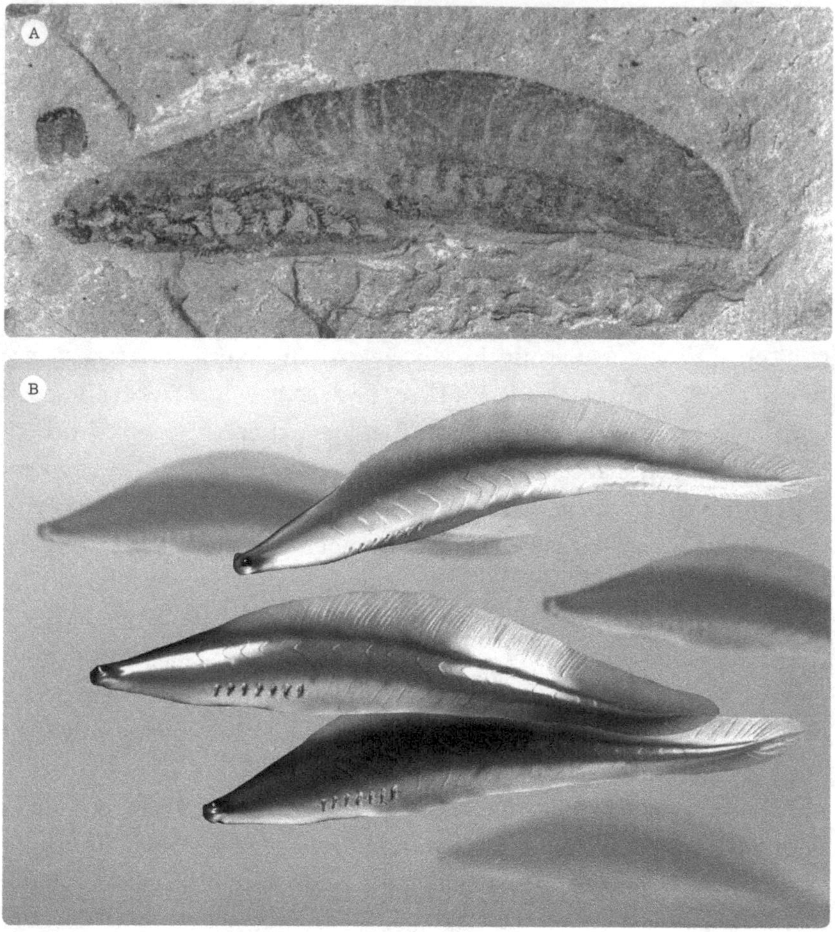

Figure 8.8 ▲

Haikouichthys: (*A*) fossil; (*B*) reconstruction of its appearance in life. ([*A*] courtesy D. Briggs; [*B*] courtesy Nobumichi Tamura)

cells) that define them as vertebrates. All we need is an animal that was soft-bodied, had most of the vertebrate features, but still lacked bony armor of any kind—and the connection is complete.

Sure enough, in 1999 a group of Chinese scientists plus Simon Conway Morris reported fossils called *Haikouichthys* (fish from Haikou) from the Early Cambrian (518 million years old) Chengjiang fauna of China (which also produced *Yunnanozoon*, the fossil lancelet). This tiny fish was barely 2.5 centimeters (1 inch) long, but its fossils preserve some remarkable features (figure 8.8). They clearly show a distinct head (unlike any lancelet), with a series of up to nine discrete gills and gill slits behind the head. There is a short notochord, and the long cylindrical body has a broad dorsal fin running down the middle of the back to the tail and a ventral fin on the base of the tail. The fins are supported by radial cartilages, as in such other jawless fish as lampreys and hagfish.

The same report describes an even more primitive fish-like fossil from the Chengjiang fauna of China. Named *Myllokunmingia*, it also appears to have a discrete head and a skull made of cartilage, five or six gill slits behind its head, a notochord down its back, and a long sail-like dorsal fin running

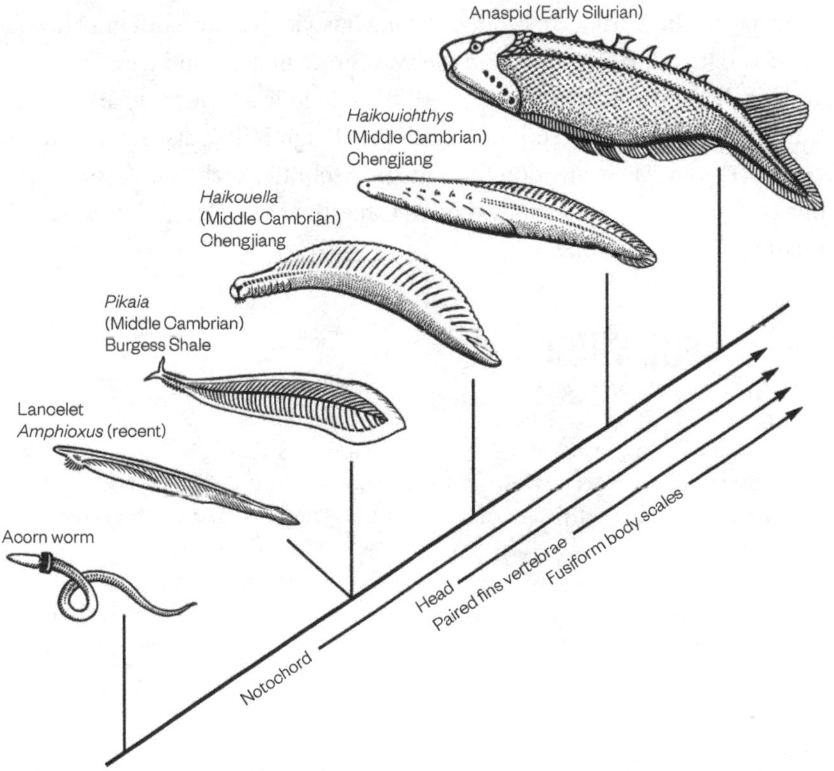

Figure 8.9 ▲

The evolutionary steps from acorn worm to lancelet to *Haikouella* and *Haikouichthys*, culminating with the bony jawless anaspids. (Drawing by Carl Buell, based on D.-G. Shu et al., "Lower Cambrian Vertebrates from South China," *Nature*, November 4, 1999; by permission of the Nature Publishing Group) from its head to the tip of its tail, with a set of paired ventral fins beneath its tail. There is only a single specimen, and it is not well preserved, so it is tough to be certain about what it is. But the available features suggest that it was an even more primitive chordate than *Haikouichthys*.

Finally, a third creature from the same Lower Cambrian beds is *Haikouella*. It is more than 20 to 40 millimeters (0.8 inch to 1.5 inches) long and is known from more than 300 specimens. It clearly has a head, a brain, gills, a notochord supporting well-developed trunk muscles running to its tail, a heart with a circulatory system, and a long dorsal fin going from trunk to tail, with small ventral fins below the tip of its tail. Some specimens show the possibility of eyes on the side of their head, a first for chordates if it is true.

In short, the Early Cambrian of China has yielded a wealth of soft-bodied chordates that are clearly on the vertebrate lineage and were more advanced than lancelets (figure 8.9). All they need is a little bony armor, and they become the armored jawless fish that Hugh Miller discovered almost 200 years ago. The transition from an invertebrate, such as an acorn worm or a tunicate, to the first unquestioned fish is now complete, with no gaps or missing fossils along the line.

SEE IT FOR YOURSELF!

None of the Chinese fossils from the Early Cambrian are on display in any museum. However, many museums have excellent displays of early fossil fish, including the American Museum of Natural History, New York; Cleveland Museum of Natural History; Field Museum of Natural History, Chicago; and National Museum of Natural History, Smithsonian Institution, Washington, D.C. The Elgin Museum in Elgin, Scotland, has the largest collection on display of fish and other fossils from the nearby Old Red Sandstone, as well as a large archive of Hugh Miller's papers, books, and notes (http://elginmuseum.org.uk/museum/collections-fossils/).

FOR FURTHER READING

Forey, Peter, and Philippe Janvier. "Evolution of the Early Vertebrates." *American Scientist* 82 (1984): 554–565.

Gee, Henry *Before the Backbone: Views on the Origin of Vertebrates*. New York: Chapman & Hall, 1997.

Long, John A. *The Rise of Fishes: 500 Million Years of Evolution*. Baltimore: Johns Hopkins University Press, 2010.

Maisey, John G. *Discovering Fossil Fishes*. New York: Holt, 1996.

Moy-Thomas, J. A., and R. S. Miles. *Palaeozoic Fishes*. Philadelphia: Saunders, 1971.

Shu, D.-G., H.-L. Luo, S. Conway Morris, X.-L. Zhang, S.-X. Hu, L. Chen, J. Han, M. Zhu, Y. Li, and L.-Z. Chen. "Lower Cambrian Vertebrates from South China." *Nature*, November 4, 1999, 42–46.

09 | THE LARGEST FISH | *CARCHAROCLES*

MEGA-JAWS

Sharks have everything a scientist dreams of. They're beautiful–God, how beautiful they are! They're like an impossibly perfect piece of machinery. They're as graceful as any bird. They're as mysterious as any animal on earth. No one knows for sure how long they live or what impulses–except for hunger–they respond to. There are more than two hundred and fifty species of shark, and everyone is different from every other one.

PETER BENCHLEY, *JAWS*

A VISIT TO SHARKTOOTH HILL

When I was growing up in southern California in the late 1950s and the 1960s, I was hooked on dinosaurs and other fossils. By the time I was a Cub Scout, I had made trips to most of the important fossil localities near my home, including the shell beds at Topanga Canyon and the mammal-bearing deposits at Red Rock Canyon, both of which date to the Miocene. But again and again, I heard stories about the legendary Sharktooth Hill near Bakersfield, where the shark teeth and marine fossils were deposited in the deep waters of the ancient Central Valley of California about 16 to 15 million years ago. Yet no one knew how anyone could go there to collect, since most of the bone bed was on private land behind fences that were marked "No Trespassing."

My career moved on to other things through the ensuing 30 years until about 1997, when I heard from colleagues that a local rancher, Bob Ernst, allowed crews of students from schools and researchers from nonprofit organizations to collect on his land. Eventually, I reached Ernst directly, and soon it was a standard stop for my Occidental College paleontology class (and one or two Caltech paleontology classes) to visit Sharktooth Hill as a class field trip. In 2002, I realized that there was a lot more research to be undertaken at Sharktooth Hill than had been accomplished. My students and I used a technique called magnetic stratigraphy, measuring the changes in Earth's magnetic field as recorded in the rocks, in order to date the beds more precisely than ever before. I collaborated with Larry Barnes of the Natural History Museum of Los Angeles County (a veteran of Sharktooth Hill since the early 1960s) and many others to identify a wide range of land mammals that had drifted out into deep water and had been entombed and then fossilized with sharks and marine mammals. All these studies have been published (mostly in 2008) with student coauthors, so our understanding of the deposit is better than ever.

A visit to the legendary bone bed is an eye-opener. First, you drive past one huge oil field after another as you travel northeast out of Bakersfield and toward the foothills of the Sierra Nevada. The oil fields around Bakersfield are still very active and among the largest in California. Eventually, you reach a turn-off from a dirt road to a ranch gate, which you must open and close using the secret code for the lock. Another mile or two on another dirt road across the low, rolling scrub- and grass-covered hills, and suddenly you see areas that have been scraped bare by a bulldozer. You jump out, grab your gear, and plunk down flat on the surface of the bone bed.

For tools, mostly you need just an awl or a similar tool to probe the soft sand, plus a whisk broom or paint brush to dust it off. Every once in a while, Bob Ernst would hire a bulldozer to come in, scrape the "overburden" of unfossiliferous rock off the top of the bone-bearing layer, and then leave it exposed for future work. Many people also wear a dust mask as well, because the soils in the area can carry the San Joaquin Valley fever (coccidioidomycosis), a fungal disease from spores in the soil that can make you very sick. On most days, you must wear a hat and loose clothing for protection against the blazing sun and slather yourself with sunscreen, and a good stadium pillow or cushion is wise, as you will sit on the hard surface for hours.

But what rewards it yields! The bone bed is made of solid bone fragments and teeth (more than 200 specimens per 1 cubic meter [35 cubic feet] of rock) and an occasional whale skull or skeleton, all surrounded by a loose sandy matrix that is relatively easy to brush away. No hard chisels or chipping away with a rock hammer required! Each scoop or probe loosens more small shark teeth. And gloves are helpful because the tips of the shark teeth are still sharp and can still cut unprotected fingers if you are careless probing through the sand. At Sharktooth Hill, the sharks may be long extinct, but they still bite!

The teeth are overwhelmingly from different types of mako sharks (*Isurus*), although teeth from some 30 other species of shark are known (figure 9.1). You find lots of loose unidentifiable bone fragments, along with badly worn vertebrae of whales, which no one saves since they are not identifiable or diagnostic. You often get the heavily calcified ear bones of whales (very distinctive to species) and, more rarely, parts of other marine mammals, which are definitely worth saving. The bone bed yields a wide range of marine mammals, from dozens of types of whales and dolphins, to various kinds of early seals and sea lions, to such strange beasts as the hippo-like extinct mammals known as desmostylians, as well as extinct relatives of manatees in abundance.

But the biggest prizes by far are the huge triangular teeth of the gigantic shark *Carcharocles megalodon*. At the Ernst Ranch, Bob let visitors keep all the other fossils they found (and allowed museums take any good whale skulls as well)—but he kept the *C. megalodon* teeth, because they are valuable on the collector's market and they paid his bills for letting people collect at his ranch and enjoy his generosity. In 2007, my good friend Bob Ernst passed away suddenly and unexpectedly, so the situation at his ranch has now changed.

The Sharktooth Hill bone bed was long a mystery: How old is it? How was it formed? How deep was the water? How did so many bones and teeth come to be concentrated in a single layer? Barnes had figured out most of the mystery long ago, and thanks to recent work by Nicholas Pyenson of the Smithsonian Institution and me, most of the questions have been answered.

First, the easier answer. Our paleomagnetic dating showed that the section of the Round Mountain Siltstone containing the bone bed dates to between 15.9 and 15.2 million years ago, so the bone bed is roughly 15.5 million years old. The microfossils in the siltstone suggest a very great water depth (at least 1000 m [3280 feet] or more).

Figure 9.1 ▶

Typical teeth from Sharktooth Hill, including one from *Carcharocles megalodon* surrounded by those from the most abundant species, the mako shark (*Isurus*). (Photograph courtesy R. Irmis/University of California Museum of Paleontology)

But why the big concentration of bones? The deep-water basin that covered the area in the Miocene apparently had an extremely low rate of sediment accumulation, because the bone bed is thought to be a lag deposit, or a long-term accumulation of bones and teeth that build up on a seafloor with almost no sedimentation. Apparently, a local geological feature trapped or diverted most of the muds and sand eroding from the land, so they were prevented from flowing down into this patch of seafloor.

Nearly all the fossils are broken or disarticulated, which indicates that the animals died and were torn up before they sank to the bottom. There they accumulated along with all the shark teeth, which are shed constantly as sharks feed. However, a few of the skeletons of whales and other marine mammals were found complete and articulated, so occasionally a carcass sank to the bottom intact (called a "whalefall") and was not broken up by scavengers. All of this bone accumulation occurred during a period known as the Middle Miocene Climatic Optimum, when warm global climates caused a huge evolutionary radiation of plankton, marine life, and especially whales all over the world. These conditions not only led to huge pods of whales feeding in the area (and sharks as well), but also contributed to the low sedimentation rates compared with those in earlier and later stages of the Miocene.

The diversity of fossils is amazing. At least 150 species of vertebrates are known, including more than 30 kinds of shark teeth, although those from mako sharks are by far the most common (see figure 9.1). There was a huge sea turtle three times larger than the living leatherback sea turtle, the largest reptile alive today. There are lots of different clams and snails in the other parts of the Round Mountain Siltstone and especially in the shallow-water Olcese Sand, which underlies it. Fossils of at least 30 species of marine mammals are in the bed.

What my colleagues and I found most surprising, however, is the diversity of land mammals that must have floated out into deeper water as carcasses, and then sunk to the bottom. As a result of more than a century of collecting, many different and mostly unidentified fossils of land mammals reside in the museum collections that Larry Barnes, Richard Tedford, Edward Mitchell, Clayton Ray, Samuel MacLeod, David Whistler, Xiaoming Wang, Matthew Liter, and I finally published in 2008 after decades of delay. They include a mastodont, two types of rhinos, tapirs, many camels and horses, deer-like dromomerycids, true cats, dogs, weasels, and the extinct "beardogs." All these mammals are already known from nearby middle Miocene beds in places such as Barstow and Red Rock Canyon in California, as well as localities all over the western United States (especially in the Plains states of Nebraska, Wyoming, and South Dakota). While I was working on this project, I carried many of the best specimens in my hand luggage as I flew from one city to another in order to identify them at local museums.

SHARK-INFESTED WATERS OF THE MIOCENE

Giant sharks, such as those in the Sharktooth Hill area, swam in seas all over the world. Their fossils are extremely abundant in the famous Lee Creek Mine in North Carolina, the Bone Valley beds in Florida, the Calvert Cliffs shell beds along Chesapeake Bay, and many other classic Miocene marine localities in the United States. They are found in Europe, Africa, and many places in the Caribbean, including Cuba, Puerto Rico, and Jamaica. The teeth of *Carcharocles megalodon* span the globe from the Canary Islands to Australia, New Zealand, Japan, and India. They have even been dredged from the deep waters of the Marianas Trench in the Pacific near the Philippines. The oldest specimens are reported from Oligocene beds about 28 million years old. They are most abundant in rocks that formed during the warmer conditions of the early to middle Miocene, but also occur in Pliocene beds (5 to 2 million years old). The youngest known specimens are dated to about 2.6 million years ago.

The problem with studying sharks is that their teeth are the only bony parts of their bodies, so most shark fossils are known from their teeth and nothing else. The rest of the "skeleton" of a shark is made of cartilage, which rarely fossilizes (figure 9.2). Sometimes, the spinal column of sharks is partially mineralized with calcite, so a few shark backbones are known, including several that belonged to *C. megalodon*. For this reason, minute details of the teeth are the basis for classifying most fossil sharks that have no living relatives. Luckily, however, we have an excellent record of the teeth of modern sharks, so their relationships can be deciphered from the abundant soft tissues. Then most shark-tooth fossils can be related to well-known living species and their relationships become clear in context.

But there is a problem with *C. megalodon* in this regard. When Louis Agassiz saw the first specimens in 1835, he assigned them to the genus *Carcharo-*

Figure 9.2 ▲

Reconstructed cartilaginous "skeleton" of *Carcharocles megalodon*, which is more than 10 meters (35 feet) long. (Photograph courtesy Dr. Stephen Godfrey, Calvert Marine Museum, Solomons, Maryland)

don, that of the modern great white shark (*Carcharodon carcharias*). The simple broad triangular shape of the tooth, along with some other features, seemed to be a good match for that of the great white shark, just scaled up much bigger. This was the prevailing opinion for many decades and the one that most specialists followed until recently. In the past decade, though, a group of shark specialists have argued that *C. megalodon* is not related to the great white shark, *Carcharodon*, but to an extinct shark, *Carcharocles*, a slightly different member of the lamniform sharks, which also include the mako sharks and several other members of that family. There are even some who argue that the giant shark is descended from the fossil shark *Otodus* and should be included in that genus. For the moment, it seems that the majority consensus among shark paleontologists favors *Carcharocles* over the other options, and this is what I will follow in this chapter. However, the chapter could just as easily be called "*Carcharodon*," and many paleontologists would not object.

A FISH *THIS* BIG!

Whatever you call it, *C. megalodon* was a mega-predator, probably the largest fish to ever swim in the oceans. It was significantly larger than the largest extant fish, the whale shark (*Rhincodon typus*), which is a gentle plankton feeder that catches its food by opening its huge mouth and gulping large volumes of water (as does the second largest shark, the basking shark [*Cetorhinus maximus*], as well as the largest whales, the baleen whales). There is some argument that the Jurassic fish *Leedsichthys* was larger, but the specimens are too incomplete to know its length for sure. Current estimates place the maximum length of *Leedsichthys* at about 16 meters (52 feet).

Once again, however, we run into problems because we have only teeth and a few calcified partial spinal columns for *C. megalodon,* so all estimates about its length must be made with assumptions of how to scale shark tooth size to body length. Complicating the estimates are the early tendency to reconstruct the jaws (not preserved, since they are cartilage) of *C. megalodon* using all the largest teeth in a collection, rather than including the smaller lateral teeth, which taper down in size along the jaws from the largest teeth in front. Thus the famous reconstruction of the jaws of *C. megalodon* once mounted in the American Museum of Natural History was probably too large, since it used only the front teeth (figure 9.3).

Given these problems, paleontologists have devised remarkably clever ways to estimate the size of *C. megalodon* (figure 9.4). The initial estimate, by Bashford Dean of the American Museum of Natural History, was based on the exaggerated jaws (see figure 9.3), and he placed the shark's length at 30 meters (98 feet). Another method compares the height of the enamel on the largest tooth in known sharks, and that gives the much smaller length of 13 meters (43 feet). In 1996, Michael Gottfried and several other shark experts looked at 73 specimens of great white sharks of known length, and derived a formula for the body length based on the largest tooth. Their largest tooth was only 168 millimeters (6.6 inches) long, which gave a total length of 16 meters. However, there are now teeth up to 194 millimeters (7.6 inches) long, which would give an estimate closer to 20 meters (66 feet). In 2002, Clifford Jeremiah tried to estimate size by the scaling of the base of the largest teeth at the root, which produced an estimate of 16.5 meters (54 feet) in length, although the largest tooth he studied was not as big as the largest known tooth. Also in 2002, Kenshu Shimada tried a different

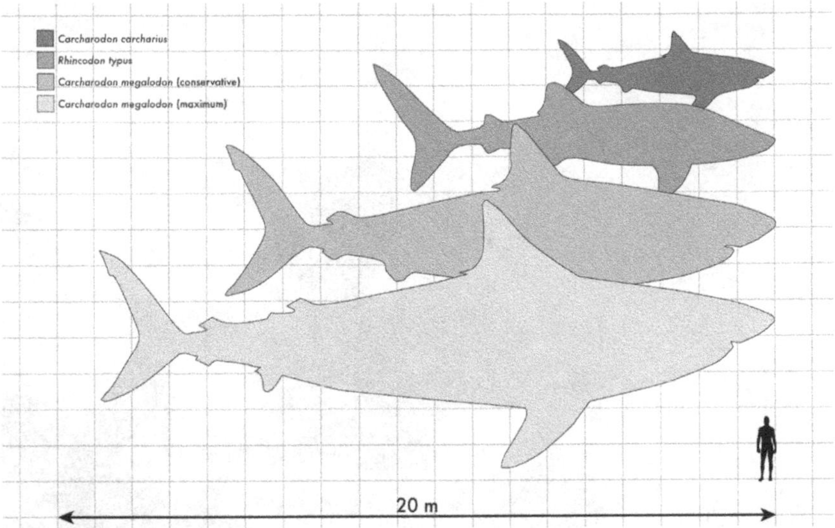

Figure 9.4 ▲
Comparison of the sizes of sharks, including the great white shark (*Carcharodon carcharias*); the whale shark (*Rhincodon typus*), the largest fish alive today; and two different size estimates of *C. megalodon*. (Drawing by Mary P. Williams)

method of scaling tooth-crown height to body length, and the largest teeth gave an estimate of 17.9 meters (59 feet). However, Patrick Schembri and Staphon Papson argued that the biggest specimens may have reached 24 to 25 meters (79 to 82 feet), almost as long as the original exaggerated estimate by Bashford Dean a century ago.

In short, there are many ways to solve the difficult problem of estimating the length of *C. megalodon*, but the consensus seems to be that they certainly reached at least 16 meters, and possibly 25 meters, in length. Even the conservative estimates are larger than the 12.7 meters (42 feet) of the largest known individuals of the living whale shark, and the 16-meter estimate of *Leedsichthys*, so no matter what method is used, *C. megalodon* was the largest fish to ever swim in the oceans.

Figure 9.3 ◄
The famous reconstruction of the jaws of *Carcharocles megalodon* by Bashford Dean at the American Museum of Natural History a century ago, using only the largest teeth. Today, it would be considered too large because it does not include the smaller side teeth. (Image no. 336000, courtesy American Museum of Natural History Library)

Figure 9.5 ▲
Life-size reconstruction of *Carcharocles megalodon*, displayed at the San Diego Natural History Museum. (Photograph by the author)

Once an estimate of length is obtained, an attempt can be made to calculate the body mass for a fish that size. Gottfried and his colleagues looked at the length-versus-body mass distribution for 175 specimens of great white sharks at various growth stages to derive a formula that predicts mass given body length. A *C. megalodon* about 16 meters long would have weighed about 48 metric tons (53 tons). A 17-meter (56-foot) *C. megalodon* would have weighed about 59 metric tons (65 tons), and a 20.3-meter (67 foot) monster would have topped off at 103 metric tons (114 tons).

Even though only teeth and a few partially mineralized backbones of *C. megalodon* have been found, the cartilaginous skeleton of this monster can be reconstructed by scaling up from the cartilage of the modern great white shark. Such a reconstruction has been done and is on display (see figure 9.2) at the Calvert Marine Museum on Solomon's Island, Maryland, a repository for many of the amazing Miocene fossils of the Calvert Cliffs along

Chesapeake Bay. Several institutions have built life-size reconstructions of *C. megalodon* in action, including the San Diego Natural History Museum (figure 9.5).

MONSTER OF THE SEAS

The sheer size of *Carcharocles megalodon* raises a question: Why did it grow so big? The most common answer seems to be that sharks were responding to the great abundance of large prey in the Miocene, especially the huge radiation of many types of whales and dolphins that developed in the early and middle Miocene. *C. megalodon* was bigger than all but the largest whales known from the same beds, so it was a true "super-predator," capable of killing and eating almost anything that swam in the Miocene oceans.

There is abundant fossil evidence of this behavior. Deep gouges and scratches that could have been produced by only the huge teeth of *C. megalodon* have been found on many fossil whale bones, suggesting that the sharks scratched the bones as they tore flesh from the carcasses. The list of whales with traces of *C. megalodon* attacks is very long, including dolphins and other small whales, cetotheres, squalodontids, sperm whales, bowhead whales, and rorquals like the fin whale and blue whale, plus seals, sea lions, manatees, and sea turtles (which were three times the size of the largest extant sea turtles). A *C. megalodon* tooth was found associated with the bitten ear bone of a sea lion. There were also several finds of *C. megalodon* teeth embedded in whale backbones, and numerous cases partially scavenged whale carcasses (especially at Sharktooth Hill) have been found surrounded by shed *C. megalodon* teeth.

Of course, this does not exhaust the list. Most sharks (especially great whites) are indiscriminate, opportunistic feeders and attack anything that moves that they can catch. This is why so many modern sharks have ocean trash (including road signs, boots, and anchors) in their stomachs when they are cut open. So *C. megalodon* certainly would have eaten smaller fish and most other sharks when it could catch them. But its large size is primarily an adaptation to attacking large prey like whales, which no other marine predator could threaten until *C. megalodon* came along.

The bite marks on one particular whale specimen about 9 meters (30 feet) long suggests how *C. megalodon* preferred to attack. The marks seem to focus on the tough bony areas (shoulders, flippers, rib cage, upper spine) rather than on the soft underbelly, which modern great whites target. This suggests that *C. megalodon* tried to crush or puncture the heart or lungs of the whale, which would have killed it quickly. This, in turn, explains why the teeth of *C. megalodon* are so thick and robust: they were adapted for biting through bone. Another common strategy focused on the flippers, since fossils of the hand bones have the highest frequency of bite marks of all. A big bite to crush, cripple, or rip off one flipper would have been sufficient to disable the prey and allow the shark to finish it off with several more bites.

The predatory behavior of these mega-sharks gives us additional clues as to why they slowly vanished over the late Miocene to Pliocene. Even though they were at the top of the food chain in the middle Miocene, by the early Pliocene there were even bigger whales that they could not attack and more large predatory whales, such as squalodontids and sperm whales. The late Miocene sperm whale *Livyatan melvillei* was truly gigantic (18 meters [60 feet] long), the largest mammalian predator ever to swim the oceans (the genus name is a homonym of "Leviathan," and the species name honors Herman Melville, the author of *Moby-Dick*). This monster could have eaten *C. megalodon* if it wanted to.

Then as the global oceans got colder during the Pliocene (especially after the Arctic ice cap formed about 4 to 3 million years ago), *C. megalodon* teeth seem to get scarcer and scarcer. When they last appear, in rocks of the late Pliocene, they are extremely rare, suggesting that a combination of the competition from very large predatory whales and the increasingly colder oceans was too much for them. Whatever the cause, they are truly extinct.

DOCU-FICTION

When cable television exploded in the 1980s, there were dozens of channels, each niche-marketed to a specific audience, whether it was golf or police procedurals or history. Unfortunately, the deregulation of the television market in the late 1980s turned them all into commercial channels that were forced to compete with one another for the best ratings, and soon their original missions were all but forgotten. Discovery Channel (originally established to broadcast science documentaries) now airs fake "documentaries" about paranormal and pseudoscientific topics. Naturally, the abandonment of its original mission to be scientific and educational extends to its relicts of science documentaries as well.

At one time, the highlight of the programming on Discovery Channel was Shark Week, when it aired nothing but documentaries about real sharks and their biology. Then in 2013, the channel broadcast a ridiculous pseudo-documentary called *Megalodon: The Monster Shark Lives*, which in 2014 was followed by *Megalodon: The New Evidence*. Both programs featured vague and scary and eerie footage, poorly lit shots, computer-graphic reconstructions, actors billed as scientists, and many "reenactments" of an alleged family's encounter with a live *Carcharocles megalodon* while on a cruise.

Only in the final few seconds of credits of either show did there appear a disclaimer that the program was entirely fiction. During their publicity appearances, the producers kept hinting that it *could* be true. Naturally, most people who watched only part of the shows or who did not see the disclaimer took them seriously, and thus many viewers believe that *C. megalodon* is still out there, lurking in the deep and waiting to get them.

Scientists and science journalists were horrified, and there was a huge backlash against Discovery Channel for airing these "docu-fictions" or "fake-umentaries" and passing them off as fact. But it was probably to no avail—*Megalodon: The Monster Shark Lives* attracted 4.8 million viewers, the most watched show in the history of the network. Count on Discovery Channel to come out with similar programs for Shark Week each year. After all, it is not on the air as a public service, as are PBS and the BBC, so it has no obligation to truth or reality. Thanks to deregulation, its only mission is to attract viewers and garner ratings for its advertisers, no matter how low it must stoop to do so.

SEE IT FOR YOURSELF!

Since Bob Ernst's death, an organization called the Ernst Quarries (www.sharktoothhillproperty.com) allows access to the bone bed to most nonprofit groups (for a nominal fee that is really worth it). The Buena Vista Museum of Natural History and Science (http://www.sharktoothhill.org/index.cfm?fuseaction=page&page_id=11) offers digging privileges to its members.

A number of museums have exhibits of fossils of *Carcharocles megalodon* or reconstructions of the shark. The jaws of *C. megalodon* are suspended from the ceiling of the Hall of Vertebrate Origins in the American Museum of Natural History, in New York, and many other fossil fish and sharks are on display. The Buena Vista Museum of Natural History and Science, in Bakersfield, California, houses the largest collection of Sharktooth Hill fossils, including jaws of *C. megalodon*. A 10.6-meter (35-foot) long reconstructed skeleton and many teeth are on display at the Calvert Marine Museum, in Solomons, Maryland. The Florida Museum of Natural History, in Gainesville, has a striking display with several reconstructed jaws of *C. megalodon* of different sizes. A life-size model of *C. megalodon* hangs from the ceiling of a gallery at the San Diego Natural History Museum, and cases of teeth are on display.

FOR FURTHER READING

Compagno, Leonard, Mark Dando, and Sarah Fowler. *Sharks of the World*. Princeton, N.J.: Princeton University Press, 2005.

Ellis, Richard. *Big Fish*. New York: Abrams, 2009.

——. *The Book of Sharks*. New York: Knopf, 1989.

——. *Monsters of the Sea: The History, Natural History, and Mythology of the Oceans' Most Fantastic Creatures*. New York: Knopf, 1994.

Ellis, Richard, and John E. McCosker. *Great White Shark*. Stanford, Calif.: Stanford University Press, 1995.

Klimley, A. Peter, and David G. Ainley, eds. *Great White Sharks: The Biology of* Carcharodon carcharias. San Diego: Academic Press, 1998.

Long, John A. *The Rise of Fishes: 500 Million Years of Evolution*. Baltimore: Johns Hopkins University Press, 2010.

Maisey, John G. *Discovering Fossil Fishes*. New York: Holt, 1996.

Renz, Mark *Megalodon: Hunting the Hunter*. New York: Paleo Press, 2002.

10 THE ORIGIN OF AMPHIBIANS | *TIKTAALIK*

FISH OUT OF WATER

What possessed fish to get out of the water or live in the margins? Think of this: virtually every fish swimming in these 375-million-year-old streams was a predator of some kind. Some were up to sixteen feet long, almost twice the size of the largest *Tiktaalik*. The most common fish species we find alongside *Tiktaalik* is seven feet long and has a head as wide as a basketball. The teeth are barbs the size of railroad spikes. Would you want to swim in these ancient streams?

NEIL SHUBIN, *YOUR INNER FISH*

FROM WATER TO LAND

Ever since Charles Darwin published *On the Origin of Species* in 1859, scientists have sought fossils that show how one crucial evolutionary transition had taken place: how fish crawled out of the water and became land-living creatures. Of course, an entire class of vertebrates, the Amphibia, are still living in that transition. Some of them spend nearly all their time in the water and rarely go out on land. Others never enter the water at all, but must live in moist habitats. Many have a mixture of the two lives.

Even before the publication of Darwin's book, some scientists noticed the similarities between amphibians and lungfish, which show many amphibian-like features (especially the lungs), but still are fish with fins. Yet the fins of lungfish and other lobe-finned fish have the same bones as those in the limbs of amphibians. But even that was not so clear-cut. The South American lungfish (*Lepidosiren paradoxa*) is so specialized that it has only tiny ribbon-like fins and swims like an eel. When it was discovered in 1837, it was thought to be a degenerate amphibian. Almost the same thing happened when Richard Owen described the African lungfish (*Protopterus*) in 1839. A staunch opponent of evolution, Owen ignored the obvious connections between the anatomy of lungfish and amphibians, and emphasized their bizarre specializations, such as the tiny ribbon-like fins. Only when the Australian lungfish (*Neoceratodus forsteri*) was discovered in 1870 was it possible to see that some living lungfish have robust lobed fins that have all the same bones as the amphibian limb. This was further confirmed when more and more primitive lungfish fossils showed that most of the lungfish had many amphibian-like features (see figure 8.2), not the bizarre specializations of the African and South American lungfish.

Still, the gap between lungfish and the earliest amphibians in the fossil record was a large and frustrating one. In 1881, Joseph F. Whiteaves described *Eusthenopteron foordi*, probably one of the best transitional fossils. Unfortunately, his description was only two paragraphs, had no illustrations, and made no mention of how this fish showed amphibian-like features. *Eusthenopteron* was a large (up to 1.8 meters [6 feet] long) lobe-finned fish that was much more amphibian-like than either extant lungfish or coelacanths (figure 10.1). It is known from hundreds of beautiful specimens from a famous locality near Miguasha, on Scaumenac Bay, Quebec. Although *Eusthenopteron* still had a fish-like body, its lobed fins had all the right bones from which to build the amphibian hand and foot, and its skull had the right pattern of bones to be ancestral to the amphibian skull.

More discoveries of fossils showed that many lungfish and other lobed-fin fish had lived in the Late Devonian (385 to 355 million years ago). By the Early Carboniferous (355 to 331 million years ago), there had been a handful of unquestioned amphibians (in the nineteenth century, called by the now-obsolete names "stegocephalians" and "labyrinthodonts"), although their fossils are much more abundant in rocks of the Late Carboniferous. So where were the transitional fossils? Many Late Devonian localities with fossils of marine fish were found, but few that seemed to be from freshwater and that had much potential for yielding a fossil on the cusp between fish and amphibian.

The breakthrough came through accident and political expediency. In the 1920s, Norway and Denmark were arguing over which country owned East Greenland. Consequently, the Danish government and a foundation established by Carlsberg Brewery (the famous Danish beer maker) funded

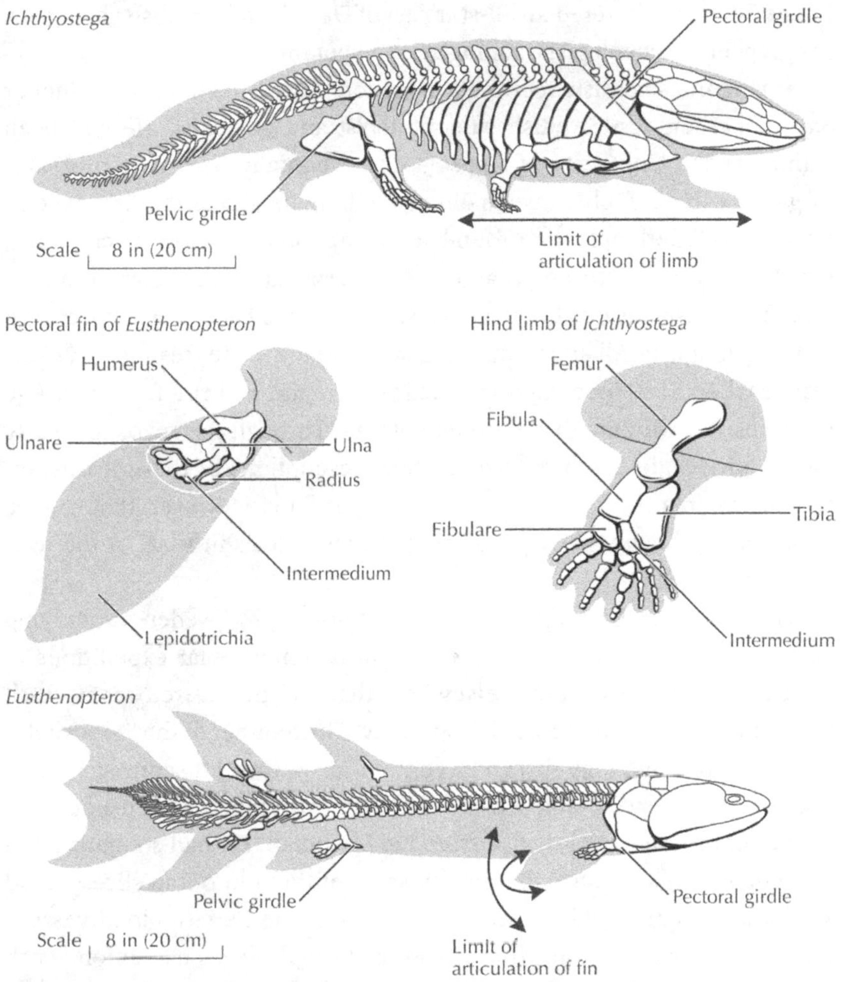

Figure 10.1 ▲

Comparison of the skeletal elements of *Ichthyostega* and *Eusthenopteron*. (Drawing by Carl Buell; from Donald R. Prothero, *Evolution: What the Fossils Say and Why It Matters* [New York: Columbia University Press, 2007], fig. 10.5)

a three-year expedition to East Greenland that visited the gigantic island in the summers of 1931 to 1933. The members of the expedition hoped to conduct enough scientific research and exploration in East Greenland that Danish territorial rights would be recognized, since Norway had done no exploration there. It was led by the famous Danish geologist and explorer

Lauge Koch and featured an all-star cast of Danish and Swedish geologists, geographers, archeologists, zoologists, and botanists.

Among the scientists recruited to explore East Greenland was Gunnar Säve-Söderbergh, a Swedish paleontologist and geologist. He had been trained at the University of Uppsala and eventually became a professor of geology there. Only 21 years old at the time he joined the first expedition, Säve-Söderbergh soon found fossils of some remarkable creatures, which he named *Ichthyostega* and *Acanthostega*, as well as more primitive lobe-finned fish like *Osteolepis*, which was much like *Eusthenopteron*, as well as many lungfish. All apparently had swum in the same fresh- or brackish waters when East Greenland was near the tropics and the Devonian Age of Fishes was winding to a close (chapter 8). Through the 1920s and early 1930s, Säve-Söderbergh published short descriptions of these fossils, intending to do a much more detailed analysis later. However, that chance never came, though, because he died of tuberculosis in 1948, at the relatively young age of 38.

Säve-Söderbergh was part of a larger tradition in Sweden of studying early fossil fish. Because the Swedes had mounted polar expeditions to Greenland, Spitsbergen, and elsewhere that had discovered many fossil fish, they soon became a Swedish specialty. The founder of the "Stockholm school" of paleontology (based largely at the Swedish Museum of Natural History) was the venerable Erik Stensiö, who was famous for his detailed studies of armored jawless fish from the Devonian. He had so many good specimens at his disposal that he cut some of them into thin slices (serial sectioning) so he could examine the details of the nerves, blood vessels, and other internal anatomy that are normally invisible in description of fish fossils. Today, high-resolution X-ray computed tomography allows paleontologists to make a "CAT-scan" of a solid fossil without slicing it up and destroying it for other uses.

After Säve-Söderbergh's death, his Greenland fossils were studied by Stensiö's successor, Erik Jarvik. He had accompanied Säve-Söderbergh on some of the later trips to Greenland, and then returned to collect more fossils. Jarvik was a careful, methodical worker, never one to rush to publish. He spent years slicing up specimens of *Eusthenopteron* to see the details of the internal anatomy of its skull. He worked on Säve-Söderbergh's *Ichthyostega* fossils for *50 years*, finally releasing his detailed publication about them in 1996, when he was 89 years old! The profession of vertebrate paleontology is legendary for scientists sitting on important fossils for years without publishing anything for the rest of us to see, but Jarvik takes the cake as one of the slowest workers of all. Although Jarvik's research was important and his descriptive work was impressive, he proposed many odd notions about different fossil groups that no other paleontologists considered to be plausible. He died in 1998, at the ripe old age of 91.

Since Jarvik's complete description of *Ichthyostega* did not appear until 1996, Säve-Söderbergh's original reconstructions of the fossils were the only well-documented "fishibian" from the 1920s until the 1980s. Thus *Ichthyostega* became the archetypal transitional fossil between *Eusthenopteron* and early amphibians (see figure 10.1). Like amphibians, it had four legs with toes, rather than the lobed fins of its ancestors. However, its forelimbs were not strong enough to do much walking, and the most recent analyses suggest that it could move only by short hops, dragging its more flipper-like hind limbs behind. The forelimbs and, especially, the hind limbs were much better adapted for use in the water, where they propelled the animal along (as newts and salamanders swim). *Ichthyostega* had robust ribs with flanges that would help support its chest cavity and lungs out of the water, but they were not capable of the rib-assisted breathing found in many amphibians. The other amphibian-like feature was its long flat snout with eyes directed upward and its short braincase; *Eusthenopteron* had a more fish-like cylindrical skull, with a short snout and a long braincase, eyes facing sideways, and big gill covers. Other than the limbs and the bones of its shoulder and hips, however, *Ichthyostega* was really fish-like. It still had a large tail fin, as well as many fishy features of the skull, such as large gill covers, hearing adapted for water, and a lateral-line system (canals on the face used to sense motion and currents in the water).

In the 1980s, the locus of research on "fishibians" shifted from Sweden to Cambridge University, where Jenny Clack, Per Ahlberg, Michael Coates, and others were active in collecting more fossils and redoing the work of the "Stockholm school" paleontologists. As Clack describes it:

> In 1985, I began to think about the possibility of an expedition to East Greenland, at the instigation of my husband Rob. Along the trail, I met Peter Friend of the Earth Sciences Department across the road in Cambridge, who had been leader of several expeditions to the part of Greenland in which I was interested. It turned out that he'd had a student, John Nicholson, who'd collected a few fossils as part of his thesis work on the sediments of the Upper

> Devonian of East Greenland between 1968 and 1970. Peter retrieved these specimens from a basement drawer and also showed me John's notebook from his 1970 expedition. John's note that on Stensiö Bjerg, at 800 metres [2625 feet], *Ichthyostega* skull bones were common was startling, and portentous. The fossils that he'd collected fitted together to make a single small block of three partial skulls and shoulder girdle bits—not of *Ichthyostega*, but of its at that time lesser known contemporary, *Acanthostega*. Peter suggested I get in touch with Svend Bendix-Almgreen, Curator of Vertebrate Palaeontology in the Geological Museum in Copenhagen. The Danes still administered expeditions by geologists to the National Park of East Greenland, where the Devonian sites are located, so he would be the person to start with in my attempts to mount an expedition there. Peter also suggested I contact Niels Henricksen of the Greenland Geological Survey (GGU). By sheer coincidence, and great good fortune, the GGU had a project in hand in the very place where I needed to go, and their last season there was the summer of 1987. With funds from the University Museum of Zoology and the Hans Gadow Fund in Cambridge and the Carlsberg Foundation in Copenhagen, I, my husband Rob, my student at the time, Per Ahlberg, and Svend Bendix-Almgreen and his student Birger Jorgenson arranged a six-week field trip in the care of the GGU for July and August of 1987. Using John Nicholson's field notes, we eventually pinned down the locality from which the *Acanthostega* specimens had come, and then the exact in-situ horizon that had been yielding them. It was in effect, a tiny, but very rich, *Acanthostega* "quarry."

The discovery of much more complete specimens of *Acanthostega* was a big breakthrough. In 1952, Jarvik named *Acanthostega*, based on poor material that received little study. But all the new fossils that Clack and her group collected in the late 1980s and the 1990s made *Acanthostega* much more complete and informative than the original *Ichthyostega* material (figure 10.2). In most respects, the smaller *Acanthostega* was much more fish-like than *Ichthyostega*. Unlike those of *Ichthyostega*, the limbs of *Acanthostega* would not have allowed it to crawl on land—it lacked wrists, elbows, or knees. Instead, its limbs were only capable of only paddling and pulling it through obstacles underwater. Even more surprising, it had as many as seven or eight fingers on its hands, not the standard five fingers that most vertebrates have! *Acanthostega* had a much larger fin on its tail than did *Ichthyostega*, and its ribs were too short to support its body on land and allow

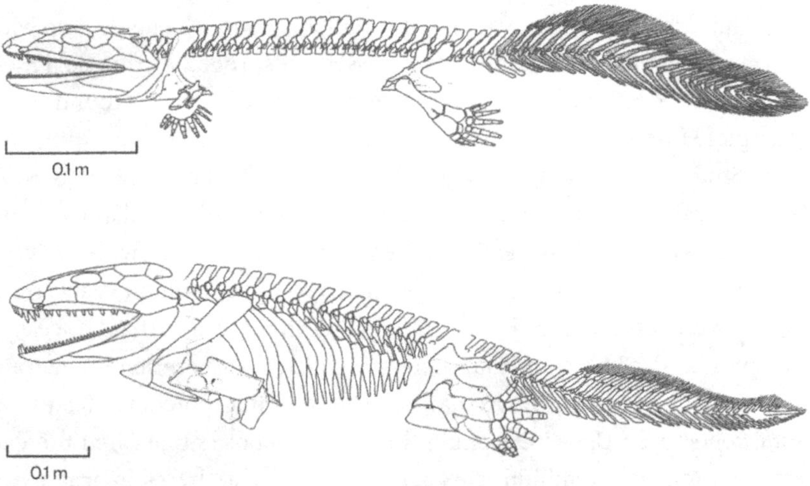

Figure 10.2 ▲

Comparison of the skeletons of *Ichthyostega* (*top*) and *Acanthostega* (*bottom*). (Drawing courtesy M. Coates, based on research by M. Coates and J. Clack)

it to breathe without the support of water. Yet it also had a few advanced amphibian-like features: its ear could hear in air as well as in water, and it had strong bones in its shoulder and hip region, four limbs with toes, and a neck joint that allowed it to rotate its head. By contrast, a fish has no "neck" that allows rotation—it must turn the entire front half of its body to change direction or snap at prey.

YOUR INNER FISH

Jenny Clack's work revitalized the research on the fish-amphibian transition, and soon many other paleontologist were getting into the act. One of them was an eager and enthusiastic young scientist named Neil Shubin. He was educated in paleontology as an undergraduate at Columbia University and the American Museum of Natural History in New York, where I was a graduate student at the time. There we met in 1980, and together we worked on the evolution of the horse *Mesohippus*, his first research project that was published. He went on to earn a doctorate at Harvard, studying the evolutionary and developmental mechanisms that dictate how amphibian limbs and toes form. His first job was teaching anatomy to medical students at the University of Pennsylvania in Philadelphia, where he hooked up with Ted Daeschler of the Academy of Natural Sciences. Together, they searched road cuts of Devonian red beds across Pennsylvania until they found some incomplete fossils of fish and "fishibians."

But Shubin was looking for bigger fish to fry. As he describes in his book *Your Inner Fish*, he and Daeschler knew that they had to find rocks older than 363 million years (such as the East Greenland rocks that had yielded *Ichthyostega* and *Acanthostega*), but younger than 390 to 380 million years (from which have been recovered most of the lobe-finned fish that are ancestral to amphibians). Shubin and Daeschler predicted that there should be transitional fossils more primitive than *Acanthostega* but more advanced than *Eusthenopteron* in Upper Devonian freshwater deposits that filled the gap between 380 and 363 million years ago. They looked at the geologic maps in the first edition of the legendary historical geology textbook *Evolution of the Earth* (1971) by Robert H. Dott Jr. and Roger Batten. When they studied the map of Upper Devonian outcrops, they saw three likely candidates: eastern Pennsylvania (where they were already working), East Greenland (already collected by the Danes and Swedes and by Clack's group), and Ellesmere Island in the Canadian Arctic (which no one had studied). Further study of published geological survey reports showed that these outcrops were Upper Devonian, between 380 and 363 million years in age, and the right rock type to preserve freshwater fish and amphibian fossils. These rocks turned out to be about 375 million years old.

By the late 1990s, Shubin and Daeschler and their crew had all the permits and equipment, as well as funding for supplies and helicopter time to take them into and out of the region. Running a major expedition to this harsh region is no picnic! Researchers need a full complement of Arctic gear, especially cold-weather clothing for protection against the freezing summer temperatures and rugged tents that can stand up to hurricane-force winds and provide warmth and shelter during the frequent storms. In addition to rock picks, shovels, and other standard tools for collecting, they carried rifles because polar bears were a serious threat.

Starting in 2000, they made short trips of a few weeks at the peak of the summer to Ellesmere Island, with poor results in the first few years because the rocks were marine, not freshwater, in origin. Finally, they found the freshwater fossiliferous rocks they had been seeking. In 2000, they found what they called Bird Quarry, which by 2003 had yielded abundant fragmentary fish fossils. In 2004, they dug 3 meters (10 feet) below the surface level of the quarry and discovered *Tiktaalik*, the fossil that made all the hardships worthwhile. Shubin and his colleagues picked the name *Tiktaalik*, which in Inuktitut, one of the Inuit languages, means "burbot," a freshwater fish of the region. It took two more years before the fossils were properly prepared for study, and all the descriptions and analyses were ready, so *Tiktaalik* was announced in two papers published in 2006, with the description of the hind limbs appearing in 2012.

More than 10 individuals of *Tiktaalik* have been recovered, ranging in length from 1 to 3 meters (3.3 to 10 feet) (figure 10.3). Even better, the best specimen of *Tiktaalik* is nearly complete, with just portions of the hind limbs and tail missing, although the hind limbs are known from other specimens. As one would expect for a specimen that is 12 million years older than *Ichthyostega* or *Acanthostega*, *Tiktaalik* is more fish-like in many ways. Its lobed fins had all the elements ancestral to the amphibian limb, but still had fin rays, rather than toes. It had fish-like scales, a combination (as do most of the "fishibians") of both gills (shown by the gill-arch bones) and lungs (shown by the spiracles in its head), and a fish-like lower jaw and palate. But unlike any fish, it had amphibian features, too: a shortened, flattened skull with a mobile neck; notches in the back edge of the skull for the eardrums on the back of the skull; and robust ribs and limbs and shoulder and hip bones. Like *Acanthostega*, its fins were not strong enough or flexible enough to allow it to drag itself across land for very far or walk with its belly off the ground; instead, they were probably used to paddle in shallow water and to support the animal so it could see above the surface. Like the other "fishibians" (and many modern amphibians, especially newts and salamanders), it probably spent most of its time in water, hunting on the margins of the streams in which it lived.

As Robert Holmes wrote in *New Scientist*:

> After five years of digging on Ellesmere Island, in the far north of Nunavut, they hit pay dirt: a collection of several fish so beautifully preserved that their skeletons were still intact. As Shubin's team studied the species they saw to their excitement that it was exactly the missing intermediate they were looking for. "We found something that really split the difference right down the middle," says Daeschler.

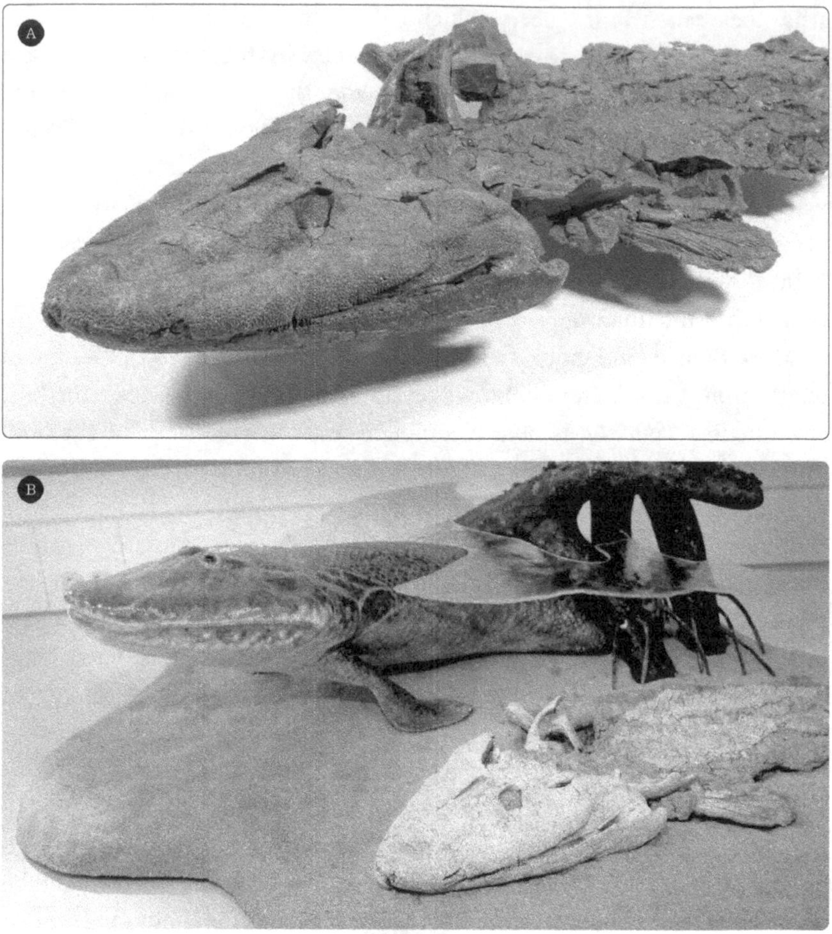

Figure 10.3 ▲

Tiktaalik: (*A*) skeleton; (*B*) reconstruction of its appearance in life. (Courtesy N. Shubin)

And Clack commented, "It's one of those things you can point to and say, 'I told you this would exist,' and there it is."

The search for even more transitional fossils continues. But one thing is clear: making the transition from water to land is not the gigantic leap that paleontologists and biologists thought it was for more than a century. You need look no further than the huge radiation of the ray-finned fish (Actinopterygii), which include 99 percent of the fish in fish tanks, fish markets, and big aquariums. Except for lampreys, hagfish, sharks, rays, lungfish, and coelacanths, all extant fish are ray-finned fish. They do not have the robust bones of the lobe-finned fish, but long thin rods of bone or cartilage to support their fins.

Ray-finned fish have found a number of ways to use their flimsy fins to move about on land. For example, mudskippers live half in and half out of the water, propped up in hallow mudflats or mangrove roots and using their front fins to crawl slowly on the air-water interface (figure 10.4). The "walking catfish" is a major pest in the southeastern United States because it can wriggle across land from one pond to another to find food or escape from a drying pool. The climbing perch can also drag itself across land in search of better pools and can even crawl up trees. Many fish, such as gobies and sculpins, adapted for tide-pool life spend part of their time in the air during low tide, and have modified their front fins for crawling along and for pushing up against rocks. Other mostly aquatic fish have modified their front-fin rays into "fingers" that can be used to dig into the surface underwater and pull the fish forward.

None of these groups of ray-finned fish are closely related to one another, so all these adaptations for land life evolved completely independently. Clearly, there are strong pressures and big advantages for fish to exploit land habitats (even if for only minutes to hours), and they have found different solutions to what was once thought to be an insoluble problem. Thus

Figure 10.4 ▲
Mudskipper feeding on worms on a mudflat in Japan. (Photograph by Alpsdake; from Wikimedia Commons)

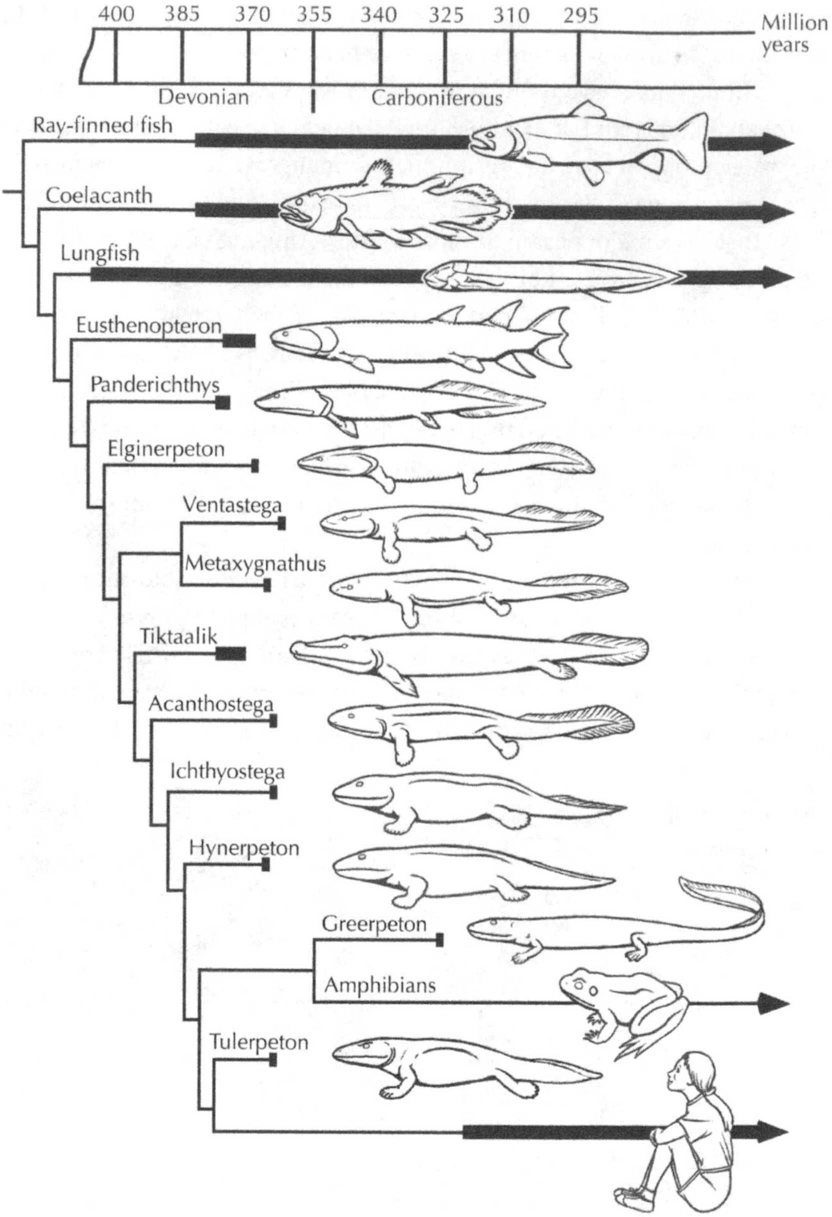

Figure 10.5 ▲

The evolution of amphibians from fish. (Drawing by Carl Buell; from Donald R. Prothero, *Evolution: What the Fossils Say and Why It Matters* [New York: Columbia University Press, 2007], fig. 10.6) the gradual changes in lobe-finned fish to become first semi-aquatic and then fully terrestrial animals are not the near-impossibility that scientists once imagined.

Recently, a group of scientists led by Emily Standen published a study that showed just how easy it is for a fish to leave the water. Their experiment focused on a very primitive bony fish, the bichir (*Polypterus*) of Africa, which is distantly related to such primitive ray-finned fish as the sturgeon and the paddlefish. Its fins are not unlike those of the earliest lobe-finned fish, and thus it is almost like a link between lobe-finned and ray-finned fish. The researchers raised bichirs on land, rather than in their normal watery habitat (they are good air breathers). Sure enough, after a few generations of breeding, their fins became more robust and better suited for crawling on land through a mechanism called developmental plasticity, which allows animal bodies to modify themselves during embryonic development to adapt to new challenges. As Standen pointed out, developmental plasticity may explain not only why so many kinds of ray-finned fish have adapted to crawling on land or in water, but also the mechanisms that allowed lobe-finned fishes to do the same.

Thus we now have a continuous sequence of "fishibians," from unquestioned fish-like creatures (such as the lobe-finned fish), through intermediates like *Tiktaalik* and *Acanthostega* and *Ichthyostega*, to animals that are even more amphibian-like (figure 10.5). Anyone who cannot imagine how fish crawled out of water and became land animals need only look at these incredible fossils to see the answer.

SEE IT FOR YOURSELF!

To my knowledge, fossils of *Ichthyostega* and *Acanthostega* are housed in only the University Museum of Zoology, Cambridge University, and the Naturhistoriska riksmuseet, in Stockholm, where a few specimens are on display.

Several museums in the United States display replicas of the skeleton and reconstructions of *Tiktaalik*, including the Academy of Natural Sciences of Drexel University, Philadelphia; Field Museum of Natural History, Chicago; Museum of Comparative Zoology, Harvard University, Cambridge, Massachusetts; and Museum of Natural History and Science, Cincinnati. Some of the best displays of lobe-finned fish fossils and early amphibians are at the American Museum of Natural History, New York.

FOR FURTHER READING

Clack, Jennifer A. *Gaining Ground: The Origin and Early Evolution of Tetrapods.* Bloomington: Indiana University Press, 2002.

Daeschler, Edward B., Neil H. Shubin, and Farish A. Jenkins Jr. "A Devonian Tetrapod-like Fish and the Evolution of the Tetrapod Body Plan." *Nature*, April 6, 2006, 757–773.

Long, John A. *The Rise of Fishes: 500 Million Years of Evolution.* Baltimore: Johns Hopkins University Press, 2010.

Maisey, John G. *Discovering Fossil Fishes.* New York: Holt, 1996.

Moy-Thomas, J. A., and R. S. Miles. *Palaeozoic Fishes.* Philadelphia: Saunders, 1971.

Shubin, Neil. *Your Inner Fish: A Journey into the 3.5-Billion-Year History of the Human Body.* New York: Vintage, 2008.

Shubin, Neil H., Edward B. Daeschler, and Farish A. Jenkins Jr. "The Pectoral Fin of *Tiktaalik roseae* and the Origin of the Tetrapod Limb." *Nature*, April 6, 2006, 764–771.

Zimmer, Carl. *At the Water's Edge: Macroevolution and the Transformation of Life.* New York: Free Press, 1998.

11 | THE ORIGIN OF FROGS | *GEROBATRACHUS*

"FROGAMANDER"

Theories pass. The frog remains.

JEAN ROSTAND, *INQUIÉTUDES D'UN BIOLOGISTE*

"MAN, A WITNESS OF THE FLOOD"

In the early eighteenth century, scholars were still divided over the origin and nature of fossils and offered many explanations for the presence of these strange objects found in rocks. The word "fossil" comes from the Latin term *fossilis* (obtained by digging), so anything dug out of rocks (including crystals, concretions, and many other nonbiological objects) were originally called fossils. Some scientists thought that fossils were works of the devil, placed in rocks to confuse the faithful and spread doubt. Others argued that they grew in rocks under the influence of mystical "plastic forces" (*vis plastica*) or that some creatures had crept into crevices, been crushed, and died, leaving their skeletons encased in stone. Only a minority of scholars connected the fossilized shells of clams and snails to their modern descendants.

Many fossils were simply unrecognizable at the time because they looked like no extant creature. The strange triangular objects known as "tongue stones" (*glossopetrae*) were thought to have fallen from the sky and to have magical properties, including the ability to heal snake bites and detoxify poisons. But in 1669, the Danish doctor Niels Steensen (known to us by his Latinized name, Nicholas Steno) saw "tongue stones" in the mouth

Figure 11.1 ▲

Johann Scheuchzer's "*Homo diluvii testis*," displayed at the Teylers Museum in Haarlem, Netherlands. (From Donald R. Prothero, *Bringing Fossils to Life: An Introduction to Paleobiology*, 3rd ed. [New York: Columbia University Press, 2013], fig. 1.4)

of a shark and realized that they were teeth. Most people thought that ammonites were the remains of coiled snakes because the chambered nautilus would not be discovered until the early nineteenth century. The stem pieces or columnals of crinoids were believed to be stars that had fallen from the heavens.

In particular, the Bible still influenced ideas about fossils. In 1726, for example, the Swiss naturalist Johann Scheuchzer described a fossil as "the bony skeleton of one of those infamous men whose sins brought upon the world the dire misfortune of the Deluge." It was a large skeleton, about 1 meter (3.3 feet) long from the head to the hip bones; had a skull and arms and a backbone; and had been found in the rocks. Therefore, it must be a human who had died in Noah's flood. Scheuchzer named it *Homo diluvii testis* (Man, a witness of the Flood) (figure 11.1). But in 1758, the pioneering naturalist Johannes Gesner disagreed, believing it to be a catfish! Then in 1777, Petrus Camper argued that it was a lizard. In 1802, Martin van Maur bought the specimen for the Teylers Museum in Haarlem, where it still resides. In 1836, it was formally named *Andrias scheuchzeri*, which translates to "Scheuchzer's image of man."

The mistake was not rectified until almost a century after Scheuchzer first described it. After Napoleon annexed the Netherlands, the specimen found its way to Paris, where the great Baron Georges Cuvier, the founder of vertebrate paleontology and comparative anatomy, got to work on it. He prepared the skeleton in the slab to better expose the bones and found much more detail than had been visible originally, especially in the arms. In addition, he had spent his professional life studying comparative anatomy, so he knew at once that it was not a human skeleton. A few comparisons, and Cuvier realized that it was not even a primate or a mammal—but a gigantic salamander!

Such gigantic salamanders are not extinct. Two species in Japan and China are even larger than Scheuchzer's fossil (figure 11.2). The Chinese giant salamander is almost 2 meters (6.6 feet) long and can weigh as much

Figure 11.2 ▲
The Chinese giant salamander. (Photograph courtesy Luke Linhoff)

as 36 kilograms (80 pounds)! It is placed in the same genus as Scheuchzer's fossil, but is named *Andrias davidianus*. It lives in rocky hill streams and lakes with clear water, usually found in forested regions, as well as at altitudes of 100 to 1500 meters (330 to 4920 feet). The Japanese species is named *Andrias japonica*, is slightly smaller than the Chinese giant salamander, and inhabits a similar environment. Both species are endangered, since their habitats are being destroyed and such large aquatic animals need a lot of territory to survive. In addition, they are being poached for traditional Chinese medicine, which is already driving rhinoceroses, tigers, pangolins, and many other animals to extinction as well.

LIVING ON BOTH SIDES

In chapter 10, we saw how amphibians arose from lobe-finned fish in the Late Devonian. But how did they evolve into the familiar groups of living amphibians, especially the frogs, toads, and salamanders? Once again, the fossil record has produced some amazing specimens that show the stages of this evolutionary history.

The word "amphibian" comes from the Greek term *amphibion* (living on both sides)—that is, both in water and on land—and "living on both sides" is one of the distinguishing features of amphibians. Most have the ability to thrive in both environments, as long as they can get moisture. Desert toads have adapted to adapt to a world with almost no water and eke out an existence underground, keeping cool and moist. However, most amphibians still need moist places in which to lay their eggs and complete their life cycle (although a handful actually give birth to live young and skip the egg stage altogether).

The living amphibians are tremendously diverse, with over 5700 known species. More than 4800 of them are frogs and toads, but only 655 are salamanders and newts. In addition, there are about 200 species in a third group of amphibians: the apodans, or caecilians. The legless apodans burrow underground mostly in tropical soils of South America, Africa, and Asia. They have tiny eyes that can sense light and dark, and some have eyes at the tip of sensory tentacles, but most are blind. To the nonspecialist, they look almost like giant earthworms.

Amphibians range enormously in size, from the tiny New Guinean frog *Paedophryne amanuensis*, which is only 7.7 millimeters (0.3 inch) long, to the

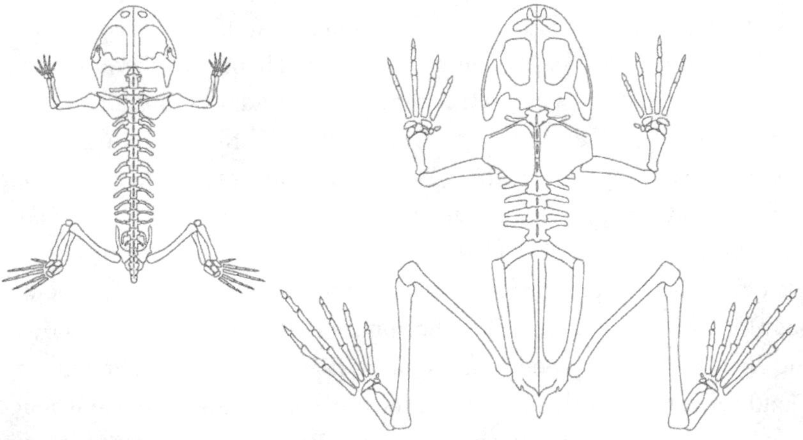

Figure 11.3 ▲

Comparison of the skeleton of *Triadobatrachus* (*left*) with that of a modern frog (*right*). Although they look superficially similar, *Triadobatrachus* was much more primitive than any modern frog in having many trunk vertebrae, small simple hips rather than an elongate hip structure, small fore- and hind limbs that did not allow it to jump, a slightly longer tail, and a much more primitive skull. (Drawing by Mary P. Williams)

huge Chinese giant salamander. Salamanders and newts retain the simple elongate body form, with a long tail and four simple limbs, of the most primitive amphibians (such as *Tiktaalik*, *Ichthyostega*, and *Acanthostega* [chapter 10]).

Frogs are the most spectacularly divergent from this ancestral body plan of all the living amphibians. As anyone who has dissected a frog in high-school biology class knows, they are truly unique in their body design (figure 11.3). Although adult frogs and toads have no tail, their larvae (tadpoles) hatch with a tail that is resorbed into their body as they mature. The head of frogs is short, with a blunt broad snout that allows them to open their mouth wide as they capture food (often using a long sticky tongue). Their very long muscular hind legs enable them to make huge leaps (both to catch prey and to escape predators) as well as swim with great power. The trunk of the frog skeleton is also short, with tiny stumpy ribs and very elongated hip bones to support the hind-leg muscles. Since frogs cannot use their ribs for breathing, they use an inflatable pouch in their throat that can pump air in and out (as well as make a variety of sounds). Frogs range tremendously in size, from the tiny New Guinean frog to the Goliath frog, which is more than 300 millimeters (12 inches) long and weighs 3 kilograms (7 pounds). It is so big that it eats birds and small mammals, as well as insects.

If the Goliath frog were not impressive enough, in 1993 a group of scientists working in the Upper Cretaceous rocks of Madagascar found the fossil of an even bigger frog. After 15 years of fitting all the pieces together (including most of the skull from 75 fragments), they published a description of it in 2008. They named it *Beelzebufo ampinga* (devil's toad). The genus name is a composite of Beelzebub (Lord of the Flies), another name for the devil, and *Bufo*, the genus of common toads; the species name is Malagasy for "shield." It was a ceratophrynine, a member of the group known as the "horned toads" of South America, so this family once extended across Gondwana, which included most of the present-day Southern Hemisphere. Its most remarkable feature was its size. Based on the nearly complete skeleton, it was 40 centimeters (16 inches) long and weighed 4 kilograms (9 pounds)—one-third again as large as the Goliath frog! It had a very large head and a wide mouth, and it is speculated that it could eat even baby dinosaurs, which roamed Madagascar at the time.

RICHES OF THE RED BEDS

This is just a glimpse of the range of size and diversity of living amphibians. What about their fossil ancestors? Starting with "fishibians" (chapter 10), there was a huge evolutionary explosion of different kinds of amphibians during the Carboniferous (355 to 300 million years ago) and Permian (300 to 250 million years ago). Most belong to three major groups that are extinct, but they were once the largest and most dominant animals on land until reptiles took over that role in the Early Permian.

By far the best place to collect Early Permian amphibians and contemporaneous land animals are the red beds of northern Texas, especially in the area around Wichita Falls and Seymour (and across the state line in Oklahoma). These incredible fossil deposits were discovered by the pioneering paleontologist Edward Drinker Cope in 1877. Working with just a horse and wagon and one or two local helpers, he found the ground literally covered with fragments of bone, along with skulls and skeletons. He collected a full wagonload in just a few days, thus beginning the long tradition of American paleontologists collecting in these rich deposits, and shipped them back to Philadelphia for study.

Almost every paleontologist who has published on the evolution of early reptiles and amphibians has collected in the red beds of Texas, including the giants of the field whose name every paleontologist knows well: Samuel Wendell Williston of the University of Kansas (in the 1890s) and the University of Chicago (until his death in 1918), Alfred S. Romer of the University of Chicago (in the 1920s) and Harvard (until the 1970s), and Everett "Ole" Olsen of the University of Chicago (and later UCLA).

The conditions for collecting are no picnic. The area is blazing hot in the summer, with windstorms that blow red dust into everything: food, beverages, equipment, and eyes and other sensitive areas. The groundwater is as hot as tea and nasty tasting, filled with pink mud and alkali, so those who drink too much of it get kidney stones. Once they find a good locality, collectors have to dig in deep and hunker down, trying to keep cool and avoid breathing the dust.

But the rewards are worth it! The most common animal in the red beds is the fin-backed, tiger-size predator *Dimetrodon*, familiar from dinosaur plastic toy sets and children's dinosaur books (chapter 19). However, *Dimetrodon* was not a dinosaur, but a very early member of the lineage that gave rise to the mammals, known as synapsids or "protomammals" (once called mammal-like reptiles, although synapsids were not reptiles). Most specimens reached 2 to 4 meters (7 to 14 feet) in length, weighed up to 270 kilograms (600 pounds), and had spines 1.2 meters (4 feet) tall on their back to support their fins. They were the top predator of their time, feeding on smaller fin-backed synapsids like the herbivore *Edaphosaurus*, as well as a variety of primitive true reptiles, such as the lizard-size *Captorhinus*, which was closely related to turtles.

But the synapsid and reptile denizens of the Texas red beds are only a tiny part of the story. Even though *Dimetrodon* ruled the planet in the Early Permian, amphibians reached their acme of size and diversity, and many of them were top predators that competed for food in this harsh landscape.

WHEN AMPHIBIANS RULED THE WORLD

The most abundant and impressive of the three groups of Late Paleozoic amphibians was the temnospondyls (formerly, labyrinthodonts). Most resembled fat crocodiles, with long trunks and tails as well as strong limbs that sprawled out to the sides. Unlike crocodiles, however, they had huge flattened skulls with eye sockets that pointed upward, and rows of sharp conical teeth arrayed around their large snouts. The head of some specialized temnospondyls known as archegosaurs superficially resembled that of crocodiles, with a long narrow snout. One of them was *Prionosuchus*, from the Pedro do Fogo Formation in Brazil, which dates to the Middle Permian (270 million years ago). *Prionosuchus* lived in lagoons and rivers, and had not only a crocodile-shaped body, but a long very narrow snout that was specialized for catching fish and other aquatic prey, as does the gavial (or gharial). If it was truly 9 meters (30 feet) long, as some claim, *Prionosuchus* was the largest amphibian that has ever lived—and larger than any living crocodile as well—although others argue that the estimates of the tail and body are too long, and it may have been only 5 meters (16 feet) in length.

The earliest temnospondyls were only about 1 meter (3.3 feet) long, but by the Permian, they were among the largest land creatures the planet had ever seen. One of the commonest fossils in the Early Permian red beds of Texas is that of *Eryops*, a big temnospondyl known from numerous complete skeletons (figure 11.4*A*). It had a sprawling body more than 2 meters (6.6 feet) long, with a robust tail and limbs, and a skull well over 60 centimeters (2 feet) long in big individuals! *Eryops* was one of the largest terrestrial animals of the Early Permian, capable of hunting prey both in water and on land. The slightly more primitive *Edops*, also from Early Permian red beds of Texas, had an even longer skull and thus was even larger than *Eryops*.

By the Late Permian, the large terrestrial temnospondyls had retreated to a completely aquatic lifestyle, possibly due to competition from all the large predatory synapsids on land at the time. Temnospondyls managed to survive the worst mass extinction in Earth history at the end of the Permian (250 million years ago). They straggled on into the Triassic (250 to 200 million years ago), when they were common in the swamps and lake deposits of places like the Petrified Forest in Arizona. These last temnospondyls had weak legs that would not have supported them on land, flattened heads with eyes that looked upward only, and huge flat bodies that were adapted to living in shallow water and feeding on aquatic prey.

Figure 11.4 ▶

Early amphibians: (*A*) the temnospondyl *Eryops*; (*B*) reconstruction of the lepospondyl *Diplocaulus*; (*C*) the anthracosaur *Seymouria*. ([*A* and *C*] courtesy Wikimedia Commons; [*B*] courtesy Nobumichi Tamura)

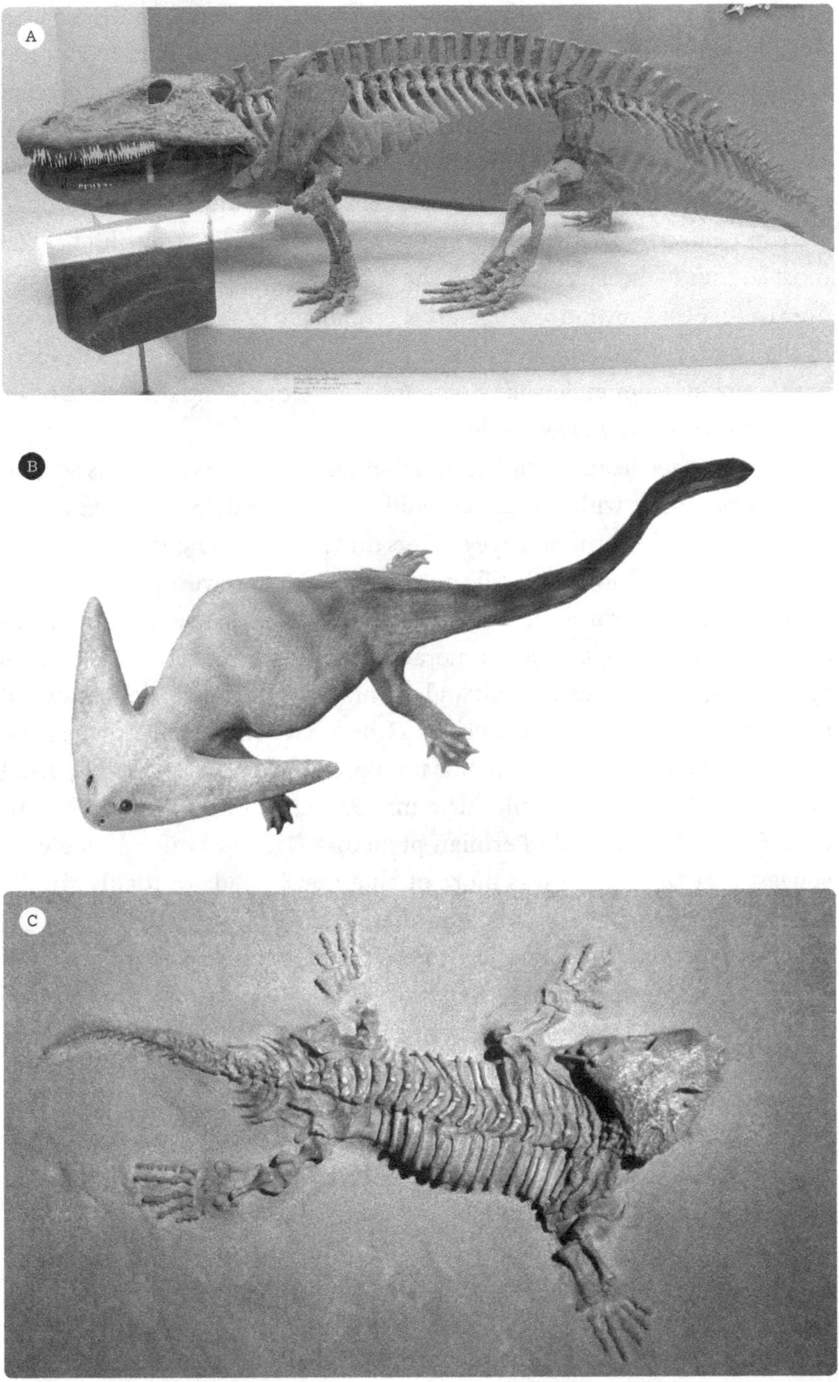
A
B
C

The second group of extinct amphibians was the lepospondyls, which lived from the Early Carboniferous to the Early Permian, but only in Europe and North America. Most were smaller than the temnospondyls that lived alongside them and had long salamander-like bodies with tiny legs, suggesting that they were mainly aquatic. Some, such as the aistopods, lost their legs entirely and looked like aquatic snakes. Others, the microsaurs, were more lizard-like in body form, with deep skulls and strong limbs. The most famous of the lepospondyls is the strange-looking *Diplocaulus* (see figure 11.4*B*). Best known from the Early Permian red beds of Texas, it was one of the largest of the lepospondyls, reaching a length of 1 meter (3.3 feet), with a stocky salamander-like body. It had armor plating over most of its body and strong, wide jaws.

But it was the head of *Diplocaulus* that was truly bizarre. It was shaped like a boomerang, with a flattened skull from each side of which extended a large flattened "horn" and eye sockets that pointed straight up. The function of these odd "horns" is still controversial. Some have argued that they were used as a hydrofoil, allowing *Diplocaulus* to swim smoothly in an up-and-down motion with the boomerang head shape providing lift. But its body was relatively weakly built and did not have the robust bones needed to support strong swimming muscles. Others have suggested that the head shape would have made it difficult for a predator to eat *Diplocaulus* head first, since the "horns" would have made the head too wide to swallow, even for the largest Early Permian predators. The upward-pointing eyes suggest that *Diplocaulus* was more of an ambush predator that lay in the bottom of streams and ponds, and then lunged forward and upward to catch its prey with its strong jaws, possibly stunning it with a blow from its "horns." The most likely hypothesis, however, is that the "horns" were analogous to the horns and antlers of antelope and deer. Males use their horns and antlers primarily as a display structure to advertise their strength and dominance while trying to find mates. That the growth of these "horns" can be traced through their younger stages and that there seem to have been both robust males and smaller-horned females appear to make this hypothesis most likely.

The third group of extinct amphibians is known as the "anthracosaurs," a wastebasket group for all the more advanced amphibians that are on the lineage leading to reptiles (see figure 11.4*C*). The Texas red beds are full of some amazing ones, including the 3-meter (10-foot) long, hippo-size herbivore *Diadectes*, and the extremely reptile-like *Seymouria* (named after Seymour, Texas, in the heart of the red beds).

FINDING THE "FROGAMANDER"

The giants of the mid-twentieth-century rush to the Texas red beds (such as Romer and Olson) are gone now, but their students continued to visit and collect important fossils. Some of the foremost successors were Robert Carroll of the Redpath Museum in Montreal (a student of Romer at Harvard), Robert Reisz (the first student of Carroll, now at the University of Toronto), the late Nicholas Hotton of the Smithsonian Institution (a student of Romer and Olson at Chicago), and the late Peter Vaughn (a student of Romer who trained many paleontologists during his career at UCLA, along with Olson). The current generation of paleontologists, intellectual grandchildren of Romer and Olson, have been making many important discoveries.

During an expedition to the Seymour area in 1994, undertaken by the Smithsonian and led by Hotton, the crew was working a locality nicknamed Don's Dump Fish Quarry. They found many fossil fish and a number of amphibians, but there was no time to clean all the fossils and do a detailed study in the field. According to the story, Hotton recognized the importance of one particular fossil (found by Peter Krohler, a curatorial assistant at the Smithsonian) and kept it in his pocket with a slip of paper on which was written "Froggie." But Hotton died in 1999 and never got the chance to study it or publish it.

Five years later, a group of younger scientists retrieved the unstudied specimen from the collections and spent countless hours finishing the preparation on it to completely expose the fossil (which was only partly visible when Hotton had it). Finally, in 2008, Hotton's "Froggie" was described and published, 14 years after it was found. The authors of the paper included Jason S. Anderson of the University of Calgary (a student of both Carroll and Reisz), plus Robert Reisz, Stuart Sumida of California State University, San Bernardino (a student of Vaughn), and Nadia Fröbisch of the Museum für Naturkunde in Berlin (a student of Carroll). They named it *Gerobatrachus hottoni* (Hotton's ancient frog), although the press labeled it the "Frogamander" as it spread the news of the discovery.

The specimen itself is a nearly complete skeleton only 11 centimeters (4.3 inches) long, found lying on its back with some of the hip region, tail, and

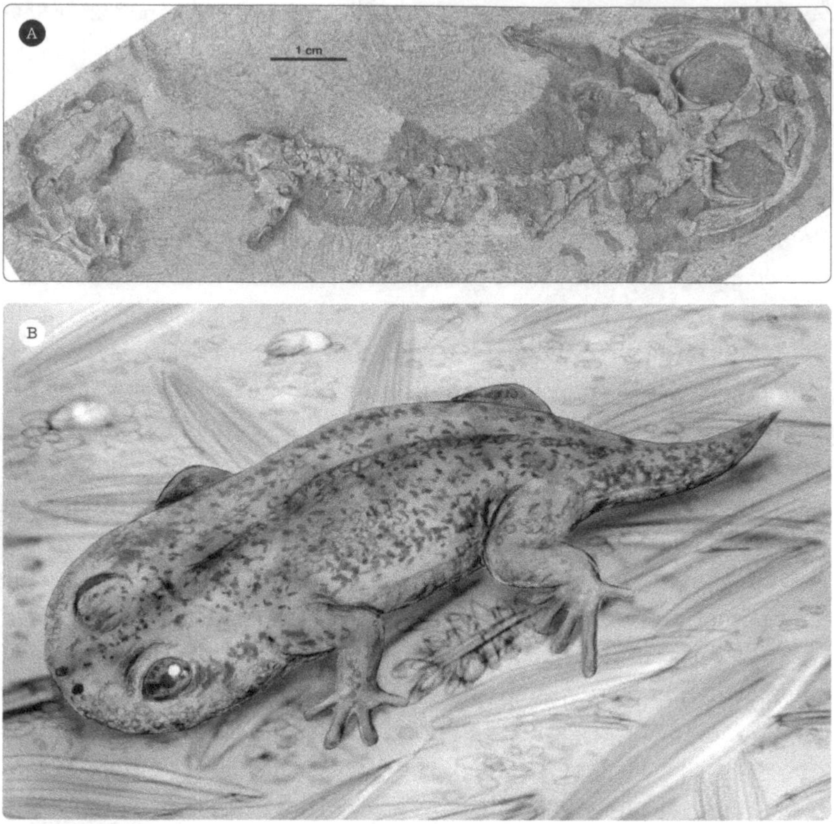

Figure 11.5 ▲

Gerobatrachus hottoni: (*A*) the only fossil; (*B*) reconstruction of its appearance in life. ([*A*] courtesy Diane Scott and Jason Anderson; [*B*] courtesy Nobumichi Tamura)

shoulder bones missing (figure 11.5*A*). What first catches your eye when you see the fossil is the combination of a salamander-like body with a broad flat frog-like snout (hence the nickname "Frogamander"). It has many other anatomical features of the skull and skeleton typical of frogs, especially the large eardrum. Most important, its teeth are attached to the jaw on tiny pedestals with a distinct base (*pedicellate* teeth), a feature unique to the living amphibians and just a few other extinct amphibians.

Fossils that do not fit into modern groups, but are squarely between them, are true transitional fossils, sometimes called (improperly) "missing links." *Gerobatrachus* is the perfect transitional fossil linking frogs and salamanders. The oldest known salamander is *Karaurus sharovi*, from the Late Jurassic (about 150 million years old) of Kazakhstan. The oldest known frog is *Triadobatrachus massinoti*, from the Early Triassic (240 million years old) of Madagascar (figure 11.6; see figure 11.3). *Triadobatrachus* looks similar to living frogs, with its broad snout and long webbed feet, except that it had a long trunk region with 14 vertebrae in its spinal column; all modern frogs have shorter trunks with four to nine vertebrae. It still had a short tail that was not lost, even in adults, unlike any living frog. Its hind legs were larger than those of any salamander, but nowhere near the large muscular legs of all modern frogs, so *Triadobatrachus* could swim strongly but not jump. All these features, and many others, make *Triadobatrachus* the perfect transitional fossil between modern frogs and more primitive forms like *Gerobatrachus*, the "Frogamander."

At 290 million years old, *Gerobatrachus* is much older than any member of the frog or the salamander lineage, and it is so primitive in its features that it cannot be called either a frog or a salamander. It contributes to the evidence that frogs and salamanders were not created as separate "kinds" but evolved from common ancestors, one of which could have been *Gerobatrachus*.

Figure 11.6 ▲
Reconstruction of the primitive Triassic frog *Triadobatrachus*. (Courtesy Nobumichi Tamura)

SEE IT FOR YOURSELF!

The "Frogamander" is not on display at any museum, as far as I know. However, large fossils of the Permian amphibians of Texas, including *Eryops* and *Diplocaulus*, can be seen at the American Museum of Natural History, New York; Denver Museum of Nature and Science; Field Museum of Natural History, Chicago; Museum of Comparative Zoology, Harvard University, Cambridge, Massachusetts; National Museum of Natural History, Smithsonian Institution, Washington, D.C.; and Sam Noble Oklahoma Museum of Natural History, University of Oklahoma, Norman.

FOR FURTHER READING

Anderson, Jason S., Robert R. Reisz, Diane Scott, Nadia B. Fröbisch, and Stuart S. Sumida. "A Stem Batrachian from the Early Permian of Texas and the Origin of Frogs and Salamanders." *Nature*, May 22, 2008, 515–518.

Bolt, John R. "Dissorophid Relationships and Ontogeny, and the Origin of the Lissamphibia." *Journal of Paleontology* 51 (1977): 235–249.

Carroll, Robert. *The Rise of Amphibians: 365 Million Years of Evolution*. Baltimore: Johns Hopkins University Press, 2009.

Clack, Jennifer A. *Gaining Ground: The Origin and Early Evolution of Tetrapods*. Bloomington: Indiana University Press, 2002.

12 THE ORIGIN OF TURTLES | *ODONTOCHELYS*

TURTLE ON THE HALF-SHELL

Behold the turtle. He makes progress only when he sticks his neck out.

JAMES BRYANT CONANT

TURTLES ALL THE WAY DOWN

> After a lecture on cosmology and the structure of the solar system, William James was accosted by a little old lady. "Your theory that the sun is the centre of the solar system, and the earth is a ball which rotates around it has a very convincing ring to it, Mr. James, but it's wrong. I've got a better theory," said the little old lady. "And what is that, madam?" inquired James politely. "That we live on a crust of earth which is on the back of a giant turtle." Not wishing to demolish this absurd little theory by bringing to bear the masses of scientific evidence he had at his command, James decided to gently dissuade his opponent by making her see some of the inadequacies of her position. "If your theory is correct, madam," he asked, "what does this turtle stand on?" "You're a very clever man, Mr. James, and that's a very good question," replied the little old lady, "but I have an answer to it. And it is this: The first turtle stands on the back of a second, far larger, turtle, who stands directly under him." "But what does this second turtle stand on?" persisted James patiently. To this the little old lady crowed triumphantly. "It's no use, Mr. James—it's turtles all the way down."

There are many versions of this story. Some are attributed to the philosopher Bertrand Russell; others, to the philosopher and psychologist William

James, the writer Henry David Thoreau, the famous skeptic Joseph Barker, the philosopher David Hume, or such scientists as Thomas Henry Huxley, Arthur Eddington, Linus Pauling, and Carl Sagan. They hearken back to the supposed Hindu legend of how the world was supported on the back of an enormous turtle. According to Bertrand Russell, in a lecture presented in 1927:

> If everything must have a cause, then God must have a cause. If there can be anything without a cause, it may just as well be the world as God, so that there cannot be any validity in that argument. It is exactly of the same nature as the Hindu's view, that the world rested upon an elephant and the elephant rested upon a tortoise; and when they said, "How about the tortoise?" the Indian said, "Suppose we change the subject."

All these renditions relate to the problem of infinite regress ("turtles all the way down") without offering any explanation of what supports the turtle at the very bottom. This debate about ultimate causes has been going on for centuries.

But this story is also a metaphor for a different question: If we follow the fossil record of turtles back in time, what would we find at the beginning? What kind of animal was not quite a turtle, yet a transitional form that was closer to turtles than to anything else? How could a creature have been "half a turtle"? This is a common taunt of creationists when they try to distort the fossil record. They point to most fossil turtles, for example, and claim that they are "just turtles" or are "all within the turtle kind," not a form that links turtles to other reptiles. Even when they are presented with the anatomical features that show the earliest turtles had very primitive features not found in any later turtle, it's "just a turtle." They cannot imagine a creature that has features of "half a turtle." How can a fossil have "half a turtle shell" when most turtles need both top shell (carapace) and bottom shell (plastron) to protect their bodies?

Fortunately, for many years culminating with an amazing discovery in 2008, the fossil record has yielded specimens that show most of the steps between a generic reptile and a true turtle.

TRANSITIONAL TURTLES

Before we reach the turtle at the bottom of the stack, let's look at the evolutionary history of turtles. Even though most people think that all turtles

Figure 12.1 ▲

The cryptodires (*top*) pull their neck into an *S*-curve in the vertical plane, and retract their head completely inside their shell; the pleurodires, or "side-necked turtles" (*bottom*), fold their neck sideways and pull their neck and head under the front lip of their shell. (Drawing by Mary P. Williams)

look alike, there are 455 genera and more than 1200 species of turtles. Many of them are endangered due to human poaching, habitat destruction, and the pet trade. Within the constraints of their highly specialized armored bodies, they have adapted to a wide variety of lifestyles, including the fully marine sea turtles, the freshwater turtles, and the terrestrial tortoises.

All modern turtles belong to two distinct groups. The familiar pond turtles, sea turtles, and land tortoises are members of the cryptodire (hidden neck) turtles. There are more than 250 species of cryptodires. They are easy to recognize because when they pull their head under the front lip of their shell, the neck coils back on itself, with the neck vertebrae folded in a vertical plane just below the front of the carapace (figure 12.1). From the outside, it looks as though they have pulled their head straight into the core of the shell. In addition to this distinctive head movement, all cryptodires have other specializations of the head and jaw muscles, discovered not long ago by Eugene Gaffney of the American Museum of Natural History.

The second group is the pleurodire (side-necked) turtles. When they pull their head into their shell, the neck folds sideways in the horizontal plane, and the head and upper neck are tucked in just under the lip in the front of the carapace (see figure 12.1). Side-necked turtles are a very specialized and distinctive group of turtles, with only about 17 genera and about 80 species. Most are found in the remnants of the ancient southern continent Gondwana (particularly Africa, Australia, and South America). The fossil record of pleurodires is not as good as that of cryptodires, but they were diverse across both Gondwana and the ancient northern continent Laurasia in the Cretaceous and Early Cenozoic; their restriction to the continents of the Southern Hemisphere is a more recent artifact of their reduced diversity.

Some of the side-necked turtles are truly odd, like the *matamata*, which has a very peculiar appearance (figure 12.2). It lives on the bottom of streams

Figure 12.2 ▲

The *matamata*, a living pleurodire with a reduced toothless jaw, broad mouth, and flat head. Rather than biting its prey, it sucks its prey in by opening its mouth wide and expanding its broad throat cavity. (Courtesy Wikimedia Commons)

and ponds in the Amazon and Orinoco basins of South America. Its shell is covered by bumps and ridges that disguise it. Its nostrils are extended into a long snorkel that allows it to lurk underwater with just the tip exposed. When a fish or another small prey animal swims too close to the *matamata*, it suddenly opens its broad mouth, expands its huge throat, and sucks the prey down in a flash! It cannot bite or chew with its highly reduced jaws, but must swallow the prey whole after it squeezes the excess water out of its mouth and throat with its strong neck muscles.

Most fossil turtles are relatively small, roughly in the same size range as the living ones, although there are giant tortoises on isolated islands, such as the Galápagos, west of Ecuador, and the Aldabras, in the Indian Ocean. The largest living turtles are the sea turtles, whose immense size is supported by the buoyancy of the water in which they live. Of these, the leatherback sea turtle is the biggest (and the fourth biggest of all the reptiles). Large individuals can be more than 2.2 meters (7 feet) long and weigh up to 700 kilograms (1540 pounds). The leatherback gets its name because most of its bony shell has been reduced, and the skeleton of its back is covered with only a thick tough hide. This loss of bony armor keeps the leatherback from being too dense and sinking too fast, since its skin is thick enough to deter most predators (and full-grown leatherbacks have very few predators).

In the geologic past, however, there were some true monster turtles. The largest was the sea turtle *Archelon* (Greek for "king of turtles"), which swam in the shallow inland seas of what is now western Kansas, along with such other marine reptiles as plesiosaurs, ichthyosaurs, and mosasaurs (figure 12.3). The largest specimens of *Archelon* are more than 4 meters (13 feet) long and about 5 meters (16 feet) wide from the tip of one flipper to that of the other. It weighed more than 2200 kilograms (4850 pounds). Like many sea turtles, it had just an open framework of bone on its back and four jagged plates on its belly. Like the modern leatherback, it probably was covered mostly by thick skin.

The extinct giant land turtles could not grow quite this large, but nonetheless they dwarfed any modern giant tortoises. One of the largest was *Colossochelys*, which was more than 2.7 meters (9 feet) long and 2.7 meters wide and weighed about 1 metric ton (1.1 tons) or more. Discovered in Pakistan in the 1840s, its fossils have been found from Europe to India to Indonesia and date from 10 million years ago to 10,000 years ago, the end of the last Ice Age. It would have looked like a gigantic version of the Galápagos tortoise.

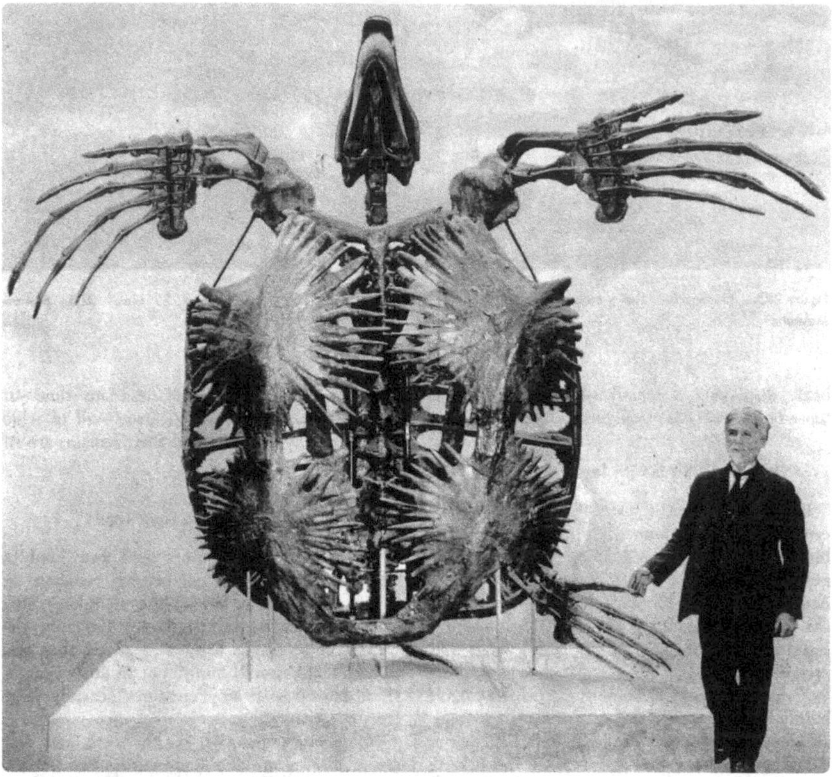

Figure 12.3 ▲
The most complete skeleton of *Archelon*, the gigantic sea turtle from the Cretaceous seas of Kansas. (Photograph courtesy Peabody Museum of Natural History, Yale University, New Haven, Connecticut)

Even bigger was *Carbonemys*, from swamp deposits about 60 million years old in Colombia. It was actually the size of a smart car, more than 1.7 meters (5.5 feet) long, and it could have eaten just about any creature it encountered, including crocodilians. It was one of the largest creatures in its world during the Paleocene—except possibly for *Titanoboa*, found in the same beds, which at 14 meters (45 feet) long was the largest snake that has ever lived (chapter 13). Like most South American turtles, *Carbonemys* was a pelomedusoid, a member of a group of side-necked turtles that is common in South America.

The largest of all land turtles was another monster from South America, the appropriately named *Stupendemys*, found in swamp beds of the Urumaco Formation in Venezuela that date to about 6 million to 5 million years ago, as well as in Brazil (figure 12.4). Like *Carbonemys*, it was a member of the pleurodire group known as pelomedusoids. It was most similar to the living Arrau turtle (*Podocnemis expansa*), except that it was much larger. As the name says, its size was truly stupendous: its shell was more than 3.3 meters (11 feet) long and 1.8 meters (6 feet) wide.

These extreme examples give a small indication of the huge evolutionary diversification of turtles and tortoises. The next question is: Which turtles are lower in the stack, and thus more primitive, than any of the members of the extant cryptodires and pleurodires?

Figure 12.4 ▲

The shell of *Stupendemys*, displayed at the Himeji Science Museum in Hyogo, Japan. (Courtesy Wikimedia Commons)

THE FIRST LAND TURTLE

The oldest land turtle known, and the oldest known turtle until 2008, is *Proganochelys*. Although it was not large (only about 1 meter [3.3 feet] long), it was an extremely primitive member of the turtle clan (figure 12.5)—more primitive than either the cryptodire or the pleurodire branch. It had a long neck that was covered with armored spikes and thus could not retract into the shell. *Proganochelys* is known from a number of complete or nearly complete skeletons, originally found in the Upper Triassic beds of Germany, dating to 210 million years ago, but later discovered in Greenland and Thailand as well.

To someone who does not know anatomy or zoology, it looks like just any other turtle. A closer look reveals that *Proganochelys* was very different from any subsequent turtle. Even though it had a carapace, there were far more plates in its upper shell, especially around the margin of the shell and protecting the legs, than in that of any later fossil turtle or living turtle. It had a long tail with a hard spiky outer sheath, terminated by a tail club. It lived alongside some of the first dinosaurs, so it had many large predators to contend with. Its skull was much more like those of primitive reptiles than of any living turtles. Although it had a beak, as do modern turtles, it still had teeth in its upper palate, so it was the last turtle with teeth of any kind. The combination of both beak and teeth suggests that it was omnivorous, eating both live prey and some plants. Since its neck could not retract, it could not pull its big armored head under its shell for safety, as do pleurodires and cryptodires.

So we come to the oldest known land turtle. *Proganochelys* was clearly a turtle with a shell, even though in most other aspects it was just a primitive reptile and very different from any later turtle. For the longest time, creationists dismissed it as "just a turtle" and said that it was impossible to imagine a turtle without its shell. Then, in 2008, the questions about the origin of turtles were finally answered.

TURTLE ON THE HALF-SHELL

For decades, Chinese paleontologists had been working on a very important fossil locality, the Guanling Biota, near the village of Xinpu, in Guizhou Province (in south-central China, just west of Hong Kong and one province

Figure 12.5 ▲

Proganochelys, the earliest land turtle: (*A*) fossil specimen and shell; (*B*) reconstruction of its appearance in life. ([*A*] courtesy Wikimedia Commons; [*B*] courtesy Nobumichi Tamura) north of the Vietnamese border). The black shales of the Wayao Member of the Falang Formation were deposited in the Nanpanjiang Basin during the Late Triassic (about 220 million years ago). This basin was bordered by uplands on three sides, but the embayment opened toward the southwest, where it was an extension of the ancient Tethys Seaway, which once stretched from the Mediterranean to Indonesia. The black shales are typical of deposits that formed in deep, stagnant waters, allowing very little scavenging or decay, so the quality of the preservation of fossils they contain is amazing. Even though the water was deep and low in oxygen, land was not far away, as indicated by the presence of fossilized driftwood and terrestrial animals. Some of these creatures probably swam in the margins of this sea or in the deltas that drained into the Nanpanjiang Basin.

Over the years, the Guanling Biota has yielded amazing fossils of marine reptiles as well as marine invertebrates (especially ammonites and huge "sea lilies," or crinoids) that document the changes in the oceans during the Late Triassic. The reptiles include nearly complete skeletons of "fish lizards," or ichthyosaurs, up to 10 meters (33 feet) in length (chapter 15), as well as of mollusc-eating reptiles known as placodonts and a group of marine reptiles known as thalattosaurs. At one time, 17 genera were named from this fossil assemblage, but recent work has reduced this list to eight genera and species.

Along with all these newly discovered species of marine reptiles, Chinese scientists found a very interesting collection of fossils that they published in 2008. Based on a complete skeleton and many partial skeletons, they named it *Odontochelys semitestacea* (toothed turtle with half a shell). One could not ask for a better transitional fossil between turtles and other reptiles (figure 12.6). In answer to the puzzle "how could turtles have evolved from no shell to a full shell?" *Odontochelys* provides the answer. It had a full bony shell on its belly (plastron), but on its back were only robust ribs and no shell at all! In other words, the transition from no shell to full shell is to form the plastron first, but not the carapace. It is truly a "turtle on the half-shell."

Figure 12.6 ▶

Odontochelys: (*A*) the best of the known fossils, showing an incomplete carapace on its back (*left*), but a complete plastron on its belly (*right*); (*B*) reconstruction of its appearance in life. ([*A*] courtesy Li Chun; [*B*] courtesy Nobumichi Tamura)

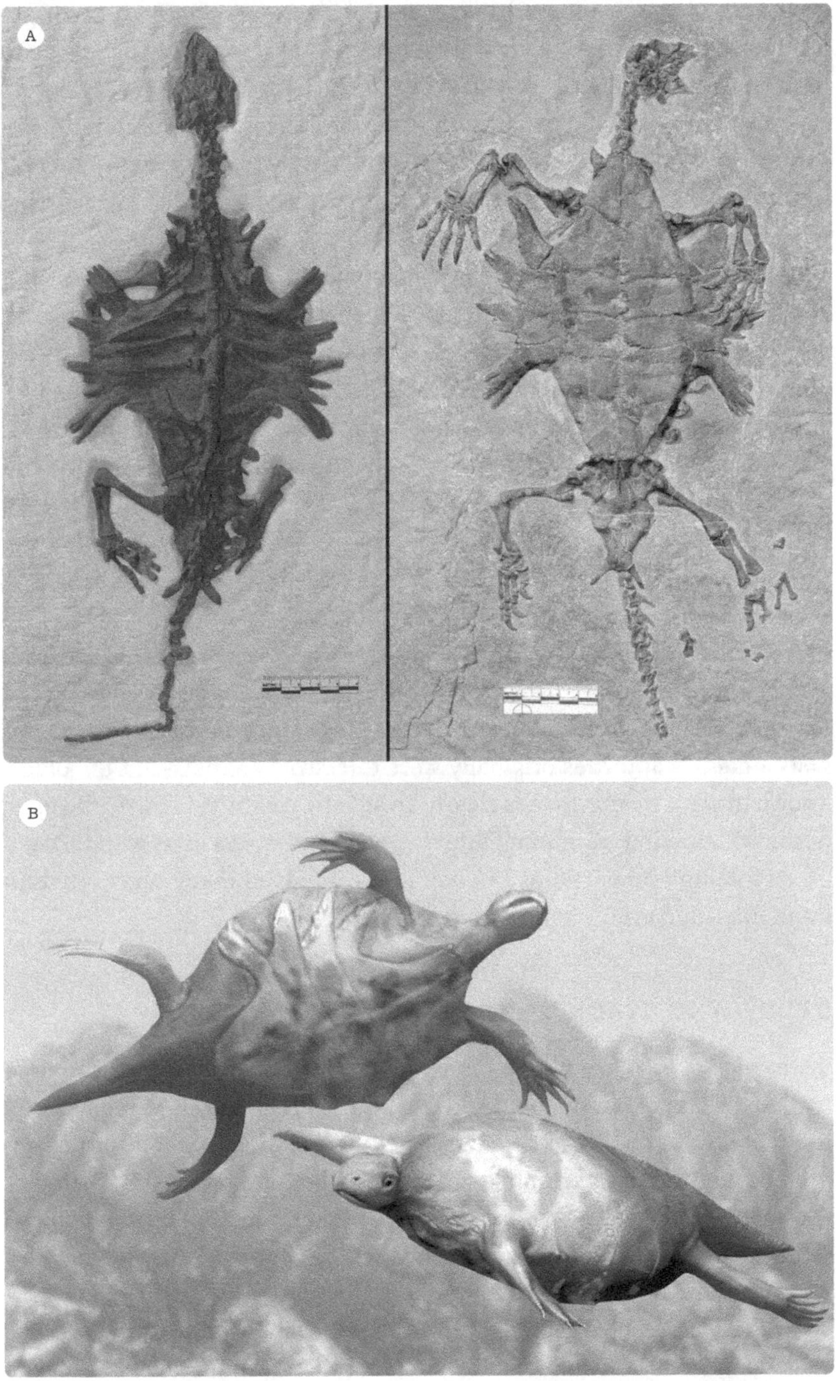
A
B

In addition to this remarkable trait, *Odontochelys* had another intriguing feature: a full row of teeth on the rim of its jaws, the last such turtle to have teeth rather than the toothless beak of all later turtles. Once again, we can see the evolutionary transition from reptiles with teeth in their jaws; through the "half turtle" *Odontochelys*, with normal reptilian teeth, and *Proganochelys*, with no teeth in its jaws but some on its palate; to later turtles, which have no teeth.

Odontochelys resolves another long-standing debate as well. For decades, some paleontologists argued that the turtle carapace comes from small plates of bone developed from skin (osteoderms) that become fused, while others contended that the carapace evolved mostly from the expansions of the back ribs. *Odontochelys* shows that the latter position is correct, since it had broadly expanded back ribs that were beginning to develop and connect into a shell, and there are no osteoderms on top of or embedded between the ribs. This is confirmed by embryological studies of turtles that track the development of the carapace from the developmental changes in the back ribs; no osteoderms are involved.

Yet another question was answered by *Odontochelys*: In what environment did turtles first evolve? Most of the later turtle fossils, such as that of *Proganochelys*, come from deposits formed on land, so many paleontologists argued that turtles originally were terrestrial animals. But the oldest known turtle, *Odontochelys*, is clearly an aquatic creature, living in the open ocean and possibly swimming into the rivers and deltas in its world. Based on its forelimb proportions, *Odontochelys* resembled many turtles that inhabit small and even stagnant bodies of water.

BELOW THE STACK OF TURTLES

With *Odontochelys*, we have a fossil that is truly transitional between nonturtle reptiles and undoubted turtles. But where among the branches of the reptiles did turtles come from? The traditional idea is that turtles are members of the most primitive group of reptiles, the anapsids, which lack the specialized openings in the back of the skull found in most advanced reptiles. This view has been around for almost a century and is still the most widely accepted.

In the past 20 years, though, it has been challenged by a new source of data: molecular sequences of DNA and proteins found in all reptiles. A number of such studies have placed the turtles within the Diapsida, the

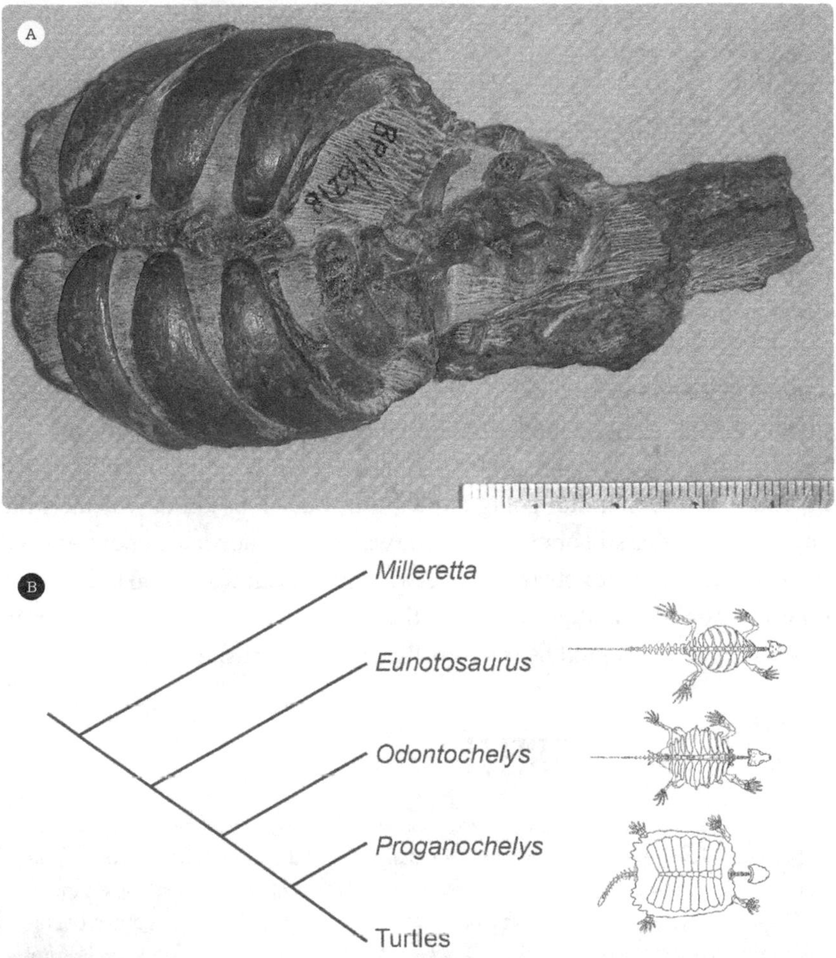

Figure 12.7 ▲

Eunotosaurus, a primitive Permian reptile with flared ribs that suggest the earliest stage of turtle evolution: (*A*) partial specimen, showing the distinctive flange-like ribs, which make a partial shell; (*B*) family tree of *Eunotosaurus* and other primitive turtles, showing the transition from reptiles to turtles. ([*A*] courtesy B. Rubidge, Evolutionary Studies Institute, University of the Witswatersrand, Johannesburg, South Africa; [*B*] Redrawn out of copyright by E. Prothero, originally from Tyler R. Lyson et al., "Transitional Fossils and the Origin of Turtles," *Biology Letters* 6 [2010])

group that includes lizards and snakes, plus crocodiles and birds. Some studies classify turtles with the lizards or in the crocodile–bird cluster.

The most recent analysis by a group at Yale University and the Universität Tübingen in Germany, however, makes a strong case for turtles being the most primitive group of living reptiles. They point to *Eunotosaurus*, a fossil from South Africa first described in 1892 by Harry Govier Seeley (figure 12.7). This creature is fairly common in beds that date to the Middle Permian (about 270 million years old), although complete skeletons with good skulls are rare. *Eunotosaurus* looked mostly like a large fat lizard, except for some key features of the skeleton. The most striking of these were the greatly expanded, broad flat back ribs, which almost connected with each other to form a complete shell. This, along with many other anatomical features, convinced many scientists that turtles come from the lineage of primitive reptiles. They argued that the molecular analyses are fooled by a problem known as long-branch attraction, whereby an isolated group that diverges early from the family tree often ends up with genetic patterns that falsely place it in the wrong group.

Thus the search is still going on, with the questions about which reptile gave rise to turtles still open. This is the way science normally operates until the evidence becomes clear and overwhelming (as it was when *Odontochelys* was first published). Stay tuned—the way this story is going, a different answer may be accepted by the time this book is published!

SEE IT FOR YOURSELF!

Odontochelys is not on display in any museum, as far as I know, but the American Museum of Natural History, in New York, has many of the other fossil turtles on display, including *Colossochelys*, *Stupendemys*, and *Proganochelys*. The Yale Peabody Museum of Natural History, in New Haven, Connecticut, has the biggest and most famous specimen of *Archelon*, and specimens are in the American Museum of Natural History and in the Naturhistorisches Museum in Vienna. Other museums with replicas of *Stupendemys* are the Himeji Science Museum in Hyogo, Japan, and the Osaka Museum of Natural History. The Museum für Naturkunde Stuttgart displays some of the original German material of *Proganochelys*.

FOR FURTHER READING

Bonin, Franck, Bernard Devaux, and Alain Dupré. *Turtles of the World*. Translated by Peter C. H. Pritchard. Baltimore: Johns Hopkins University Press, 2006.

Brinkman, Donald B., Patricia A. Holroyd, and James D. Gardner, eds. *Morphology and Evolution of Turtles*. Berlin: Springer, 2012.

Ernst, Carl H., and Roger W. Barbour. *Turtles of the World*. Washington, D.C.: Smithsonian Institution Press, 1992.

Franklin, Carl J. *Turtles: An Extraordinary Natural History 245 Million Years in the Making*. New York: Voyageur Press, 2007.

Gaffney, Eugene S. "A Phylogeny and Classification of the Higher Categories of Turtles." *Bulletin of the American Museum of Natural History* 155 (1975): 387–436.

Laurin, Michel, and Robert R. Reisz. "A Reevaluation of Early Amniote Phylogeny." *Zoological Journal of the Linnean Society* 113 (1995): 165–223.

Li, Chun, Xiao-Chun Wu, Olivier Rieppel, Li-Ting Wang, and Li-Jun Zhao. "An Ancestral Turtle from the Late Triassic of Southwestern China." *Nature*, November 27, 2008, 497–450.

Orenstein, Ronald. *Turtles, Tortoises, and Terrapins: A Natural History*. New York: Firefly Books, 2012.

Wyneken, Jeanette, Matthew H. Godfrey, and Vincent Bels. *Biology of Turtles: From Structures to Strategies of Life*. Boca Raton, Fla.: CRC Press, 2007.

13 THE ORIGIN OF SNAKES | *HAASIOPHIS*

WALKING SERPENTS

Then the Lord God said to the woman, "What is this you have done?" The woman said, "The serpent deceived me, and I ate." So the Lord God said to the serpent, "Because you have done this, Cursed are you above all livestock and all wild animals! You will crawl on your belly and you will eat dust all the days of your life. And I will put enmity between you and the woman, and between your offspring and hers; he will crush your head, and you will strike his heel."

GENESIS 3:13–15

GOODNESS, SNAKES ALIVE!

If ever there were creatures in the animal kingdom that provoke strong reactions in people, it is snakes. They are among the most hated and feared of all the animals, yet most snakes are actually beneficial to humans because they kill rodents and other pests. But many people have a strong, often irrational fear of most snakes that can become a true, paralyzing phobia (ophidiophobia). Their cold stare with unblinking eyelids, their flicking tongues, and their slithering about without legs are unnerving to many people.

Surely, however, the biggest factor for the nearly universal fear of snakes is that some are venomous. In Australia, the 10 most common snakes are extremely dangerous, so this fear is justified. A high percentage of snakes in tropical Africa and Asia are venomous as well. In the United States, though, the only common venomous snakes are rattlesnakes, copperheads, and cottonmouths. And they are greatly outnumbered by the harmless ones that we routinely slaughter. Most people do not allow a snake to live, let alone get close to one or try to study and understand it. The exceptions, of course, are the people who love all of natural history, especially those whose fascination with reptiles leads to a serious interest (and perhaps even a career) in herpetology.

Probably because of our long evolutionary history of living with dangerous snakes, snakes have long had a big impact on human culture, often being featured in myths and legends. In ancient Egypt, the cobra adorned the crown of Pharaoh, while Medusa, a Gorgon of Greek mythology, had snakes on her head instead of hair. Hercules had to kill the Lernean Hydra by cutting off its nine snake heads, each of which grew a new head as soon as it was severed. The Greeks also revered snakes in medicine, so the symbol of healing, the caduceus, is a staff with two entwined snakes. Snakes are worshipped in the Hindu and Buddhist religions. For example, the neck of the Hindu god Shiva is wrapped by cobras, and Vishnu is depicted as sleeping on a seven-headed snake or within a snake's coils. In addition, snakes were an important part of Mesoamerican mythology and religion as well. The Chinese have long revered snakes, as well as eaten them in their cuisine as a delicacy. One of the twelve signs of the Chinese zodiac is the snake. And, of course, in Genesis 3:1–16, the serpent in the Garden of Eden tempts Eve with the fruit of the Tree of Knowledge of Good and Evil. There are even modern Christian groups that practice snake-handling as a form of worship (and most handlers are bitten and eventually die).

Whatever your personal feelings about snakes, they are clearly one of the most successful and diverse groups of animals on Earth. Despite their highly specialized, predatory lifestyle (none eat anything but live prey), more than 2900 species are clustered in 29 families and dozens of genera. They are found from the Arctic Circle in Scandinavia to Australia in the south, and on every continent except Antarctica. They live as high as 4900 meters (16,000 feet) in the Himalayas and below sea level in warm coastal waters from the Indian Ocean to the western Pacific. Many islands have no snakes (Hawaii, Iceland, Ireland, New Zealand, most of the South Pacific), but not necessarily because St. Patrick or anyone else drove them out. More likely, it was impossible for snakes to reach these islands from the nearest mainland, even when sea level dropped during the last peak glaciation and most land mammals were able to walk to distant islands. Some of these islands (such as Ireland) were almost completely under sheets of ice, while others (such as Hawaii) were just too remote.

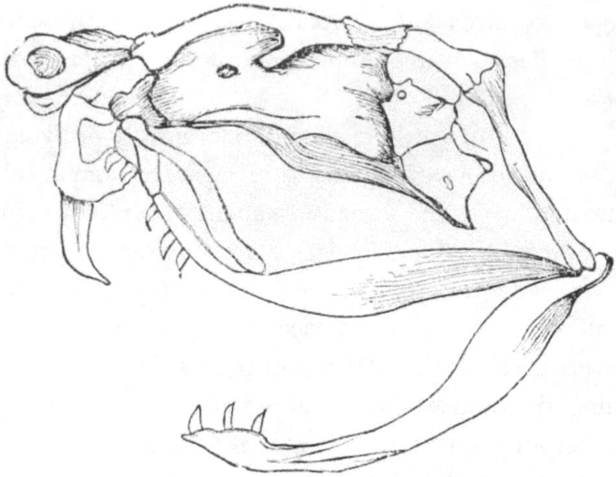

Figure 13.1 ▲
Skull of a snake, showing the delicate struts of bone. (Courtesy Wikimedia Commons)

Snakes have many remarkable features, some of which are unique to them. Their skull is composed of a series of small bony struts linked by highly elastic ligaments and tendons (figure 13.1). Thus they can stretch and wrap their entire head around a prey animal and then slowly ratchet the jaws up the body until it is completely swallowed. Meanwhile, they can hold their breath until the prey is past their throat. For weeks, they slowly digest their meal whole. During this time, snakes are often in torpor and in hiding while the difficult process of digestion of a solid unchewed carcass takes place. The bulge of the prey can be seen moving through their body as digestion proceeds.

Although some snakes have good eyesight, the majority can see only a blurry image of their surroundings and tend to be better at tracking movement; a few are blind. Instead of eyesight, most snakes flick their forked tongue to "taste" the smells in the air, using the Jacobsen's organ on the roof of the mouth to "taste" the scents brought in by the tongue. In addition, many snakes have heat-sensing pits on their snouts that allow them to detect the presence of warm-blooded animals (both predators and prey). Snakes have lost their external ears, and most of them "hear" by feeling vibrations alongside their lower jaw. (That is one of the reasons that "snake charming" is bunk. Since the snake must keep its lower jaw to the ground in order to hear, when it rears up, it is responding to the movements of the "snake charmer," not to the sound of the flute.)

Behind the skull are almost 200 to 400 vertebrae. In contrast, humans have only 33 vertebrae, and most animals with tails have about 50. The attached ribs make up nearly the entire body of snakes. The ribs are covered by a criss-crossing truss of muscles that allow snakes to control their movements, as well as propel themselves along with a variety of sinuous motions. The body consists mostly of a very elongated trunk region (rib cage) and a short tail. Inside the long body usually are two lungs, with the left lung being highly reduced (or sometimes absent) due to the limited space in the narrow body. All the other paired organs, such as the kidneys and gonads, are staggered along the length of the body. The most primitive snakes (especially the boas and their relatives) retain vestiges of their hip bones and thighbones, which no longer function as limbs but serve in courtship and sexual combat. These vestigial bones demonstrate the ancestry of snakes in four-legged animals.

Snakes show an enormous range of size for such a restrictive body plan. The smallest is the Barbados threadsnake, only about 10 centimeters (4 inches) in length, which could curl up easily on a dime. Most snakes are about 1 meter (3.3 feet) long, big enough to subdue their normal prey of rodents and other small mammals and birds (and, occasionally, other snakes). At the other extreme are the reticulated python and the anaconda, two huge boa constrictors. The anaconda is a specialized swimmer that drags its prey underwater as it crushes the air out of it. It can reach 6.6 meters (22 feet) in length, and up to 70 kilograms (154 pounds) in weight. The reticulated python is not as heavy, but can be a bit longer, reaching 7.4 meters (24 feet). Both of these snakes are so large that they can swallow big prey, such as goats, sheep, small cattle, and capybaras. But they are nothing compared with the giants of the past.

The recently discovered *Titanoboa*, from deposits in Colombia that date to the Paleocene (60 to 58 million years ago), shatters the records held by living snakes like the anaconda (figure 13.2). Although it is known from hundreds of vertebrae and part of the skull, the size of these bones is so enormous that the entire snake is estimated to have reached about 15 meters (50 feet) in length, as long as a school bus, and weighed about 1135 kilograms (2500 pounds). *Titanoboa* lived about 5 million years after the giant dinosaurs

A
B

vanished. In the tropical swamps of Colombia, it lived alongside gigantic crocodilians and turtles as well as other huge reptiles. Their gigantic size was probably due to the absence of large mammalian predators, which were yet to evolve, or large dinosaurs. Thus the niche for giant predator was occupied by such reptiles as snakes, crocodiles, and turtles.

Titanoboa broke the previous record held by *Gigantophis*, a monster snake in the extinct Gondwana family Madtsoiidae, from beds in Egypt and Algeria that date to the Eocene (40 million years ago). *Gigantophis* reached 10.7 meters (35 feet) in length, still much longer than the largest anaconda or reticulated python. Another huge snake of the family Madtsoiidae was *Wonambi*, an inhabitant of Australia during the last Ice Age. It reached 6 meters (20 feet) in length, one of the largest reptiles and biggest predators that Australia has ever seen. Its head, however, was small, so it could not have eaten the rhinoceros-size wombat relatives called diprotodonts or the gigantic kangaroos of the Ice Age in Australia, but most other game was within reach. It died out about 50,000 years ago, along with the bulk of the Australian megafauna of marsupial mammals.

WHENCE THE SERPENT?

Snakes are a marvel of adaptation and success, and have been so ever since the dinosaurs vanished from the planet. But where did they come from? How does a four-legged reptile turn into a snake? Where are the transitional fossils that demonstrate this evolution?

Actually, becoming legless is the simplest part of the transformation. It has happened in many different groups of four-legged animals, all independently evolved. The examples of leglessness among reptiles include not only the snakes, but an entire group of living reptiles called the amphisbaenians, as well as several groups of lizards, including some skinks, the Australian flap-footed lizards, "slow worms," and "glass lizards." Among amphibians, the apodans, or caecilians, developed worm-like bodies, while the sirens have only stunted forelimbs and no hind limbs. In addition, at

Figure 13.2 ◀

Titanoboa: (*A*) Jonathan Bloch comparing a large vertebra of *Titanoboa* with a much smaller vertebra of an anaconda; (*B*) life-size reconstruction of its appearance in life. ([*A*] photograph by Jeff Gage/Florida Museum of Natural History; [*B*] photograph courtesy Smithsonian Institution)

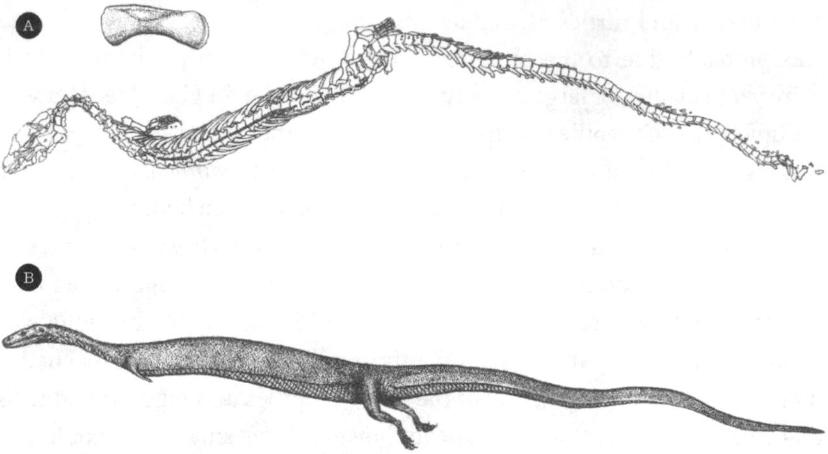

Figure 13.3 ▲

The transitional fossil *Adriosaurus*, which had tiny forelimbs but fully functional hind limbs: (*A*) skeleton; (*B*) reconstruction of its appearance in life. (Courtesy M. W. Caldwell)

least two extinct groups of amphibians, the aistopods and lysorophids, became limbless as well. Nearly every one of these animals is a burrower, so the loss of limbs appears to aid in digging through the ground or soft mud. There is a simple reason why losing all the limbs is so easy. The development of the limb buds and, eventually, the limbs is controlled by a specific set of Hox genes and of Tbx genes, so all it takes is for those genes to shut off the commands to develop limbs, and the limbs vanish.

Nonetheless, finding a fossil snake caught in the act of losing its limb seems to be extremely unlikely. Most snakes do not fossilize, since they are built of hundreds of delicate vertebrae and ribs that are usually broken and disassociated, and only a handful of snakes are known from partial or complete articulated skeletons. The vast majority of fossil snakes are known from only a few vertebrae, so the diagnostic characteristics of these creatures must come from little details of the spinal column.

Despite all these obstacles, the prehistoric record has produced a remarkable set of fossils that document the transition from four-legged lizards to legless snakes. The first stage is represented by a number of fragmentary fossils from the Jurassic. Then there is a fossil known as *Adriosau-*

rus microbrachis, which was found in 2007 in rocks in Slovenia that date to the middle Cretaceous (about 95 million years ago) (figure 13.3). Its name means "Adriatic lizard with small arms." *Adriosaurus* was an extremely slender, long-bodied marine lizard that had fully functional forelimbs but vestigial, nonfunctional hind limbs.

Next came a wide variety of snakes that had lost their forelimbs, but still had tiny nonfunctional hind limbs. For example, *Najash rionegrina* was a burrowing land snake described in 2006 from the Candeleros Formation in Argentina and dating to about 90 million years ago. (Nahash is a biblical Hebrew name for the serpent in the Garden of Eden.) *Najash* still had pelvic bones, the vertebrae that attach to the pelvis, and vestigial hind limbs that retained the thighbone and shin bone.

Even more specialized and snake-like are a series of extraordinary fossil snakes from the Late Cretaceous marine rocks of Israel and Lebanon. The most complete of these fossils is *Haasiophis terrasanctus* (figure 13.4). Its name means "Haas's snake from the Holy Land," after the Austrian paleontologist Georg Haas, who found the locality and was describing the fossil before he died in 1981. *Haasiophis* was discovered in the limestones of the Ein Yabrud locality in the Judean Hills, near Ramallah on the West Bank, and is about 94 million years old. It is a nearly complete skeleton, missing only the tip of its tail, and is about 88 centimeters (35 inches) long. The skull and most of the vertebrae look much like those of the other primitive snakes. But the hind limbs are still present and very tiny, including the thighbone, both tibia, and part of the feet. Unlike the hind limbs of *Najash*, the hip bones of *Haasiophis* are tiny and are no longer attached to the spinal column, so they are completely vestigial and useless. *Haasiophis* and many other marine snakes of the Cretaceous apparently had a vertical fin or paddle-shaped tail, as do living sea snakes.

A slightly larger snake from the Ein Yabrud locality is *Pachyrhachis*, described by Haas in 1979. Although its fossils are less complete than those of *Haasiophis*, it also has tiny vestigial hind limbs on its 1-meter (3.3-foot) long body. Its ribs and vertebrae are very thick and dense, which would have helped it in diving in the Cretaceous seas.

A third snake from the marine limestone of the Middle East is *Eupodophis descouensi*, which was found in rocks about 92 million years old in Lebanon (not far from Ein Yabrud) (figure 13.5). (Its genus name means "good limbed snake," and its species honors the French paleontologist

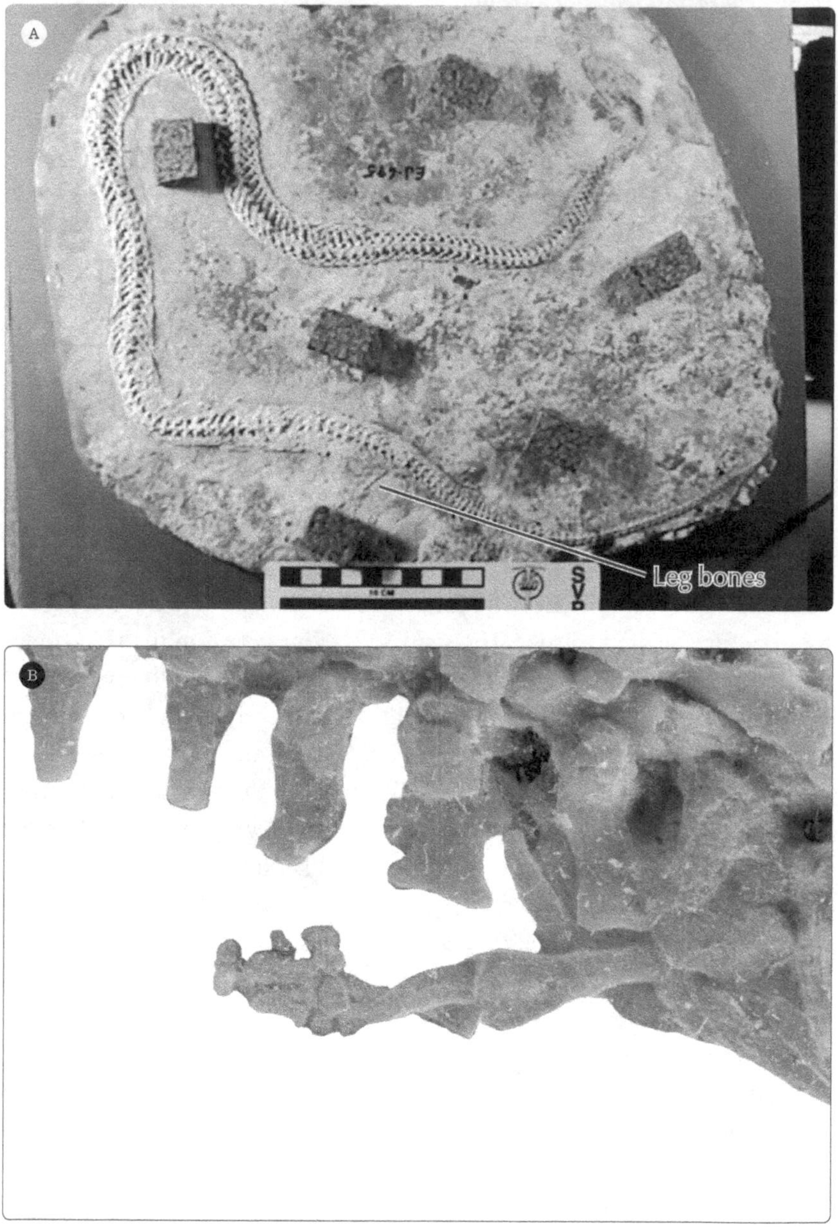

Figure 13.4 ▲

The two-legged snake *Haasiophis*: (*A*) complete articulated skeleton with the vestigial hind limbs preserved (the large dark blocks are cork spacers to protect the specimen from anything stacked on top of it); (*B*) detail of the vestigial hind limbs. (Courtesy M. Polcyn, Southern Methodist University)

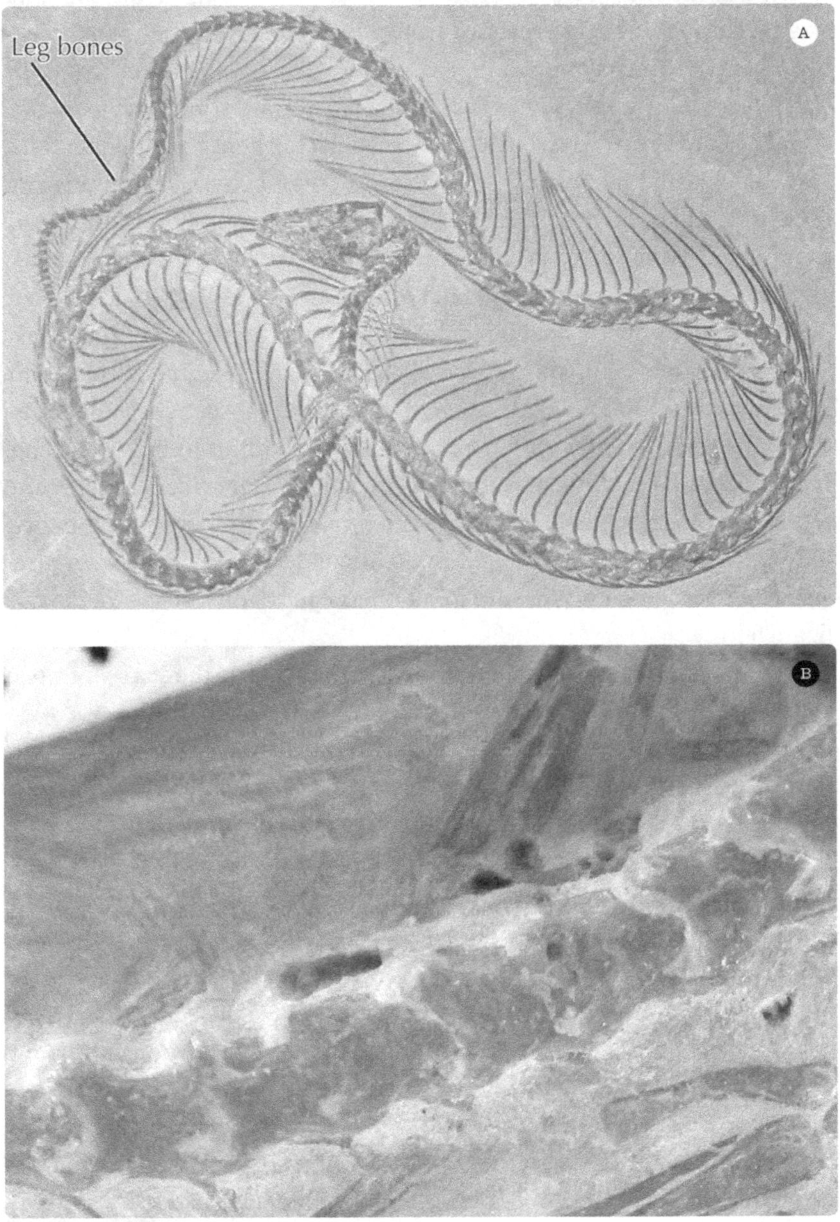

Figure 13.5 ▲

The two-legged snake *Eupodophis*: (*A*) complete skeleton with the vestigial hind limbs preserved; (*B*) detail of the spine, showing the vestigial hind limbs. (Courtesy M. W. Caldwell)

Didier Descouens.) It was 85 centimeters (34 inches) long, about the same size as *Haasiophis*, but its limbs are even more reduced and tiny than those of the other two Cretaceous two-legged snakes: *Haasiophis* and *Pachyrhachis*.

Thus not only the vestigial hind limbs of several extinct marine snakes from the Late Cretaceous, but also the vestigial hip bones and thighbones—sometimes with tiny "spurs" projecting from their bodies—of primitive extant snakes like the boas and their relatives, are mute but powerful testimony of the evolution of snakes from creatures with legs.

But from what ancestor did snakes descend? The earliest ideas were proposed in the 1880s by the pioneering paleontologist and herpetologist Edward Drinker Cope, who noticed that snakes have many anatomical similarities to the monitor lizards, such as the goanna of Australia and the Komodo dragon of Indonesia (and even more similarities to the Cretaceous marine lizards known as mosasaurs). The anatomical evidence still seems to support the relationship of snakes and monitors, although recent molecular data do not but are ambiguous. Some molecular sequences do place snakes closest to monitor lizards, but others put them outside any extant lizard family.

The view that snakes lost their legs when they took to the sea seems to be supported by the many fossils of marine snakes from the Cretaceous of the eastern Mediterranean (Slovenia, Israel, Lebanon). According to this scenario, the loss of external ears and the fused transparent eyelids of snakes make sense as adaptations for swimming, rather than for burrowing.

Another school of thought argues that snakes evolved from burrowing lizards, not swimming lizards, like the earless burrowing monitor *Lanthanotus* of Borneo. To proponents of this idea, the clear eyelids of snakes would protect the eyes against the abrasion of grit while burrowing, and the absence of external ears would keep dirt out of the ear region. The terrestrial adaptations of *Najash* are consistent with this view, although *Najash* is slightly later in time than the marine snakes *Haasiophis*, *Pachyrhachis*, and *Eupodophis*. The most primitive of all known snakes, however, is *Coniophis*, which had the head of a lizard but a body like a snake, although the fossil is too incomplete to determine what limbs it may have had. Nevertheless, it was terrestrial, not marine. Yet the aquatic lizard *Adriosaurus* is an even more primitive snake relative, and it had four limbs and swam in the ocean.

Thus the mystery of the nearest relatives of snakes is still unsolved. This is how science marches on, and controversies like this are essential for the scientific process so that we continually scrutinize the evidence and keep our options open. No matter how this debate is resolved, the fact that many fossils exhibit features of the transition from four-legged to two-legged to legless shows that snakes did evolve from four-legged ancestors.

SEE IT FOR YOURSELF!

A life-size model of *Titanoboa* is displayed at the University of Nebraska State Museum, Lincoln.

FOR FURTHER READING

Caldwell, Michael W., and Michael S. Y. Lee. "A Snake with Legs from the Marine Cretaceous of the Middle East." *Nature*, April 17, 1997, 705–709.

Head, Jason J., Jonathan I. Bloch, Alexander K. Hastings, Jason R. Bourque, Edwin A. Cadena, Fabiany A. Herrera, P. David Polly, and Carlos A. Jaramillo. "Giant Boid Snake from the Paleocene Neotropics Reveals Hotter Past Equatorial Temperatures." *Nature*, February 5, 2009, 715–718.

Rieppel, Olivier. "A Review of the Origin of Snakes." *Evolutionary Biology* 25 (1988): 37–130.

Rieppel, Olivier, Hussan Zaher, Eitan Tchernove, and Michael J. Polcyn. "The Anatomy and Relationships of *Haasiophis terrasanctus*, a Fossil Snake with Well-Developed Hind Limbs from the Mid-Cretaceous of the Middle East." *Journal of Paleontology* 77 (2003): 536–558

KING OF THE FISH-LIZARDS

She sells sea-shells on the sea-shore.
The shells she sells are sea-shells, I'm sure.
For if she sells sea-shells on the sea-shore
Then I'm sure she sells sea-shore shells.

SHE SELLS SEASHELLS BY THE SEASHORE

In the late eighteenth century, the seaside town of Lyme Regis on the Dorset coast of southern England was a popular summer tourist destination for the rich and fashionable to splash in the waves and enjoy the cool sea breezes. Gathering shells was a popular hobby, as was collecting fossils and other curiosities. No one thought of fossils as anything more than quaint objects found in the rocks, suitable for naming and labeling, but not revealing anything not already known from the book of Genesis. No one knew about dinosaurs yet (not discovered until the 1820s and 1830s) or about most of the extinct life on the planet. Indeed, most people (especially scholars) denied that extinction could even happen, because God looked after even the humblest sparrow and would not allow even one of his creations to die out. As Alexander Pope's poem *An Essay on Man* (1733) put it, "Who sees with equal eye, as God of all, / A hero perish, or a sparrow fall." By 1795, a humble British surveyor and canal engineer named William Smith began to notice that fossils occurred in a definite sequence across all of Britain, but it was another 20 years before his discovery began to be understood.

In Lyme Regis, a poor cabinetmaker named Richard Anning and his wife, Molly, were trying to eke out a living. He and his wife had many children, but nearly all died in infancy, as was common in those days of poor medicine and deadly childhood diseases with no cure. Their oldest daughter died at age four when her clothes caught on fire. Five months after this tragedy, in 1799, Mary Anning was born. When she was 15 months old, lightning struck and killed three women in the village, but did not harm baby Mary, who was being held in the arms of one of the women. Mary had only limited schooling in church, where she learned to read and write, but education for working-class women was rare in those days. As soon as Mary was old enough, she joined her father and her older brother, Joseph (the only other surviving sibling), on their trips to collect fossils along the sea cliffs in the Lower Jurassic (210 to 195 million years ago) Blue Lias beds. They were full of "snake-stones" (ammonites), "devil's fingers" (belemnites), "devil's toenails" (the oyster *Gryphaea*), and "verteberries" (vertebrae). Many of the townspeople collected fossils to sell to rich tourists during the summer as a supplement to their meager incomes during the hard years of the early nineteenth century. In addition, the Anning family encountered further discrimination, since they were Dissenters from the Anglican Church and thus shut out of many parts of life.

Tragedy struck again in November 1810 when Richard Anning, suffering from the effects of tuberculosis and dangerous falls while fossil collecting on the cliffs, died at the age of 44. Molly, Joseph, and Mary (then only 11 years old) were obliged to collect fossils full time in the hope of earning a small income. The first lucky find happened a year later, when Joseph discovered an amazing skull more than 1.2 meters (4 feet) long embedded in the rock and chiseled it out; Mary later found the rest of the skeleton. First identified as a "Crocodile in Fossil State" because of its long snout, it was bought and sold by a series of wealthy collectors.

In 1814, the specimen was described by Everard Home, but he could make no sense of it (figure 14.1). He classified it as a fish because it was aquatic and had vertebrae that resembled those of fish, but he recognized its many reptilian features as well. He considered it to be a "missing link" on the "great chain of being" between fishes and reptiles. However, he was not implying that one evolved from the other, an idea that was still 40 years in the future. Then in 1819, Home decided that it was a link between a lizard and the salamander *Proteus*, so he named it Proteo-Saurus.

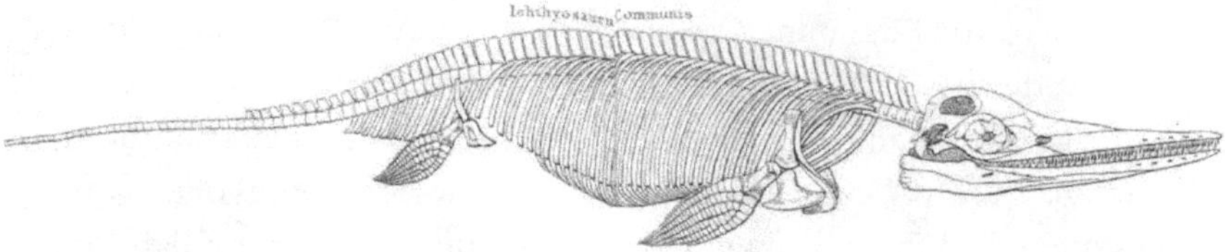

Figure 14.1 ▲
William Conybeare's illustration of the first known fossil of an ichthyosaur. (From William Conybeare, "Additional Notices on the Fossil Genera Ichthyosaurus and Plesiosaurus," *Transactions of the Geological Society of London*, 2nd ser., 1 [1822])

In 1817, Charles Dietrich Eberhard Konig (born Karl Dietrich Eberhard König), curator of the Department of Natural History at the British Museum, informally called the fossil *Ichthyosaurus* (from the Greek words for "fish" and "lizard") because he realized that it has features of both fish and lizards. By May 1819, he was able to purchase it for the British Museum, where it still resides. In 1822, British geologist William Conybeare formally described and named the specimen, along with many others, as *Ichthyosaurus*, making the name valid for all later fossils of this kind (and eliminating the need to use the name Proteo-Saurus).

Meanwhile, Joseph Anning reduced his collecting when he became an apprentice to an upholsterer, and Mary had to support the family with her fieldwork. Most of her best finds were made during the stormy winter months, when the waves caused the cliffs to erode and create fresh exposures of fossils. They were also the most dangerous months, since the cliff could collapse at any time or the waves could sweep away collectors if they misgauged the time of low tide. As the Bristol *Mirror* said of her in 1823:

> This persevering female has for years gone daily in search of fossil remains of importance at every tide, for many miles under the hanging cliffs at Lyme, whose fallen masses are her immediate object, as they alone contain these valuable relics of a former world, which must be snatched at the moment of their fall, at the continual risk of being crushed by the half suspended fragments they leave behind, or be left to be destroyed by the returning tide: —to her exertions we owe nearly all the fine specimens of Ichthyosauri of the great collections.

Anning had several close calls. In October 1833, she barely escaped being buried alive in a landslide that killed her constant companion, a black-and-white terrier named Tray (figure 14.2). As she wrote to her friend Charlotte Murchison later that year, "Perhaps you will laugh when I say that the death of my old faithful dog has quite upset me, the cliff that fell upon him and killed him in a moment before my eyes, and close to my feet . . . it was but a moment between me and the same fate."

But her perseverance was rewarded. In 1823, she found the first complete specimen of a long-necked plesiosaur (see figure 15.5), which further baffled the British scientific community. A year later, she discovered the first fossil of a pterosaur known outside Germany. She collected numerous fossil fish that were described by other scientists, as well as many ammonites and other molluscs. She found evidence of an ink sac in the bullet-shaped shells known as belemnites, proving that were the fossils of an extinct squid. She realized that what people had been calling "bezoar stones" were actually fossil feces, which William Buckland later published as his own idea and called them coprolites. Even though she had received little education, she read every scientific paper she could find and often hand-copied them (including the detailed illustrations). In 1824, Lady Harriet Silvester wrote of her:

> The extraordinary thing in this young woman is that she has made herself so thoroughly acquainted with the science that the moment she finds any bones she knows to what tribe they belong. She fixes the bones on a frame with cement and then makes drawings and has them engraved. . . . It is certainly a wonderful instance of divine favour—that this poor, ignorant girl should be so blessed, for by reading and application she has arrived to that degree of knowledge as to be in the habit of writing and talking with professors and other clever men on the subject, and they all acknowledge that she understands more of the science than anyone else in this kingdom.

By 1826, when Anning was only 27, she had saved enough money to open her own shop, where nearly all the famous geologists and paleontologists called to visit her and buy fossils. They included Louis Agassiz, William Buckland, William Conybeare, Henry De la Beche, Charles Lyell, Gideon Mantell, Roderick Murchison, Richard Owen, and Adam Sedgwick, among others. American collectors established their own museums with her fossils, and royalty from several countries bought her best specimens.

Figure 14.2 ▲

The only known portrait of Mary Anning, shown with her rock hammer, collecting bag, heavy garments, and terrier Trey, who was killed in a landslide while she was collecting. (Courtesy Wikimedia Commons)

Figure 14.3 ▲

An ichthyosaur and two plesiosaurs battling off the coast of Lyme Regis, painted by Henry De la Beche in 1830. One of the first depictions of a known prehistoric scene, it is considered to be the first piece of art in a genre now called paleoart. Lithographs of this scene were sold to raise money for Mary Anning. Plesiosaurs and ichthyosaurs were well known in the early nineteenth century, but dinosaurs had not yet been discovered. (From Henry De la Beche, *Duria Antiquior—A More Ancient Dorset* [London, 1830])

Despite their high opinion of Anning, however, the British gentlemen who had founded the discipline of modern geology in the early nineteenth century did not accept her as their equal because of her low social class and Dissenter religious opinions. She later converted to the Church of England to remove this obstacle. All her amazing specimens were described by the rich gentlemen who bought them, with little or no credit given to her collecting or preparation of the fossils. None of her ideas reached print during her lifetime, since there were no opportunities for her to publish for herself. She died in 1847 of breast cancer, but by that time members of the British geological community had come to appreciate her importance. They raised money to help her during her final months and paid her funeral expenses, installed a stained-glass window in her honor in her church, and eulogized her at their meetings (an honor reserved only for members). She was even the subject of an article by Charles Dickens. According to many people, the

Figure 14.4 ▲
"Awful Changes," a satirical cartoon drawn by Henry De la Beche in 1830, showing Professor Ichthyosaurus lecturing on the strange creature from the previous creation (the human skull). Charles Lyell was one of the last geologists to deny the reality of extinction and to believe that the history of Earth was cyclic, with extinct species returning in a later incarnation. Eventually, Lyell had to concede that the fossil record shows directional change and that extinct species never return. (From Henry De la Beche, *Duria Antiquior—A More Ancient Dorset* [London, 1830])

tongue-twisting poem "She sells seashells by the seashore" was written about her as well.

Today, Anning is recognized as not only the first and greatest woman paleontologist, but also one of the pioneers of paleontology. Her discoveries transformed the view of the world she had been born into. By the 1830s, people had begun to think hard about the implications of the extinct ichthyosaurs and plesiosaurs, and to talk about a dark deadly "antediluvian world" in whose seas these monsters had swum (figure 14.3). A few years later, dinosaurs were added to this scenario. Even though in the 1820s and earlier, Baron Georges Cuvier had shown that mammoths and mastodonts and giant ground sloths must be extinct, it was the enormity of the extinction of such large animals as ichthyosaurs and plesiosaurs that finally made scientists reconsider their literal interpretations of Genesis. They looked at the bizarre antediluvian world with horror, especially given the baleful glare of the ichthyosaur eyes. Some, like Lyell, argued that Earth has gone through cycles and extinct animals have reappeared (figure 14.4). But eventually, most scholars were forced to reject the idea of a perfect creation with no change or the reality of Noah's flood. Without intending to, Mary Anning, a devout and humble woman who was only making a living by collecting and selling fossils, laid the foundation for an enormous revolution in scientific thinking before she died at the young age of 47.

THE "FISH-LIZARDS"

Mary Anning's discoveries opened the door to the world of an amazing group of animals: the ichthyosaurs. Dinosaurs were known in the early nineteenth century merely from fragments of teeth and jaws, and thus were poorly understood until complete skeletons were found in the 1880s (chapter 17). In contrast, fossils of ichthyosaurs were often found complete or nearly complete. This allowed naturalists to quickly ascertain that these creatures were indeed reptiles, yet with a body form that closely mimicked that of dolphins and whales, thanks to convergent evolution. Most ichthyosaurs had a long narrow pointed snout and many sharp conical teeth for catching swimming prey. Most had large eyes, apparently for seeing in murky water, and the eyes of some species were protected by a ring of small bones around the pupil of the eyeball called a *sclerotic ring.* The bones of Later Jurassic ichthyosaurs show signs of decompression sickness, demonstrating that they were deep divers that often suffered the effects of holding their breath for a very long time and of nitrogen being released from their blood as they rose from the deep waters.

The head of ichthyosaurs merged with their body, as in many aquatic animals that are streamlined for full-time swimming. Recent estimates put their fastest speeds at 2 kilometers (1.2 miles) an hour, a bit slower than the fastest living dolphins and whales. Their dolphin-like body sported a dorsal fin (analogous to those in dolphins and fish), supported by cartilage but not by bone and visible only on specimens with soft-tissue preservation.

But their hands were modified into flippers made of dozens of little discs of bone formed by the multiplication and division of finger bones into many tiny parts. Their hind feet were modified into much smaller paddles (lost altogether in whales and dolphins), apparently not used much for propulsion during swimming. The rear of the body tapered into a fish-like tail with flukes aligned in the vertical plane, so ichthyosaurs swam with a side-to-side motion of the tail part of the body (as do most bony fish).

The last vertebrae of the spine bent downward sharply in a "kink" to support the lower lobe of the tail; the upper lobe was not supported by bone, only cartilage. In the early days of ichthyosaur research, scientists puzzled over this "kink" of the tail vertebrae and thought that it might be an artifact of preservation, possibly due to the drying and contraction of tendons. But Richard Owen correctly inferred that it was a product of a bi-lobed tail fluke, and his insight was confirmed in the late nineteenth century when the amazing locality of Holzmaden in Germany was found. At this site, fossils are preserved with dark outlines of their soft tissues. This allowed paleontologists to see the nature of the upper lobe of the ichthyosaur tail for the first time, as well as the outline of the dorsal fin (which is not supported by bone, so it is usually not visible).

Much is known about ichthyosaur paleobiology, since there are numerous well-preserved complete articulated skeletons, often with soft-tissue outlines and stomach contents. Most ichthyosaurs are thought to have fed like dolphins and whales, rapidly catching swimming prey (squids and belemnites, ammonites, fish, and the like) with their long toothy beaks, and this is confirmed by preserved stomach contents. Some early ichthyosaurs had blunt crushing teeth for eating molluscs, while others had toothless bills and were thought to have fed by suction (as do many fish). In many specimens, we find evidence that they ate smaller ichthyosaurs. A number of predators were willing to attack them, leaving scars on their faces and bones. Some ichthyosaurs had a short lower jaw and an upper jaw modified into a long sword that they may have used for slashing at schools of fish to disable some prey (as do swordfish and sailfish).

Early on, scientists speculated about how such a completely aquatic animal could have moved on land, especially to lay eggs on a beach (as do sea turtles), considering that its flippers were not large enough to allow it to drag itself out of the water and across the sand. Then some specimens from Holzmaden that showed a baby emerging tail first from what could have been the birth canal of the mother confirmed what scientists had guessed all along: ichthyosaurs gave birth to live young, following internal fertilization, and never laid eggs on land (just as dolphins and whales bear their young in the sea, nurturing them to their first breath).

In short, ichthyosaurs show amazing convergent evolution on the dolphin body plan, yet they were reptiles, not mammals, and displayed many fundamental differences from mammals. But where did these highly specialized creatures come from?

THE ORIGIN OF ICHTHYOSAURS

As we do for plesiosaurs (chapter 15), we have an excellent series of transitional fossils that show how ichthyosaurs originated (figure 14.5). First there is *Nanchangosaurus* from the Early Triassic of China. It has a normal reptilian body, except that it shows the longer snout seen in all ichthyosaurs. When it was first described, paleontologists were not even sure in what group to classify it, since it has so many primitive reptilian features, but the precocious skull points toward ichthyosaurs.

Then comes *Utatsusaurus* from the Early Triassic of Japan. It has the more streamlined, torpedo-like body form of ichthyosaurs, yet the hands and feet are primitive and not yet modified into flippers. It has the long ichthyosaur snout, but there is no downward kink in the tail vertebrae, just a gentle bend that suggests small lobes on the upper fluke. Next is *Chaohusaurus* from the Early Triassic of China, which has a fully ichthyosaurian skull, with short snout, simple teeth, and large eyes. But its robust limbs are just beginning to show the signs of developing into the typical paddles of ichthyosaurs, and there is only a slight kink in the tail for the upper-fluke lobe.

Even more specialized is *Mixosaurus* from the Middle Triassic of Germany and other places (see figure 14.5). It is a classic transitional fossil, halfway between advanced ichthyosaurs and their more primitive ancestors. It has the fully dolphin-like body, long snout, large eyes, and dorsal fin of most ichthyosaurs. The hands and feet are clearly modified into flippers, but the number of finger and toe bones has not yet multiplied. The tail shows an even better developed downward bend than that of *Chaohusaurus*, with just a small lobe on the upper fluke. The Late Triassic *Californosaurus* is even more specialized, with an even more modified front paddle and the first sign of reduction of the hind paddle, along with a sharper downward bend

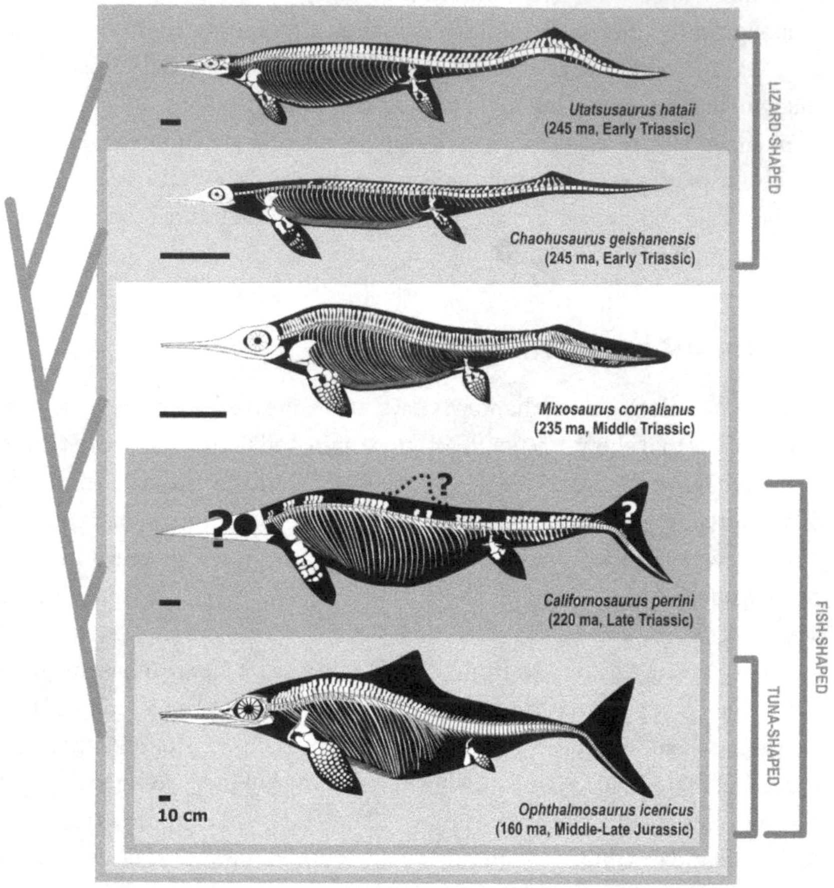

Figure 14.5 ▲

The evolution of ichthyosaurs from more primitive reptiles during the Triassic. (© Ryosuke Motani)

in the tail. It presumably had an upper fluke on its tail fin, although it is not well enough preserved to tell.

All these intermediate forms gradually acquired the standard body plan of the fully advanced Jurassic ichthyosaur, such as *Ophthalmosaurus* (see figure 14.5): long toothy snout, small skull with huge eyes protected by sclerotic rings, completely streamlined body with a dorsal fin, large front flippers with extra finger bones, small hind flippers with extra bones as well, and the sharp downward kink of the tail vertebrae that indicates the fully symmetrical upper and lower lobes of the tail. This was the kind of creature that Mary Anning first brought to light in 1811, and now it can be traced back to reptiles that barely look like ichthyosaurs at all.

WHALE-REPTILES OF THE TRIASSIC

So far, we have looked at the normal range of ichthyosaurs, most of which were no more than 3 to 5 meters (10 to 16 feet) in length. But there were some whale-size ichthyosaurs as well. The most impressive of these was *Shonisaurus*.

The original specimens of *Shonisaurus* come from one famous locality, the Berlin-Ichthyosaur State Park in south-central Nevada (figure 14.6). It is located in West Union Canyon, at an altitude of 2133 meters (7000 feet) in the Shoshone Mountains, about a six-hour drive north from Las Vegas or a three-hour drive east from Reno—literally, in the middle of nowhere. The state park incorporates not only the fossil site, but also the mining town of Berlin, which is now a ghost town. Early miners in the area knew about the fossil ammonites and clams, and some saw the huge bones as well. Some of them used ichthyosaur bones to build hearths! In 1928, Siemon Muller of Stanford University recognized the bones as belonging to ichthyosaurs, but he did not have the resources to collect or study them.

After another 24 years passed, Margaret Wheat of Fallon, Nevada, collected some of the long-neglected fossils and sent them to Charles L. Camp of the University of California Museum of Paleontology in Berkeley. Camp was interested, so in 1953 he visited the site and decided to undertake a serious excavation and study of the fossils. After that visit, Camp wrote in his field notes:

> Si Muller says he found these ichthyosaur remains in 1929 and 30 and tried to sell the idea to us then and subsequently—Mrs. Wheat told me about them last September and said that the vertebrae were very large (up to 1 ft. in dia.) and 21 pounds weight. . . . We went up the south facing slope . . . and looked over the material exposed by Mrs. Wheat's broom. . . . It is a series of six or more vertebrae in hard limestone, and much more float below. These are monster vertebrae — larger than any ichthyosaur vertebrae hitherto known and later in age then the Middle Triassic Cymbospondylus (Leidy).

Figure 14.6 ▲

Berlin-Ichthyosaur State Park, near Gabbs, Nevada: (*A*) entrance plaza with life-size bas-relief of *Shonisaurus*; (*B*) bone bed, showing the huge area of bones, within the main building. (Photographs courtesy Lars Schmitz)

Figure 14.7 ▲

Shonisaurus: (*A*) mounted skeleton, displayed at the Nevada State Museum in Las Vegas; (*B*) reconstruction of its appearance in life. ([*A*] photograph by the author; [*B*] courtesy Nobumichi Tamura)

During the summers of 1954 through 1957, Camp plus Samuel E. Welles and the museum crew undertook to work the site in earnest (with a second effort by Camp from 1963 to 1965). They managed to excavate one nearly complete skeleton, which is on display at the Nevada State Museum in Las Vegas (figure 14.7*A*). But they left most of the bed intact, as it had been found, but with the bones cleaned and prepared so they can be seen much more clearly than when they were mostly buried.

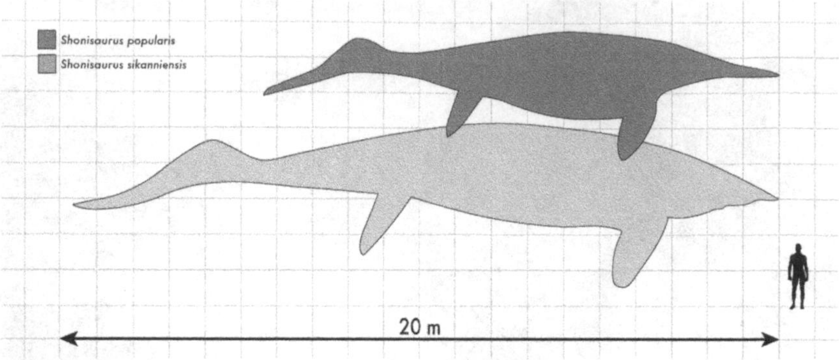

Figure 14.8 ▲
Comparison of the sizes of the ichthyosaurs *Shonisaurus popularis* and *S. sikanniensis*. (Drawing by Mary P. Williams)

While Camp and his team were working, they saw several blinding flashes and heard the booms of nuclear tests from the Nevada Test Range, only 240 kilometers (150 miles) to the south. As Camp wrote after the May 15, 1955, blast of 28 kilotons:

> The 14th big atom went off this morning at 5, 200 miles away. I sat up in bed and saw a violet-pink flash lasting a fraction of a second. About 15 min. later a low grumbling thunderous roar came in like thunder shaking the earth a little. This came in two or three crescendos. About 3–5 min. later a more subdued noise like far away growling of lions came through the air without quite so much force.

Fortunately, the radioactive fallout from these tests blew to the east and not to the north, so the paleontologists were never contaminated. Camp lived to the age of 82, dying in 1975 of cancer—but apparently never exposed to high levels of radiation. But the residents of St. George, Utah, were not so lucky.

The concentration of skeletons is staggering, with at least 40 individuals represented. Camp originally thought that they had been stranded by low tide, like beached whales, but a later study by Jennifer Hogler demonstrated that this portion of the Luning Formation (Upper Triassic [about 217 to 215 million years old]) is a deep-water deposit. Thus the reason that so many ichthyosaur carcasses sank to the bottom but were not disturbed is still a mystery. The absence of encrusting invertebrates, the relatively undisturbed bones, and the complete nature of the skeletons suggest that the very deep stagnant water could not support any scavengers or many other organisms, for that matter.

Shonisaurus was the size of a large whale, roughly 15 meters (almost 50 feet) in length. It had a long toothless snout (except when it was young), suggesting to some scientists that it did not swim fast to catch prey (figure 14.8; see figure 14.7*B*). Rather, it inhaled its meals as they swam by or, like most large whales and the whale sharks, may have fed more on plankton than on large animals. It had a deep, round body and relatively long pectoral and pelvic fins, made entirely of the huge round finger elements that result when finger bones turn into a flipper (hyperphalangy). There was apparently no dorsal fin, and like many other Triassic ichthyosaurs, it had only a small upper lobe on its tail, with just a slight downward turn of the tip of the tail vertebrae, not the sharp kink seen in ichthyosaurs of the Jurassic.

Camp took a long time to complete his study of the fossils, finally publishing his results in 1976. He named the creature *Shonisaurus* after the Shoshone Mountains and Native Americans of the area and gave it the

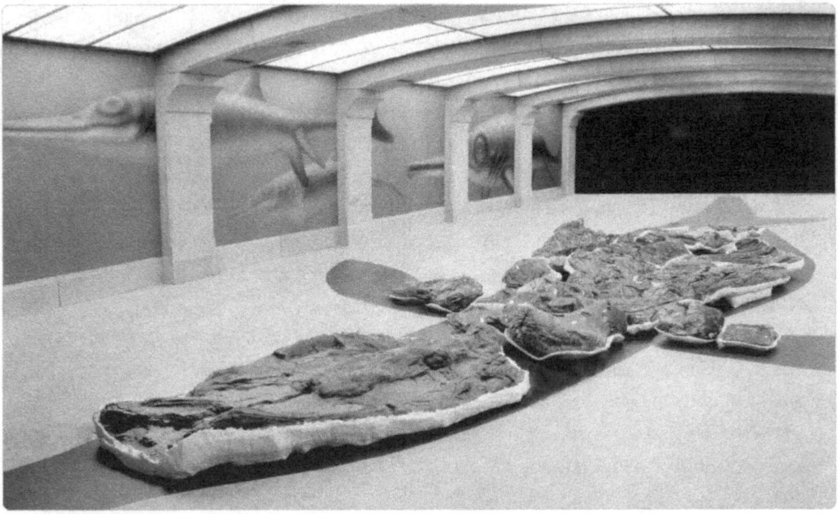

Figure 14.9 ▲

The giant British Columbia *Shonisaurus* (or *Shastasaurus*?), displayed at the Royal Tyrrell Museum in Drumheller, Alberta. (Photograph courtesy Royal Tyrrell Museum, Drumheller, Alberta)

trivial species name *popularis* (common). In the late 1950s, Wheat, Camp, and Welles realized that this gigantic creature deserved to be recognized as Nevada's State Fossil. It is certainly spectacular, unique to Nevada, and more charismatic than most of the more typical fossils found in Nevada. After many decades of lobbying, the Nevada state legislature officially recognized it in 1984.

In 2004, the late Betsy Nichols found and described an even bigger shonisaur from the Upper Triassic (210 million years old) Pardonet Formation in British Columbia (figure 14.9). Named *Shonisaurus sikanniensis*, it reached more than 21 meters (70 feet) in length, larger than most living whales! It, too, had a long toothless snout; a large, deep body with long narrow pectoral and pelvic fins; no dorsal fin; and a tail fluke with only a small lobe on the top. Since its discovery, *S. sikanniensis* has been reassigned by some to *Shastasaurus*, the genus of a much smaller ichthyosaur known from the Late Triassic of California. However, the most recent analysis, in 2013, supports the original opinion that it is a huge species of *Shonisaurus*.

SEE IT FOR YOURSELF!

Mary Anning's original fossils are displayed at the Natural History Museum in London, and many others are in the Sedgwick Museum of Earth Sciences, Cambridge University. Many museums in the United States feature excellent specimens of ichthyosaurs, including the American Museum of Natural History, New York; Carnegie Museum of Natural History, Pittsburgh; Field Museum of Natural History, Chicago; and National Museum of Natural History, Smithsonian Institution, Washington, D.C. In Germany, many museums display fossils of ichthyosaurs from Holzmaden, including the Museum für Naturkunde (Humboldt Museum), Berlin; Naturmuseum Senckenberg, Frankfurt; Paläontologisches Museum München; and Staatliches Museum für Naturkunde, Stuttgart.

The Berlin-Ichthyosaur State Park can be reached from Nevada State Highway 361 to Gabbs, first turning east on Highway 844 to Grantsville, and then east on the gravel road to the site. The almost complete skeleton of *Shonisaurus popularis* is at the Nevada State Museum in the Springs Preserve Park in Las Vegas. The giant *Shonisaurus sikanniensis* can be seen at the Royal Tyrrell Museum, Drumheller, Alberta.

FOR FURTHER READING

Callaway, Jack, and Elizabeth L. Nicholls, eds. *Ancient Marine Reptiles*. San Diego: Academic Press, 1997.

Camp, Charles L. *Child of the Rocks: The Story of Berlin-Ichthyosaur State Park*. Nevada Bureau of Mines and Geology Special Publication 5. Reno: Nevada Bureau of Mines and Geology, with Nevada Natural History Association, 1981.

Ellis, Richard. *Sea Dragons: Predators of Prehistoric Oceans*. Lawrence: University Press of Kansas, 2003.

Emling, Shelley. *The Fossil Hunter: Dinosaurs, Evolution, and the Woman Whose Discoveries Changed the World*. New York: Palgrave Macmillan, 2009.

Hilton, Richard P. *Dinosaurs and Other Mesozoic Animals of California*. Berkeley: University of California Press, 2003.

Howe, S. R., T. Sharpe, and H. S. Torrens. *Ichthyosaurs: A History of Fossil Sea-Dragons*. Swansea: National Museum and Galleries of Wales, 1981.

Wallace, David Rains. *Neptune's Ark: From Ichthyosaurs to Orcas*. Berkeley: University of California Press, 2008.

15 | THE LARGEST SEA MONSTER | *KRONOSAURUS*

TERROR OF THE SEAS

There were no real sea serpents in the Mesozoic Era, but the plesiosaurs were the next thing to it. The plesiosaurs were reptiles who had gone back to the water because it seemed like a good idea at the time. As they knew little or nothing about swimming, they rowed themselves around in the water with their four paddles, instead of using their tails for propulsion like the brighter marine animals. (Such as the ichthyosaurs, who used their paddles for balancing and steering. The plesiosaurs did everything wrong.) This made them too slow to catch fish, so they kept adding vertebrae to their necks until their necks were longer than all the rest of their body. . . . There was nobody to scare except fish, and that was hardly worthwhile. Their heart was not in their work. As they were made so poorly, plesiosaurs had little fun. They had to go ashore to lay their eggs and that sort of thing. (The ichthyosaurs stayed right in the water and gave birth to living young. It can be done if you know how.)

WILL CUPPY, *HOW TO BECOME EXTINCT*

OCEANS OF THE OUTBACK

Today, the Australian outback is a semi-desert, with the dry scrub extending for hundreds of kilometers. The rare rains come as torrential downpours, and then dry billabongs (water holes) rapidly fill up. Most of the plants are adapted to growing quickly during the few weeks of wet conditions and then surviving drought for most of the year. Tall gum trees (*Eucalyptus*) cast some shade, but they are constantly dripping sap as well and shedding both their long narrow leaves and their long strips of bark. The entire ecosystem is adapted to drought. The plants burn fiercely during the now more frequent wildfires that torch the highly inflammable sap-saturated vegetation.

The animals of the outback are equally adapted to dry conditions, from the largest herbivores, the kangaroos, to the burrowing wombats and the koalas living in the gum trees.

It is hard to imagine this parched landscape any other way, but the rocks beneath much of Australia provide evidence of a very different environment. They are limestones deposited in shallow seaways that drowned much of Australia and most other continents as well. During the middle part of the Age of Dinosaurs (Early Cretaceous [about 125 to 100 million years ago]), Earth had a global greenhouse climate. Huge submarine volcanic eruptions from superplumes in the mantle pumped enormous volumes of carbon dioxide into the atmosphere. The high concentrations of greenhouse gases in the atmosphere made the planet much warmer than ever before. Scientists estimate that carbon dioxide was possibly as high as 2000 parts per million (ppm), compared with over 400 ppm today. Naturally, ice does not last on such a warm planet, so there were no polar ice caps, no glaciers in the mountains, no ice anywhere. (Sadly, a number of recent dinosaur movies seem to be unaware of this fact, showing snowy mountains in their background scenes.)

In addition, the major continents were rapidly moving apart after having been united into the super-continent Pangaea. This rapid seafloor spreading not only pumped greenhouse gases into the atmosphere, but had other effects as well. When seafloor spreading is rapid, the mid-ocean ridge has much more total volume, since it is hot and more expanded than when spreading is slow. In contrast, a slower-spreading ridge has a longer time to cool, so it sinks steeply away from the ridge crest and is less thick. The expanded ridge volume made the ocean basins shallower, displacing water to the only place it could go—onto the continents. Also contributing to the shallower water and the sea-level rise were the buildup of gigantic plateaus of lava from the submarine volcanoes and the expansion of the increasingly warmer water (the latter a factor in the rise of global sea level today).

As a result, shallow seas drowned nearly all the continents in the Early Cretaceous. Some had been submerged by the Late Jurassic, when the global greenhouse conditions had begun. Not only was Australia mostly under water, but so was most of Europe. The shallow seas covering Europe were full of new forms of plankton, a group of tiny algae called coccolithophorids. As these planktonic algae died, their minuscule calcite shells sank to the seafloor, accumulating and solidifying into huge volumes of rock that we know as chalk. These chalky seas are exposed not only in famous places like the White Cliffs of Dover, but also across northern France, Belgium, and Holland.

North America, too, had a huge shallow marine seaway that ran across what is now the Great Plains. It connected the Gulf of Mexico with the warm Arctic Ocean. Nearly all the Plains states and provinces—from Texas and Oklahoma to Kansas and Nebraska to South and North Dakota to Alberta and Saskatchewan—are covered with immense areas of shallow marine Cretaceous shales and limestones and chalk. At the Niobrara Chalk beds of western Kansas, you will be able to collect a huge number of marine fossils, including those of giant marine reptiles, enormous fish and sea turtles (see figure 12.3), and a wide spectrum of invertebrates from ammonites to clams more than 1.7 meters (5 feet) across.

SEA MONSTER OF THE OUTBACK

But no one knew this more than a century ago. In 1899, a man named Andrew Crombie discovered a scrap of bone with six conical teeth near his home in Hughenden, in Queensland, Australia. This fragment eventually made its way to the Queensland Museum, where in 1924 the director of the museum, Heber Longman, named it *Kronosaurus queenslandicus* (the genus name for Kronos and the Greek for "lizard," and the species name in honor of where it was found). Kronos (or Cronus) was one of the Titans in Greek mythology. He overthrew his parents, Uranus and Gaia, and then ate all but one of his children so they could not overthrow him. His wife, Rhea, protected her newborn child, Zeus, and fooled Kronos by getting him to swallow the Omphalos Stone, wrapped in swaddling clothes. Eventually, Zeus conquered Kronos and forced him to vomit up his other children, who became the other Greek gods and goddesses. Zeus then sent Kronos to prison in Tartarus. Clearly, Longman wanted to evoke the titanic size of the specimen in its name. Eventually, scientists from Queensland Museum returned to Crombie's original site and found more material, including a partial skull, of *Kronosaurus*.

The mention of this huge specimen spurred the Museum of Comparative Zoology at Harvard University to mount an expedition to the area. William E. Schevill, a young graduate student in paleontology who had finished his undergraduate education at Harvard in 1927, led the six-man team in late

1931. Described as a very strong man when he undertook this expedition in his mid-twenties, Schevill carried a 3-kilogram (7-pound) sledgehammer to break limestone, and he could throw it into the air and catch it as he walked. (Schevill became an expert in whale echolocation and communication based at the Woods Hole Oceanographic Institution.) The men were instructed to collect any sort of natural history specimens for the museum. As Thomas Barbour, the director, put it, "We shall hope for specimens of the kangaroo, the wombat, the Tasmanian devil, and the Tasmanian wolf." The team returned to Harvard a year later with more than 100 specimens of fossil mammals and many thousands of insects.

After the original Harvard crew returned to the United States, Schevill, who remained in Australia, recruited some locals to undertake an expedition to explore the Lower Cretaceous beds around Richmond and Hughenden. According to Australian paleontologist John Long, Schevill asked the Australian Museum if it wanted to participate, but it showed no interest, and the Queensland Museum had no funds for the undertaking or personnel who could help.

In 1932, the team reached the Grampian Valley and Hughenden, where they found the snout of a small *Kronosaurus*. Then they heard from the owner of a station ("ranch" in Australian lingo), Ralph William Haslam Thomas, that there were some huge bones on his 8100-hectare (20,000-acre) property, called Army Downs. They apparently had been lying in the ground for years, but were too heavy to move or collect. At best, people could only break off a tooth or two with a hammer and chisel. Thus no one had taken an interest in the bones until the Harvard crew arrived. The men set up camp under a large *Bahunia* tree and regularly hunted for fresh meat. One afternoon, a local family visited them to see if they needed some fresh beef. They replied, "No thanks, we're right for meat." They had been living on kangaroo meat fried in emu fat, followed by a strong cheese and treacle.

The bones were encased in thick, hard limestone nodules, so the team had to use dynamite to excavate most of them. Schevill's assistant, nicknamed the "Maniac," was the expert in dynamiting the bones out of the ground and into more manageable pieces for transport. Most of the bones at the surface had been weathered and destroyed, so only those that lay deep in the nodules remained. Parts of the back of the skull were missing, along with most of the spine and the bones of the ribs, pelvis, and shoulder.

Figure 15.1 ▲

Mounted skeleton of *Kronosaurus*, with Alfred Romer's wife, Ruth, for scale, as displayed at the Museum of Comparative Zoology, Harvard University. (Photograph courtesy Ernst Mayr Library, Museum of Comparative Zoology, Harvard University)

Eventually, the men packed 86 wooden crates of fossils weighing over 5.5 metric tons (6 tons), which were shipped back to Boston on the steamship *Canadian Constructor* on December 1, 1932. Then the heavy blocks encased in plaster jackets were sent to the preparation labs in the museum basement, where Harvard's preparators (including "Dinosaur Jim" Jensen and Arnie Miller) began to slowly work on them. The thick limestone nodules had to be chiseled away slowly but steadily, and some parts of the specimen had to be jackhammered to break the tough rock.

The skull was prepared first, but there was no impetus to do the incredibly difficult work of cleaning the rest of the skeleton. Then in 1956, a rich donor expressed an interest in the fossil because of his family's history of chasing and sighting sea serpents. He gave the museum enough money so the preparation of the rest of the skeleton was able to be finished in three years. In 1959, the nearly complete skeleton of *Kronosaurus* was put on display at Harvard (figure 15.1). Ralph Thomas, now 93 years old, was invited to Harvard for the dedication ceremony to see his fossil on display, 27 years after he had first shown it to the museum crew. Thomas and Schevill had a tearful reunion, because each thought that the other had died during World War II.

Today, there is a small local museum in Richmond, Queensland, called Kronosaurus Korner. In front of the museum is a life-size concrete replica of *Kronosaurus* as it may have appeared in the Early Cretaceous (figure 15.2).

Figure 15.2 ▲
Kronosaurus Korner in Richmond, Queensland, Australia. (Photograph courtesy Kronosaurus Korner)

Since its discovery in Australia, *Kronosaurus* has been found in one more place: Colombia. In 1977, a peasant farmer from Monoquirá turned over a huge boulder while he was tilling his field. When he looked at it later, he realized that it had a fossil in it. He alerted the scientific organizations in Colombia, and they began to excavate it. It turned out to be a nearly complete skeleton of *Kronosaurus*, one of the best fossils ever found in Colombia. Paleontologist Oliver Hampe described it in 1992 as a new species, *Kronosaurus boyacensis*.

KING OF THE SEA MONSTERS

Kronosaurus was truly an amazing creature. It had a skull almost 3 meters (10 feet) long (figure 15.3), with the front paddles reaching 3.3 meters (11 feet) in length and a total length of about 12.8 meters (42 feet). However, a recent study has suggested that in reconstructing the missing parts, the preparators may have put in too many vertebrae. Its total length may have been closer to 10 meters (33 feet). The specimen at the Museum of Comparative Zoology covers the entire wall of one gallery and takes your breath away when you first see it (see figure 15.1)! According to the account by his son, "Dinosaur Jim" Jensen mounted it to the wall with a series of curtains and other tricks that virtually hide the iron rods and supports he welded into place. He intended to make the specimen appear to be floating in the air or

Figure 15.3 ▲
Reconstruction of the head and body of *Kronosaurus*. (Courtesy Nobumichi Tamura)

water as a living, swimming creature, and that is indeed the illusion that the mount creates.

Kronosaurus was one of the largest members of a group of marine reptiles known as plesiosaurs, which includes two branches: the pliosauroids and the plesiosauroids. All plesiosaurs had a similar basic construction, other than their heads and necks. They were active swimmers that rowed their way across the Cretaceous seas using their huge front and back flippers. Plesiosaurs had a huge shoulder and hip girdle made of several bony plates on their belly for anchoring their powerful swimming muscles. Between the girdles was a mesh of belly ribs (*gastralia*) that gave their abdomens additional strength and support. In many specimens, smooth stones were found where the stomach had been inside the rib cage, suggesting that plesiosaurs swallowed stones to provide ballast. Also in the stomachs of the specimens from Queensland were fossils of their meals, which prove that *Kronosaurus* ate marine turtles and smaller plesiosaurs. Fossils of huge ammonites and giant squid lay in the same beds, and they almost certainly were food for such a gigantic predator. In addition, the plesiosaur *Eromangasaurus*, also from the same beds, has large bite marks on its skull, suggesting an attack by *Kronosaurus*.

Viewers of the popular television series *Walking with Dinosaurs* may have seen a large plesiosaur from Europe called *Liopleurodon*. The creature was animated as a monster more than 25 meters (82 feet) long, preying on dinosaurs and every other form of life during the Jurassic. In this size range, it approaches the size of the largest whales, including the blue whale (figure 15.4).

Sadly, as most paleontologists know, such television specials often get their facts wrong in the service of a more dramatic story. Having consulted on, and appeared on, numerous documentaries about prehistoric animals, I know this all too well. No matter what I say to the scriptwriters and producers, they override it to tell a more exciting story. Once the script goes to the

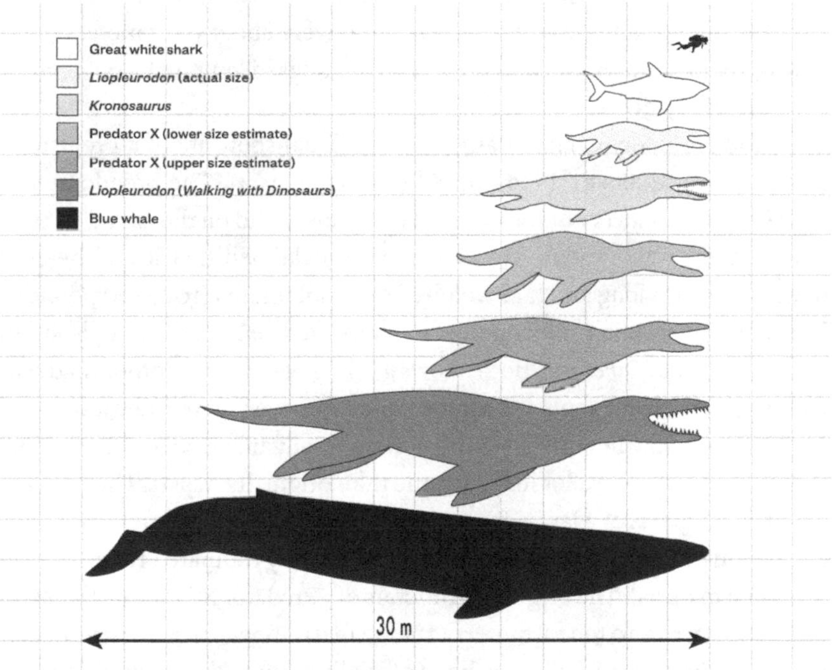

Figure 15.4 ▲
Comparison of the sizes of the plesiosaurs *Liopleurodon* and *Kronosaurus* with those of the great white shark (*Carcharodon carcharias*) and the blue whale (*Balaenoptera musculus*). The exaggerated size of "Predator X" and the gigantic *Liopleurodon* from television specials are also shown. (Drawing by Mary P. Williams)

animation studio, forget science! In most cases, what the animators draw is entirely imaginary. From the bones of prehistoric animals alone, we cannot reconstruct their color, and we cannot know how they moved precisely or what they sounded like. Any of the "stories" in these documentaries about how they interacted, how they behaved within their family groups, and so on come from pure imagination (guided by a bit of research into modern animals). Sadly, this is often the only part of paleontology that most of the public sees, and they are misled into thinking that paleontology is all about making catchy movies about extinct animals that show color and behavior and sounds, when none of that is based on real scientific data.

In fact, there are no complete specimens of *Liopleurodon* that suggest such a large size. Instead, there the fossils consist of mostly a few skulls and jaws, as well as other isolated bones. The largest complete skeleton, on display at the Museum für Geologie und Paläontologie in Tübingen, is only 4.5 meters (15 feet) long. New methods of estimating size from skulls suggest that the largest skulls belong to animals that were about 5 to 7 meters (16 to 23 feet) long, not even close to the size of the revised length of *Kronosaurus*, at 10 meters (33 feet).

In 2009, History Channel aired a sensational show about a prehistoric animal that it dubbed "Predator X" (see figure 15.4). The broadcast was based on the discovery of fossils of a large pliosauroid on the island of Svalbard in the Arctic Ocean. The documentary claimed that it had been 15 meters (50 feet) long and had weighed 5000 kilograms (100,000 pounds). The same misleading information was repeated in 2011 in an episode of the series *Planet Dinosaur*. Both shows got huge publicity in other media as well, since the claim about "the largest predator ever" gets attention.

Sure enough, when the specimens were finally unveiled and described, they turned out to be a lot less extreme than originally hyped. They consist of only a few parts of a jaw, a few vertebrae, and parts of a flipper. Sure, they are big, but the size of an animal cannot be reliably estimated from such incomplete material. The original promoters of "Predator X" revised their size estimate down to 10 to 12.8 meters (33 to 42 feet), about the same size as *Kronosaurus*. "Predator X" has now been officially named *Pliosaurus funkei*, and we were all put in a funk ourselves at the disappointment after the buildup.

Only *Kronosaurus* is completely known enough to reliably estimate its length and size. The rest is pure speculation and media hype until a much more complete large pliosaur is found.

LONG NECKS OF THE SEA

The other branch of the plesiosaurs is the more familiar type known as plesiosauroids, best known from the elasmosaurs. Instead of the heavy long snout and short neck of pliosauroids, such as *Kronosaurus*, plesiosauroids evolved in the opposite direction: tiny head and extremely long neck. Since Mary Anning's discovery of *Plesiosaurus*, the first known plesiosauroid (figure 15.5), many more have been found. These creatures were about as long as pliosauroids, but certainly not as heavy. Nonetheless, they were very large. Among the biggest was *Elasmosaurus*, which is known from complete specimens up to 14 meters (46) feet in length and was estimated at 2000 kilograms (4400 pounds) in weight. Unlike pliosauroids, plesiosauroids were probably not fast swimmers, but paddled slowly along using all four flippers for propulsion.

Since the discovery of fossils of plesiosauroids, paleontologists reconstructed them with a long, flexible snake-like neck and a head that could

Figure 15.5 ▲

A long-necked plesiosaur, *Rhomaleosaurus cramptoni*, found at Kettleness in Yorkshire, displayed at the Natural History Museum in London. (Courtesy Wikimedia Commons)

whip around easily in any direction, and most reconstructions still show them that way. More recent analyses of the weight of their neck and head, the limited muscles of their neck, and the constraints on the movement of the neck vertebrae show that the neck was not very flexible. These studies suggest that the plesiosauroid neck would have been semi-rigid and incapable of bending very far, more like a fishing pole than a snake neck. It also could not have been lifted out of the water in a swan-like fashion.

If the neck could not rotate and allow the plesiosauroids to snap in any direction, paleontologists have suggested methods of feeding that do not require a flexible neck. One proposal is that their long neck allowed them to lurk in deeper waters below the prey without being detected. Then they could poke their head into a school of fish or squid or ammonites and grab a meal before the shock wave of their massive body arrived to alert their prey to their movements. Their huge eyes are also consistent with this idea.

Another suggestion is that plesiosauroids were bottom-feeders, using their neck to plow through the mud of the seafloor in order to grab prey. Most plesiosauroids had long peg-like teeth that pointed forward, a common adaptation for spearing fish and other aquatic prey. Some plesiosauroids, like *Cryptoclidus* and *Aristonectes*, had hundreds of tiny pencil-like teeth that suggest they could have strained out small food items from either the plankton or the sea bottom.

Other scientists are not so sure that plesiosauroids had a semi-rigid neck. They point out that a lot of soft tissue is missing from the fossils (especially the cartilage between the vertebrae), and with so many neck vertebrae, their neck would still have been fairly flexible. The neck was certainly not as flexible as a snake's body, or capable of curling into an *S* shape, but these scientists argue that plesiosauroids could still have curled their neck into a fairly tight arc to reach prey. If so, then the elaborate behaviors suggested by the "rigid-neck" hypothesis are less likely.

The large body size, the flippers directly beneath their body, the lack of attachment of their hind limb bones to their spine, and other features of plesiosaurs make it unlikely that they could have crawled onto land or dug a hole in which to lay eggs, as do sea turtles. Still, many artists persist in showing plesiosaurs awkwardly splayed across rocks, with flippers far too short to drag their body across the surface. Their purely aquatic life was confirmed by the recent description of a plesiosaur fossil with an embryo in its body, showing that they gave birth to live young in the sea.

ORIGINS OF THE SEA MONSTERS

Where did such a remarkable group of animals like the plesiosaurs come from? Fortunately, we have an excellent fossil record of their origin from reptiles that bore no resemblance to plesiosaurs.

The oldest relative of plesiosaurs is a reptile known as *Claudiosaurus*, from rocks in Madagascar that date to the Permian (270 million years ago) (figure 15.6). It looks just like many other primitive reptiles of the Permian, except that it has certain key features of the skull and palate that earmark it as an early member of the marine reptile group, the Euryapsida, that includes both plesiosaurs and ichthyosaurs. It appears to have been partially aquatic, with no breastbone that might interfere with the swimming strokes of its forelegs. Thus it could swim with both front and back legs moving together, rather than with the alternating-foot pattern that characterizes the lizards. Its limbs are long, with really long toes that suggest webbed feet. In fact, many scientists have noted that it has the limb proportions and skeletal features of the Galápagos marine iguana.

In the Triassic (250 to 210 million years ago), there was a large group of primitive aquatic reptiles known as nothosaurs. They, too, were the size of large lizards (less than 1 meter [3.3 feet] long) and looked mostly like *Claudiosaurus*. They were already acquiring the long neck of some plesiosaurs, though, and a long fish-catching snout. In the limbs, a lot of bone had been reduced to cartilage, a common occurrence in aquatic vertebrates. In its shoulder girdle and hip bones can be seen the beginnings of the robust plate-like bones found in the limb girdles of plesiosaurs.

The final transitional fossil to plesiosaurs is a Middle Triassic creature from Germany known as *Pistosaurus*. It has a primitive skull with a simple snout, but its palate is much like that of the more advanced plesiosaurs. The rest of its body is transitional between plesiosaurs and other lizards, including long neck, deep body, well-developed belly ribs, and limbs that are intermediates between the plesiosaur paddle and the unspecialized nothosaur foot. The long bones of its hands and feet have turned into dozens of extra finger and toe bones, which became modified into simple disk-like bones in the paddles of plesiosaurs.

In short, the plesiosaurs may look strange and highly specialized, but we can trace their lineage back in time to lizards that show no indication of becoming giant sea monsters.

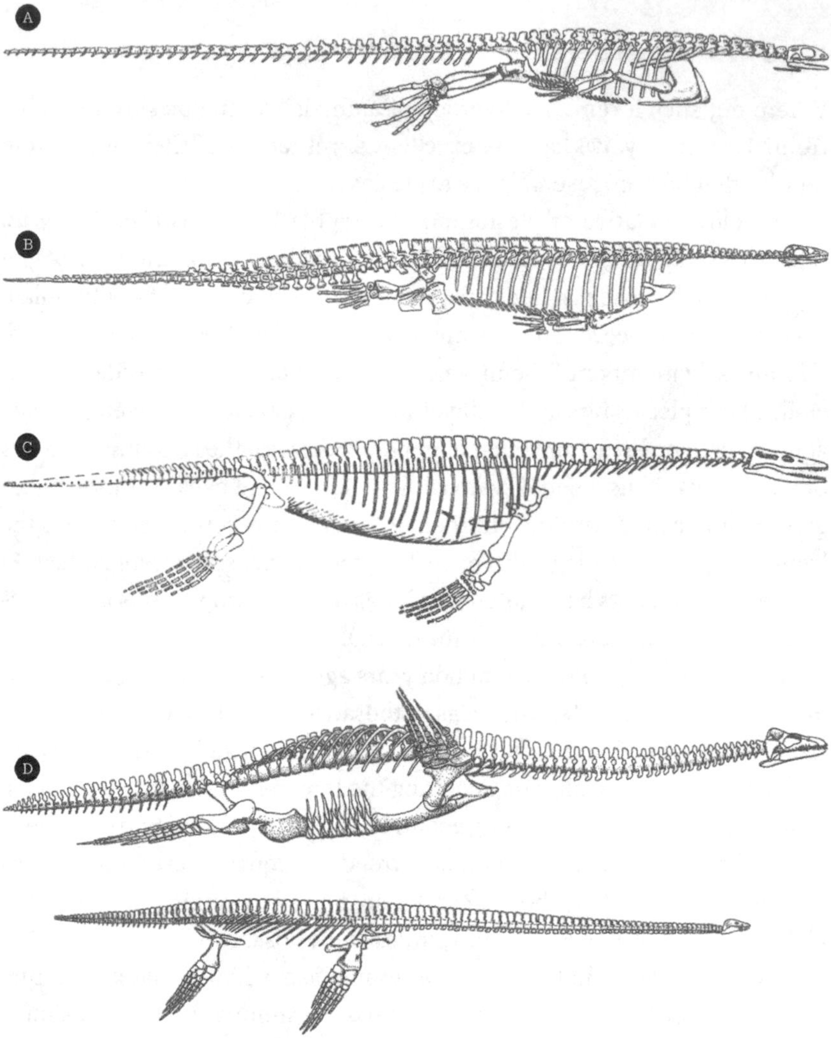

Figure 15.6 ▲

Fossils spanning the transition from reptiles distantly related to plesiosaurs to highly specialized plesiosaurs: (*A*) the primitive reptile *Claudiosaurus*, from the Permian of Madagascar, with just a few features of the Euryapsida, but still a short neck, a long tail, and relatively large hands and feet that were not yet modified into flippers; (*B*) the nothosaur *Pachypleurosaurus*, from the Triassic, with a longer neck, a stouter tail, and hands and feet modified for swimming; (*C*) the primitive true plesiosaur *Pistosaurus*, from the Triassic, with a longer neck, a longer skull, a shorter tail, and limbs partially modified into paddles; (*D*) the advanced plesiosauroids *Cryptoclidus* (*top*) and *Hydrothecrosaurus* (*bottom*), with much longer necks, smaller heads, shorter tails, and limbs fully modified into flippers. (From Robert L. Carroll, *Vertebrate Paleontology and Evolution* [New York: Freeman, 1988], figs. 12-2, 12-4, 12-10, 12-12; courtesy R. L. Carroll)

THE LOCH NESS MONSTER?

Since about the 1930s, many people have claimed that a large reptilian monster lives in Loch Ness, Scotland, and further suggesting that it is a long-surviving plesiosaur. A whole industry has been built around keeping the mystery of Loch Ness going, and a barrage of television shows try to make this myth seem plausible. As Daniel Loxton and I demonstrated, there is no possibility that a real reptilian "Loch Ness monster" exists (unless you are thinking of some unusually large fish like a sturgeon). The reasons are numerous and come from many lines of evidence:

- **BIOLOGICAL.** The climate is too cold around Loch Ness to support a large cold-blooded reptile for very long. In fact, only two species of lizards and two species of snakes live in Scotland, and Earth is currently in a relatively warm interglacial period. Basic biology shows that there cannot be just *one* Loch Ness monster, but must be a population of them, if they really have lived for the 65 million years since the plesiosaurs went extinct. If there were a population, we would routinely find plenty of bones and carcasses of them, as we do of every animal that dies in Loch Ness or any other large lake—but not a single scrap of bone has ever been found. In addition, the lake is too small and too poor in resources to support a large population of predatory reptiles. The larger the body size of an animal, the larger the home range it requires to get enough food, and Loch Ness is well below the size to support even one monster. In fact, every inch of the lake has been combed by radar and been dredged many times, so there is no chance that something big lurks in the lake that has been missed.
- **PALEONTOLOGICAL.** The fossil record of plesiosaurs is excellent, and so is the fossil record of marine vertebrates during the Age of Mammals, after the plesiosaurs went extinct. Not one bone of a plesiosaur (which are very distinctive and easy to recognize) has been found in any rocks younger than 65 million years, even though fossils of other large marine animals (sharks, whales, sea lions, manatees) routinely are unearthed in places such as Sharktooth Hill, California (chapter 9), and the Calvert Cliffs along Chesapeake Bay. Since larger fossils have a very good chance of preservation, this is conclusive evidence that plesiosaurs have been extinct for 65 million years.

◉ **GEOLOGICAL.** Loch Ness is a glacial valley that was covered by about 1.6 kilometers (1 mile) of ice only 20,000 years ago, and was ice-covered for over 2.5 million years. If the monster hid in the lake, was it locked in moving glacial ice for millions of years, as in the plot of a cheap science-fiction movie? If not, when did it arrive there? If it was hiding in other areas before entering the lake, why have no fossils been found? Besides, Loch Ness is landlocked and well above sea level, so there is no way for a large sea creature to have traveled there, especially since plesiosaurs could not crawl on land.

◉ **CULTURAL.** As Loxton and I showed, the "plesiosaur" meme about the Loch Ness monster is a recent invention. It is not found in some of the vague older reports about a mysterious creature in the water. In the legends, it was called the "water-horse," and there was nothing plesiosaur-like about it. Instead, the "plesiosaur" meme emerged from one person, George Spicer, after he saw the plesiosaur in *King Kong* in 1933. Since he and a woman, Aldie Mackay, claimed to have seen the monster, newspapers and other media have kept the phenomenon going.

In addition, numerous hoaxes have been perpetrated since the reports began, and they have fed the myth, including the "Surgeon's Photograph," the iconic image of Nessie. After the hoaxer died, it was revealed that he had photographed a toy submarine with a fake "head" stuck on top. Other deceptions included floating bales of hay covered with tarps and ropes and the "Nessie fin," which is just a grainy photograph of underwater bubbles with too much enhancement.

In short, the existence of the Loch Ness monster is completely impossible scientifically, and it has been debunked by nearly every line of evidence available. Its only support comes from vague "eyewitness reports," which are the worst possible evidence in a scientific investigation, since human eyes and brains are easily fooled. Plesiosaurs were fascinating creatures. It would be terrifying if they still swam in Earth's oceans, but, despite the persistence of the myth of the Loch Ness monster, they are truly extinct.

SEE IT FOR YOURSELF!

The skeleton of *Kronosaurus queenslandicus* is still the centerpiece of the main hall of the Museum of Comparative Zoology at Harvard University, Cambridge, Massachusetts. In Australia, the original *Kronosaurus* material is on display at the Queensland Museum, South Brisbane. The nearly complete skeleton of *Kronosaurus boyacensis* is exhibited on the very spot where it was found, and the Museo de Fosil was built over it by the people of nearby Villa de Leyva.

In Europe, fossils of plesiosaurs can be seen in many museums. In England, many of Mary Anning's discoveries from Lyme Regis are displayed at the Natural History Museum, London; and Lyme Regis Museum. The largest skull of *Pliosaurus kevani* is at the Dorset County Museum, Dorchester. Many German museums display plesiosaurs (especially from Holzmaden), including the Museum für Naturkunde (Humboldt Museum), Berlin; Naturmuseum Senckenberg, Frankfurt; and Staatliches Museum für Naturkunde, Stuttgart. The only complete *Liopleurodon* on display is at the Museum für Geologie und Paläontologie der Universität Tübingen.

In the United States, many museums display long-necked elasmosaurs, especially those from the Western Interior Seaway in the Cretaceous of Kansas, including the American Museum of Natural History, New York; Biodiversity Institute and Natural History Museum, University of Kansas, Lawrence; Denver Museum of Nature and Science; Museum of Geology, South Dakota School of Mines and Technology, Rapid City; and Sternberg Museum of Natural History, Fort Hays University, Hays, Kansas. The Natural History Museum of Los Angeles County, in Los Angeles, has an elasmosaur from the Cretaceous Moreno Hills of California, called *Morenosaurus*, suspended from the ceiling, as well as the recently described specimen of a mother plesiosaur skeleton with her embryo inside.

The Otago Museum in Dunedin, New Zealand, displays a plesiosaur found in New Zealand.

FOR FURTHER READING

Callaway, Jack, and Elizabeth L. Nicholls, eds. *Ancient Marine Reptiles*. San Diego: Academic Press, 1997.

Ellis, Richard. *Sea Dragons: Predators of Prehistoric Oceans*. Lawrence: University Press of Kansas, 2003.

Everhart, Michael J. *Oceans of Kansas: A Natural History of the Western Interior Sea*. Bloomington: Indiana University Press, 2005.

Hilton, Richard P. *Dinosaurs and Other Mesozoic Animals of California*. Berkeley: University of California Press, 2003.

Loxton, Daniel, and Donald R. Prothero. *Abominable Science: The Origin of Yeti, Nessie, and Other Cryptids*. New York: Columbia University Press, 2013.

16 | THE LARGEST PREDATOR | *GIGANOTOSAURUS*

MONSTER FLESH-EATER

I propose to make this animal the type of the new genus, *Tyrannosaurus*, in reference to its size, which far exceeds that of any carnivorous land animal hitherto described. . . . This animal is in fact the *ne plus ultra* of the evolution of the large carnivorous dinosaurs: in brief it is entitled to the royal and high sounding group name which I have applied to it.

HENRY FAIRFIELD OSBORN, "*TYRANNOSAURUS* AND OTHER CRETACEOUS CARNIVOROUS DINOSAURS"

KING OF THE TYRANT LIZARDS

Thanks to a century of publicity, *Tyrannosaurus rex* is probably the best known and most popular of dinosaurs. Discovered in the Hell Creek badlands of Montana by the legendary fossil collector Barnum Brown in 1900, it was described by the prominent paleontologist Henry Fairfield Osborn in 1905. Osborn bestowed on it the memorable name, which means "king of the tyrant lizards," and its contraction to *T. rex* is equally familiar. In fact, *Tyrannosaurus rex* is one of the few scientific names that almost everyone knows (even more than know our own genus and species, *Homo sapiens*). Brown found five skeletons altogether, and by the time the dinosaur was named and described, the American Museum of Natural History had mounted and displayed one of the spectacular skeletons (figure 16.1), the

Figure 16.1 ▶
The classic old mount of *Tyrannosaurus rex* in the American Museum of Natural History, in New York, as it appeared from about 1910 to the early 1990s. The "kangaroo" pose was based on the idea that *T. rex* was a sluggish lizard that dragged its tail. (Image no. 327524, courtesy American Museum of Natural History Library)

DINOSAURS WITH

fourth of the five that Brown found. Osborn wrote in his paper describing the first specimens that *Tyrannosaurus rex* was "the *ne plus ultra* of the evolution of the large carnivorous dinosaurs: in brief it is entitled to the royal and high sounding group name which I have applied to it."

Osborn soon got the publicity boost he wanted, when on December 3, 1906, an article in the *New York Times* on his newly announced specimens described the creature as the "most formidable fighting animal of which there is any record whatever," the "king of all kings in the domain of animal life," the "absolute warlord of the earth," and a "royal man-eater of the jungle." In another *New York Times* article, *Tyrannosaurus rex* was called the "prize fighter of antiquity" and the "Last of the Great Reptiles and the King of Them All."

Painted reconstructions by the pioneering paleoartist Charles R. Knight soon made *T. rex* the most celebrated of all dinosaurs. It has been a cultural icon ever since, appearing in every medium. Its movie credits range from the silent films *Ghost of Slumber Mountain* (1918) and *The Lost World* (1925; based on the 1912 novel by Sir Arthur Conan Doyle, creator of Sherlock Holmes); through *King Kong* (1933); to the *Jurassic Park* trilogy, the last two remakes of *King Kong*, and the film and television series *The Land Before Time* (as the "Sharptooths"). On television, it was the star of *Barney and Friends*, and it has been featured in parade floats and transformed into thousands of different items of merchandise. There was even a British rock band called T. rex. The image of the huge predator towering over other dinosaurs (and people in museum galleries) is very powerful (see figure 16.1). The late paleontologist Stephen Jay Gould says that the mounted *T. rex* skeleton at the American Museum of Natural History terrified him at age five, but also inspired him to become a paleontologist.

Naturally, a lot has been learned in the 110 years since *Tyrannosaurus rex* was first announced and described. The biggest change in our perception of the creature has been its posture. When Osborn first directed the mounting of the original fossils, the bones were put together as though *T. rex* were a big bipedal lizard, with its tail dragging on the ground (see figure 16.1). That conception of *Tyrannosaurus rex* is still reflected in the majority of toys and books and older products. But in the 1970s and 1980s, paleontologists discovered that the trackways of large predatory dinosaurs show no evidence of tail-drag marks, indicating that *T. rex* (like nearly all dinosaurs) held its tail straight out and balanced over its hips and hind legs. Many biomechan-

Figure 16.2 ▲
The modern remount of *Tyrannosaurus rex* in the American Museum of Natural History. Like all other bipedal dinosaurs, it is mounted with its body held horizontally and its tail straight out, to balance its body over its hind limbs. In the background is the old plaque mount of *Gorgosaurus*, which could not be remounted in the modern pose. (Photograph courtesy Wikimedia Commons)

ical studies showed that this was the stable posture as well. Thanks to this research (which author-screenwriter Michael Crichton followed closely), the *Jurassic Park* movies helped popularize this vision of *Tyrannosaurus rex*—as a fast-moving, intelligent predator that balanced on its hips like a horizontal beam and held its tail straight out—which is now permeating the culture and all the merchandise based on this dinosaur (figure 16.2).

Our expanding knowledge of *Tyrannosaurus rex* includes many kinds of research that could never be attempted until recently. For example, modeling of the bite force of a *Tyrannosaurus rex* suggests about 35,000 to 57,000 newtons (7900 to 13,000 pound-force) of force in the back teeth, three times more powerful than the bite force of a great white shark; 3.5 times that of the Australian saltwater crocodile; seven times that of *Allosaurus*. and 15 times that of an African lion. A more recent study increased the bite force estimates to 183,000 to 235,000 newtons (41,000 to 53,000 pound-force), equivalent to that of the largest specimens of the giant shark *Carcharocles megalodon* (chapter 9).

Its immense skull was 1.5 meters (5 feet) long, but honeycombed with many holes, pockets, and air sacs to make it lighter. The tip of its snout was *U*-shaped in horizontal cross-section, giving *T. rex* a stronger bite force than that of the predatory dinosaurs, or theropods, with *V*-shaped snouts. However, its snout was narrow compared with the wide back of its skull, so its eyes pointed forward and allowed it excellent binocular vision to get stereoscopic views and accurately estimate distances. The huge teeth (up to 30 centimeters [1 foot] long from tip of root to tip of crown) were recurved, shaped like steak knives the size of bananas, with serrated ridges to slice through flesh and reinforcing ridges on the back to strengthen them. The teeth in the front of its skull had a *D*-shaped cross-section, so they were less likely to have snapped when *Tyrannosaurus rex* bit down and pulled back.

So what did they eat? Numerous dinosaur bones have scars that could have been carved only by a tyrannosaur, and some tyrannosaurs show healed wounds from bites of other tyrannosaurs and even broken tyrannosaur teeth embedded in their faces and necks. Clearly, *T. rex* ate many kinds of dinosaurs and fought among themselves. The major argument about the diet of tyrannosaurs concerns whether they were purely predators or mostly scavengers. Like many debates, this one unnecessarily casts the two options as mutually exclusive, but nature is always more complex than oversimplified arguments. Most large mammalian predators (lions, tigers, jaguars, cougars) are both predators and scavengers. Capturing prey is so difficult that they cannot afford to be choosy, but eat carrion when they find it and hunt for fresh meat when they have no choice. There are even examples of partially consumed tyrannosaur carcasses with bite marks from other tyrannosaurs, suggesting that they were cannibalistic as well.

Tyrannosaurus rex had an *S*-shaped curve in its neck, as did most theropods. Although its neck was shorter than that of many other theropods, it was very robust and strong, allowing the creature to have tremendous power when it whipped its head around to subdue prey or rip out chunks of flesh. Its legendary tiny hands had only two functional fingers (usually portrayed as having three, a common mistake). In fact, most predatory dinosaurs had only three functional fingers, yet in the popular media many are shown with hands of five fingers. There are lots of ideas about what function such tiny arms could have had, but short, almost nonfunctional, forelimbs were typical of many of the advanced theropods, although the arms of *T. rex* were unusually small. More to the point, as the skull and jaws of theropods grew larger and larger, their arms became smaller. This suggests that they had become specialized for "power biting" and killing entirely with their massively powerful neck and jaws. The arms were just vestigial relicts that were no longer used to hold struggling prey.

At one time, tyrannosaur fossils were relatively rare, with only five known partial skeletons. But in the past 20 years, a stampede of collectors have put an enormous effort into finding them (especially after the specimen named "Sue" fetched over $8 million at auction), and now more than 50 individuals are known from partial skeletons. We have examples from all age groups—from babies to "teenagers" to young adults to very old adults—so we can see that they grew extremely rapidly until about 14 years of age, when they reached 1800 kilograms (4000 pounds). Then they added an average of 600 kilograms (1300 pounds) each year until their growth slowed as they reached maturity. As Thomas Holtz described it, tyrannosaurs "lived fast and died young," since they had rapid growth rates and high mortality rates; by contrast, mammals take longer to mature and have longer life spans as a result.

With the abundance of specimens, paleontologists have long sought to determine whether the fossils are of males or females. Many supposed differences were suggested, but most have not proved to be valid. The differences typically attributed to sexual dimorphism turn out to be mostly geographic differences between populations. One specimen, however, can definitely be sexed. In a bone of "B-rex," found in Montana, the soft tissues are relatively well preserved, including medullary tissue, which is characteristic of ovulating female birds.

Birds are descended from theropod dinosaurs closely related to tyrannosaurs (chapter 18). Although most tyrannosaur fossils are only bones, with no preservation of skin or feathers, some have skin impressions that are consistent with a feathery covering. Then, the small tyrannosaur *Dilong paradoxus* was found in the Yixian Formation of China, and it showed a coat of filamentous feathers or fluff. The discovery of *Yutyrannus huali* in China proved that feathers (mostly filamentous or downy feathers) covered nearly every part of the body. These two specimens demonstrate that tyrannosaurs should be reconstructed with a coat of down and long filamentous feathers, not with naked skin, as traditionally depicted (even *Jurassic World*, the fourth installment of the *Jurassic Park* saga, still portrays featherless dinosaurs).

Thanks to many specimens and extensive study by many paleontologists, *Tyrannosaurus rex* is by far the best known of all the predatory dinosaurs. But what other big predators were there?

OUT OF AFRICA

In the late nineteenth century, Germany was considered the leader in almost every field of scientific scholarship, especially in embryology, anatomy, and evolutionary biology. So important were its leading scientists that pioneering American paleontologists like Henry Fairfield Osborn and William Berryman Scott did postgraduate research there in lieu of the modern doctorate (which was not yet commonly awarded in the United States).

German archeologists made great strides in Egyptology, led by the legendary Karl Richard Lepsius. Heinrich Schliemann discovered and excavated the site of ancient Troy in what is now western Turkey, as well as conducted the first excavations of Mycenaean Greece. The huge Pergamon Museum in Berlin has some of the best art and artifacts from ancient Olympia, Samos, Pergamon, Miletus, Priene, and Magnesia in Greece; huge pieces of art and buildings from ancient Babylonia and Assyria; as well as the legendary bust of the Egyptian queen Nefertiti.

Paleontology was truly cutting-edge in Germany. Many of the leading scholars in paleontology and related fields from the late eighteenth into the twentieth century were German. Among the famous names were the pioneering paleobotanist Ernst Friedrich von Schlotheim (1764–1832); the legendary explorer and biologist Alexander von Humboldt (1769–1859); the early geologist Leopold von Buch (1774–1853); the embryologist and zoologist Ernst Haeckel (1834–1919), who was one of Darwin's chief supporters; Karl Alfred von Zittel (1839–1904), the author of the most widely used textbook; and the very influential paleontologist Otto H. Schindewolf (1896–1971).

Some were working on the legendary fossil beds in Germany, such as the Solnhofen Limestone and the Holzmaden Shale. But many were exploring abroad like their archeological and other scientific colleagues, especially in areas of Africa that were then German colonies. Between 1909 and 1911, for example, German East Africa (now Tanzania) was the site of huge excavations of dinosaur fossils in the Tendaguru beds by Werner Janensch, which produced the amazing complete skeleton of *Giraffatitan* (formerly

Brachiosaurus), now in the Museum für Naturkunde (Humboldt-Museum) in Berlin (see figure 17.5), as well as the stegosaur *Kentrosaurus* (with spikes on its back instead of plates) and many other unusual dinosaurs.

Another prominent German paleontologist who worked in Africa was Ernst Freiherr Stromer von Reichenbach (1870–1952). He was the leader of a famous expedition to Egypt in 1910/1911, at the same time that Jan-ensch was working in German East Africa. After two relatively unsuccessful trips out of Cairo (on one of which his colleague Richard Markgraf did discover the fossils of one of the earliest primates, *Libypithecus*, in what is now Libya), Stromer and Markgraf trekked to the far western deserts of Egypt at Bahariya Oasis, just east of the Libyan border. On January 18, 1911, Stromer finally found the bones of huge dinosaurs. In his words, he discovered

> three large bones which I attempt to excavate and photograph. The upper extremity is heavily weathered and incomplete [but] measures 110 cm [43 inches] long and 15 cm [6 inches] thick. The second and better one underneath is probably a femur [thighbone] and is wholly 95 cm [37 inches] long and, in the middle, also 15 cm thick. The third is too deep in the ground and will require too much time to recover.

He unearthed additional specimens at Bahariya over the next few weeks, but by February 1911 he had to return to Germany. Stromer spent several decades describing and publishing the fragments of some amazing dinosaurs, including the sauropod *Aegyptosaurus*, the giant crocodilian *Stomatosuchus*, and fragments of several theropods: *Bahariasaurus* and the super-predators *Carcharodontosaurus* and *Spinosaurus*.

Unfortunately, all the fossils from Bahariya Oasis were stored in the Paläontologisches Museum München (home of the Bavarian State Collection). By early 1944, Allied air raids were routinely bombing all the larger German cities, especially those with significant military targets, in preparation for the D-Day landings of June 6, 1944. The Museum für Naturkunde in Berlin (with its priceless specimens, including *Giraffatitan*, the "Berlin specimen" of *Archaeopteryx*, and many amazing ichthyosaurs from Holzmaden) had several close calls, including bombs that wiped out a railroad station next door. On the night of April 24/25, 1944, the Royal Air Force flew a huge bombing raid over Munich, and all of Stromer's fossils (as well as all the rest of the museum's invaluable and historic collections) were utterly destroyed. In one night, the work undertaken and collections amassed by biologists

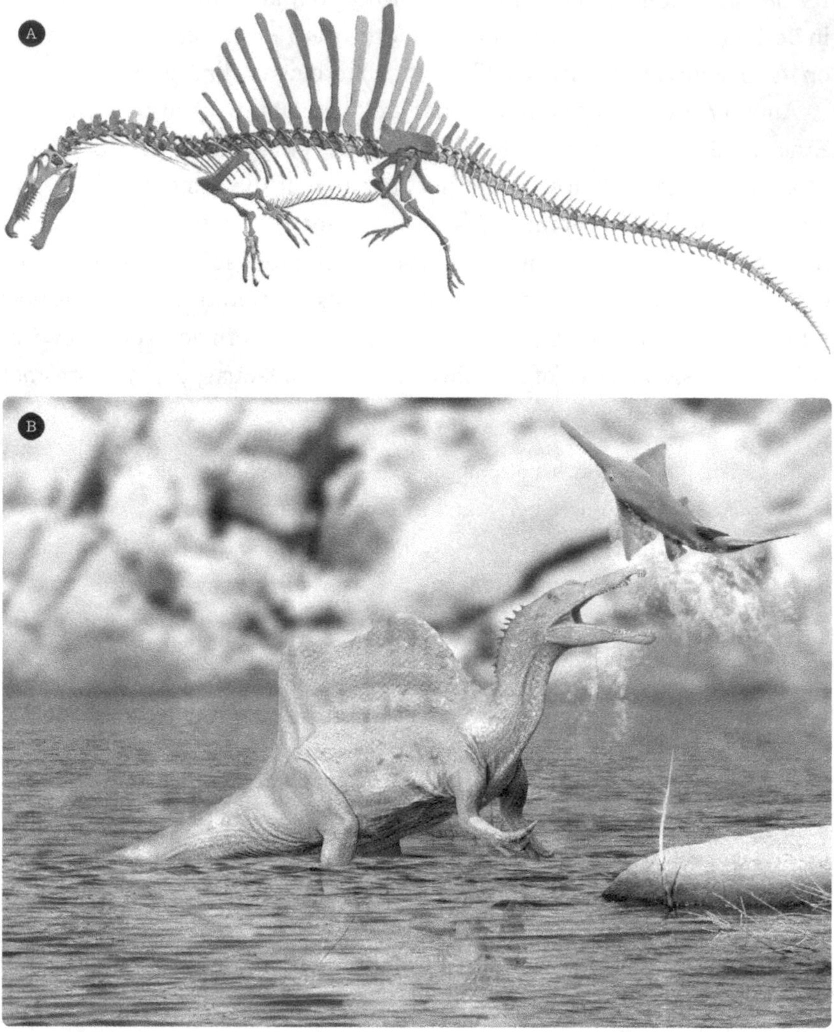

Figure 16.3 ▲

Spinosaurus: (*A*) the known bones (*dark shading*), based on the new work of Nizar Ibrahim, Paul Sereno, and others; (*B*) reconstruction of its appearance in life. ([*A*] photograph courtesy Paul Sereno and Michael Hettwer; [*B*] courtesy Nobumichi Tamura)

and paleontologists over centuries were obliterated, most with no information recording what they were and what they looked like. All that survives of Stromer's specimens are his original scientific illustrations from 1915, photographs of the exhibits, and a few fossils discovered in recent years.

By far, the most famous of the fossils from Stromer's expedition was of *Spinosaurus*. Thanks to the movie *Jurassic Park III*, *Spinosaurus* is familiar to every dinosaur fan. It was portrayed as a huge bipedal predator with a crocodile-like snout and a "sail" along its back (figures 16.3 and 16.4), so big that it beat up and devoured the smaller *T. rex*. Stromer had illustrated only a few of the huge spines that supported the "sail" along its back, plus a lower jaw, a few teeth, and some ribs and vertebrae. The spines were very long (up to 1.65 meters [5.4 feet]). Stromer described (but never illustrated) parts of the upper jaw that have been forever lost. The specimens were indeed huge (the lower jaw was 75 centimeters [30 inches] long), but so incomplete that the true size and appearance of *Spinosaurus* is really guesswork. In the late 1990s, an expedition from the University of Pennsylvania led by Peter Dodson, Matthew Lamanna, Joshua Smith, and Kenneth Lacovara returned to Bahariya Oasis. They found a few new specimens (including the sauropod *Paralititan stromeri*, whose species name honors Ernst Stromer), but not much more *Spinosaurus* material.

But several fossils of *Spinosaurus* have been discovered in Morocco and Tunisia since 1944. Recently, a group including Paul Sereno and his postdoctoral student Nizar Ibrahim found additional *Spinosaurus* material and

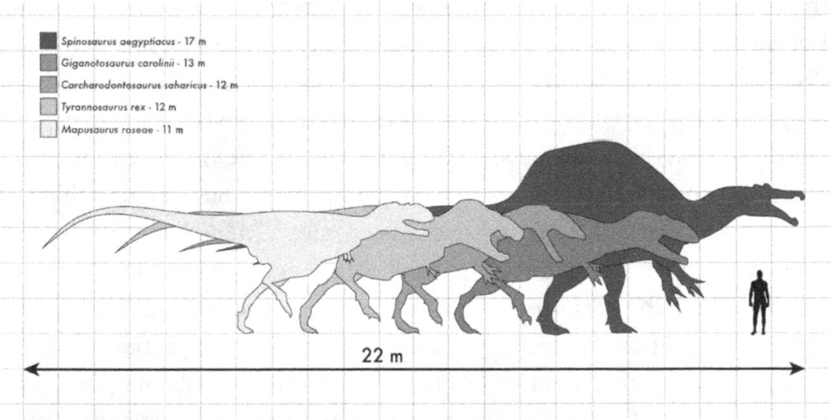

Figure 16.4 ▲

Comparison of the sizes of the major theropods: the huge size of *Carcharodontosaurus* is conjectural, since the fossils are not complete enough to reconstruct, and the depiction of *Spinosaurus* is based on the long-legged reconstruction and no longer matches the newly discovered short-legged, low-slung specimens. (Drawing by Mary P. Williams) made a big announcement of their new reconstruction. It suggests that *Spinosaurus* had a relatively long slender body and remarkably short legs and arms, nothing like the scaled-up, *T. rex*–like depiction that appears in *Jurassic Park III* (see figure 16.3*B*). Its long narrow beak was not adapted for eating other dinosaurs, but for catching fish. Its beak also has nostrils midway back up the snout, and nerve and blood-vessel channels that would have helped it sense changes in water pressure, all supportive of the idea that it lived more like a crocodile than like most bipedal theropod dinosaurs.

This physical evidence is consistent with the results of chemical studies of the bones, which show that it ate a diet of fish and other aquatic creatures. The density of its limb bones also suggests that it spent most of its time in the water. Other aquatic animals, like the hippopotamus, have very dense limb bones that serve as ballast. There is a big argument as to whether, with such stumpy limbs, it could have crawled onto land, but it certainly was not a fast land runner and not a predator that could have chased down and killed larger dinosaurs. The fingers on its hand are long and delicate, for catching smaller prey, and the bottom of its foot bones are flattened for walking on its entire foot (plantigrade), not on the tips of its toes (digitigrade), as did most dinosaurs. In addition, its long delicate fingers and toes suggest that it even may have had webbing on its hands and feet.

Finally, there is the huge "sail" on its back, formed by long extensions of the neural arch spines on top of its backbone that gave *Spinosaurus* its name. Every paleontologist has his or her own set of favorite theories to explain this feature, although most agree that it was not a true sail that could have propelled the dinosaur through the water, and it was so large and conspicuous that it would actually have made it difficult for *Spinosaurus* to have sunk below the surface and hid, as do crocodilians. Some have argued that the "sail" was a heat-gathering and -radiating device, although few other dinosaurs needed such a feature. The most common suggestion is that it was comparable to the horns and antlers of deer and antelope, mostly used for species recognition and for advertising dominance among males.

The big announcement of the newly discovered fossils of *Spinosaurus* has been greeted with some skepticism among dinosaur paleontologists because there may be some mistakes in the reconstruction of some bones (especially the hip bones). In addition, the reconstruction was based on a composite of skeletons from different individuals. Some of the bones were actually re-created as digital copies from the photographs that survive of

Stromer's original specimens and were generated by a three-dimensional printer.

In particular, the claims about the size of *Spinosaurus* must be taken with a grain of salt. Certainly, with its slimmed-down profile and short limbs, it was nowhere near as massive or heavy as the huge land predators. Ibrahim and Sereno and their colleagues claim that it was 15.2 meters (50 feet) long, but a close look at the diagram of which bones have actually been found shows that very few tailbones are among them (see figure 16.3*A*), so reconstructing the tail (and thus the dinosaur's length) is speculative at best. Many estimates of its size have been proposed, but with so little material, they are not very well constrained. In 1926, German paleontologist Friedrich von Huene (who studied the original fossils) estimated its length at 15 meters (almost 50 feet) and its weight at 6 metric tons (6.6 tons). In 1988, Gregory Paul also gave a length of 15 meters, but lowered its mass estimate to 4 metric tons (4.4 tons). But in 2007, François Therrien and Donald Henderson used newer scaling techniques and revised the estimates to 12.6 to 14.3 meters (41 to 47 feet) in length and only 12 to 21 metric tons (13.2 to 23 tons) in weight, shorter but heavier than previous estimates. If the largest *T. rex* was about 13 meters (43 feet) and about 10 metric tons (11 tons), then *Spinosaurus* was about the same size. Contrary to *Jurassic Park III*, it was probably not a giant that could toss a *T. rex* around like a toy. Since all the specimens are so incomplete, there is really no way to know.

If *Spinosaurus* cannot be conclusively shown to have been the largest predatory dinosaur, what about the other large theropod found in Africa: *Carcharodontosaurus*? It was discovered in 1924, when French paleontologists Charles Depéret and J. Savornin unearthed some huge teeth from the Lower Cretaceous Continental Intercalcaire of Algeria. The teeth resembled those of the first dinosaur ever named, *Megalosaurus*, so they christened them *M. saharicus*. In 1914, Stromer found a partial skull of this creature at Bahariya Oasis, as well as more teeth, claw bones, and assorted hip and leg bones. When he finally got around to describing this material in 1931, he renamed it *Carcharodontosaurus saharicus*, since its fossils were nothing like those of *Megalosaurus*, which had been found in England. As for its name, its huge teeth were about the size and shape of those of the great white shark (*Carcharodon carcharias*). Sadly, Stromer's fossils of *Carcharodontosaurus* were destroyed in the bombing raid over Munich in 1944 that obliterated *Spinosaurus* and the rest of his collection.

Figure 16.5 ▲
The skull of *Carcharodontosaurus* found in the Kem Kem Formation, with a human skull for scale. (Photograph courtesy Paul Sereno and Michael Wettner)

The incomplete skull seemed impressive, but so much of it was missing (and so little of the rest of the skeleton was known) that an accurate reconstruction and estimation of size seemed impossible. According to early calculations, the skull was the longest of that of any carnivorous dinosaur, but key elements were missing. When they were found, the skull length was revised down from almost 2 to 1.6 meters (6.6 to 5.2 feet). More recent measurements place the length of *Carcharodontosaurus* at 12 to 13 meters (39 to 43 feet) and its weight between 6 and 15 metric tons (6.6 and 16 tons), making it about the same size as *Spinosaurus* and a large *T. rex*. But the specimen is so incomplete that we cannot say with confidence which of these three dinosaurs was the largest (see figure 16.4).

Then Paul Sereno began a series of expeditions to the Saharan region in 1995. In the Kem Kem Formation of Morocco, near the Algerian locality where Depéret and Savornin had found the original teeth, Sereno and his crew from the University of Chicago unearthed much more complete skull material of *Carcharodontosaurus* (figure 16.5). Sponsored by the National Geographic Society, Sereno's discoveries of this and other spectacular fossils made the news. His dangerous expedition through the deadly Sahara—with its killer heat, deadly sandstorms, bad roads, and dangerous bandits and terrorists—was featured in more than one television documentary after the initial description in 1996. In 2001, Sereno's former student Hans C. E. Larsson did a detailed analysis of areas of the ear region and brain case that were unknown in other skulls. In 2007, Sereno and his student Stephen Brusatte published a description of another species, *Carcharodontosaurus iguidensis,* from the Echkar Formation of Niger. It was about the same size as the Moroccan fossils. Since Stromer's original fossils had been destroyed, Sereno and Brusatte designated the new skull from Morocco as the replacement type specimen, or neotype.

Although *Carcharodontosaurus* is built like a normal theropod in most of its known skeleton, its skull is distinctive (see figure 16.5). The roof of the skull is arched upward, and there are unusually large openings in the side of the skull, compared with those of *T. rex* and other large theropods. This made the gigantic skull much lighter despite its enormous size. *Carcharodontosaurus* had a brain about the same proportional size as that of the smaller *Allosaurus*, which is a close relative. It had a large optic nerve (and big eye openings), suggesting a strongly visual predator.

If the specimens of both *Spinosaurus* and *Carcharodontosaurus* are so incomplete that we cannot conclusively show that they were significantly larger than *T. rex*, then what is the largest predator that ever lived? It turns out that it comes not from Africa, but from another Gondwana continent in the Early Cretaceous: South America. And it is a close relative of *Carcharodontosaurus.*

BIGGER THAN *TYRANNOSAURUS REX*

Unlike *Spinosaurus* and *Carcharodontosaurus*, the dinosaur that may have been the largest of all the theropods is known from a reasonably complete skeleton. Amateur fossil hunter Rubén Dario Carolini discovered the fossils in 1993, from Lower Cretaceous beds in southern Argentina. It was named *Giganotosaurus carolinii* by Rodolfo Coria and Leonardo Salgado in 1995 (its

Figure 16.6 ▲

The best skeleton of *Giganotosaurus*, displayed at the Museo Municipal Ernesto Bachmann in Villa El Chocón, Argentina. (Photograph courtesy R. Coria)

genus name from the Greek for "big southern lizard," and its species name in honor of Carolini). Many people misread the name and mispronounce it "GIGANTO-saurus." The correct pronunciation is "GIG-a-NO-to-saur-us."

In contrast to the specimens of large theropods from Africa, *Giganotosaurus* is about 70 percent complete (figure 16.6), and includes most of the skull; the lower jaw, pelvis, and hind limbs; and most of the backbone. It is missing just the forelimbs and a few other pieces. Thus the estimate of its length comes from relatively complete skeletons and skulls, and from real limbs rather than guesswork. The largest skull and jaws were found by Jorge Calvo in 1988, and measured about 1.95 meters (6.4 feet), longer than that of any other theropod dinosaur. Like that of its close relative *Carcharodontosaurus*, its skull is slender and very lightly built, with a large arching roof and many openings surrounded by bony struts. There are roughened areas around the top of the snout and above the eyes. The back of the skull slants forward, so the jaw joints hang behind and beneath the attachment between the skull and the neck vertebrae. Given the lighter construction of its skull, *Giganotosaurus* does not seem to have had a strong bite force; apparently it was only one-third of that estimated for *T. rex*. Its broad slicing teeth, like those of sharks, in its lower jaw were much more suitable for producing slashing wounds than for biting down bulldog-style, as were the robust banana-size teeth of *Tyrannosaurus rex*. Thus, it may have attacked smaller prey, among which would have been some of the small titanosaur sauropod dinosaurs *Andesaurus*, the diplodocids *Nopcsaspondylus* and *Limaysaurus*, as well as an array of other iguanodonts, small predatory dinosaurs known as dromaeosaurs like *Velociraptor*, and many smaller animals. They are all found in the same Lower Cretaceous beds in Argentina as was *Giganotosaurus*.

Based on the limbs, the nearly complete spinal column, and the general skull and skeleton, the largest individuals of *Giganotosaurus* were about 14.2 meters (53 feet) long and weighed between 6.5 and 13.8 metric tons (7 and 15 tons). This is quite a bit longer than the largest *T. rex*, at 13 meters (42 feet) and 8 metric tons (8.8 tons). Thus *Giganotosaurus* has the best claim to have been the largest predator that ever lived.

SEE IT FOR YOURSELF!

Skeletons of *Tyrannosaurus rex* are found in many museum around the world, but the most famous ones are at the American Museum of Natural History, New York (the first one ever mounted); Carnegie Museum of Natural History, Pittsburgh (Henry Fairfield Osborn's original type specimen): Denver Museum of Nature and Science (where it was reconstructed in a "dancing" pose and hangs over the entrance lobby); Field Museum of Natural History, Chicago (which displays "Sue," the controversial specimen that cost $8 million at auction); Museum of the Rockies, Bozeman, Montana (whose curator Jack Horner has found more specimens than anyone); National Museum of Natural History, Smithsonian Institution, Washington, D.C.; Natural History Museum of Los Angeles County, Los Angeles (which has three individuals, from baby to near adult); and University of California Museum of Paleontology, Berkeley.

The new reconstruction of *Spinosaurus* was on display in the headquarters of the National Geographic Society, Washington, D.C.

In Argentina, mounted skeletons of *Giganotosaurus* are displayed at the Museo Municipal Carmen Funes, Plaza Huincul; and Museo Municipal Ernesto Bachmann, Villa El Chocón. In the United States, replicas are exhibited at the Academy of Natural Sciences of Drexel University, Philadelphia; and Fernbank Museum of Natural History, Atlanta.

FOR FURTHER READING

Brett-Surman, M. K., Thomas R. Holtz Jr., and James O. Farlow, eds. *The Complete Dinosaur*. 2nd ed. Bloomington: Indiana University Press, 2012.

Carpenter, Kenneth. *The Carnivorous Dinosaurs*. Bloomington: Indiana University Press, 2005.

Fastovsky, David E., and David B. Weishampel. *Dinosaurs: A Concise Natural History*. 2nd ed. Cambridge: Cambridge University Press, 2012.

Holtz, Thomas R., Jr. *Dinosaurs: The Most Complete Up-to-Date Encyclopedia for Dinosaur Lovers of All Ages*. New York: Random House, 2007.

Nordruft, William, with Josh Smith. *The Lost Dinosaurs of Egypt*. New York: Random House, 2007.

Parrish, J. Michael, Ralph E. Molnar, Philip J. Currie, and Eva B. Koppelhus, eds. *Tyrannosaurid Paleobiology*. Bloomington: Indiana University Press, 2013.

Paul, Gregory S. *The Princeton Field Guide to Dinosaurs*. Princeton, N.J.: Princeton University Press, 2010.

17 | THE LARGEST LAND ANIMAL | *ARGENTINOSAURUS*

LAND OF THE GIANTS

There were giants in the earth in those days.

GENESIS 6:4

GIANTS IN THE EARTH

In the early nineteenth century, the incredible world of extinct animals was mostly unknown to the public. A few large bones had been found here and there, but they usually were attributed to the biblical "giants in the earth" or otherwise dismissed and not given true scientific consideration. By 1810, Baron Georges Cuvier in France had thoroughly described the fossil mammoths and mastodonts recently discovered in Ice Age deposits in Europe and North America, and concluded that they were extinct creatures that had lived in a dark, stormy "antediluvian world," relicts of a previous creation not mentioned in the Bible. Then Mary Anning began to find the fossils of amazing marine reptiles in the Jurassic deposits of England, and soon the "antediluvian world" of huge scary ichthyosaurs and plesiosaurs dominated the imaginations of artists trying to render the prehistoric past (see figure 14.4). But none of these people had yet imagined a world dominated by dinosaurs.

There were many isolated finds of dinosaur bones before anyone finally realized that they were the remains of large extinct reptiles, instead of being misinterpreted as "dragon bones" (as the Chinese long called them) or the bones of human giants mentioned in the Bible. In 1676, a large bone was

Figure 17.1 ▲

Robert Plot's original figure of the first dinosaur ever illustrated, later called "Scrotum humanum" (actually the end of a thigh bone of a theropod dinosaur, probably *Megalosaurus*). (Courtesy Wikimedia Commons)

found in the Taynton Limestone in Stonesfield Quarry, near Oxford, which dates to the Middle Jurassic. A year later, Robert Plot, professor of chemistry at Oxford University, published *The Natural History of Oxfordshire* and illustrated it—the first time that a dinosaur bone had ever been figured in the scientific literature (figure 17.1). He correctly realized that it was the lower end of a thighbone (femur) and thought that it might be from a Roman war elephant or from a giant in the Bible.

In 1763, Richard Brookes wrote a book that republished Plot's illustration and captioned it "Scrotum humanum," since it vaguely resembles a gigantic pair of petrified human testicles. In 1970, there was a controversy as to whether the valid name of the very first described dinosaur was the unfortunate "Scrotum humanum," which would replace the younger name *Megalosaurus*. The International Commission of Zoological Nomenclature ruled that the specimen was not diagnostic enough to know for sure which dinosaur it had come from and that the name clearly had not been intended as a scientific description, since it is just two words in a caption.

Between 1815 and 1824, the Reverend William Buckland, a famous natural historian, described jaw fragments and other bones from a huge predatory "lizard" discovered near his home in Oxford, England, that he called *Megalosaurus* (Greek for "big lizard"). Then in 1825, Dr. Gideon Mantell described teeth and some other bone fragments from a huge reptile found in the Lower Cretaceous Wealden beds in Tilgate Forest, Sussex, that he called *Iguanodon* (iguana tooth).

By 1842, these and other discoveries prompted British naturalist Richard Owen to coin the word "Dinosauria" (Greek for "fearfully great lizards") to cover all these specimens. Only a year earlier, Owen had described some teeth and other huge bones from a creature found in 1825 near Chipping North that he named *Cetiosaurus* (Greek for "whale lizard"). Because the material was so incomplete, Owen thought that the fossils were those of a giant marine reptile related to crocodiles, but Mantell corrected him and suggested that they were from a giant land reptile like *Iguanodon* or *Megalosaurus*. Owen, however, did not agree and did not include *Cetiosaurus* when he named the Dinosauria in 1842. The scraps of *Cetiosaurus* were not enough to accurately reconstruct the creature at the time.

But in March 1868, some workers near Bletchingdon found the huge thighbone of a sauropod, and soon many other large limb elements and vertebrae were discovered. With these bones, it became clear that *Cetiosaurus* was not a giant crocodile, but a huge reptile that had walked on four pillar-like legs (figure 17.2). The skeleton was not complete enough to reveal the long neck and long tail that we now associate with sauropods, but it was (and still is) one of the most complete sauropods found in Europe.

It wasn't until the 1870s and 1880s that nearly complete sauropod skeletons were finally found in Colorado and Wyoming by crews working for paleontologist Othniel Charles Marsh of Yale University and teams working for naturalist Edward Drinker Cope of Philadelphia. Marsh, in particular, recovered some remarkably complete specimens from a locality called Como Bluff in south-central Wyoming. Soon each specimen he received got a name, starting with *Apatosaurus* and *Atlantosaurus* in 1877, *Morosaurus* and *Diplodocus* in 1878, and *Brontosaurus* and *Barosaurus* in 1890, while Cope named *Camarasaurus* and *Caulodon* in 1877. (Today, scientists consider *Atlantosaurus* and *Brontosaurus* to be the same as *Apatosaurus*, and *Morosaurus* to be the same as *Camarasaurus*. Only *Apatosaurus*, *Diplodocus*, *Camarasaurus*, and *Barosaurus* are still valid genera.) There were so many

Figure 17.2 ▲
Limb bones from the partial skeleton of *Cetiosaurus*, displayed at the Oxford University Museum of Natural History. (Photograph courtesy M. Wedel)

of these fossils that by 1878, Marsh could lump them (including *Cetiosaurus*), into a group he called the Sauropoda (Greek for "lizard foot"). Unfortunately, Marsh published only short papers on each of these dinosaurs, with no illustrations, so most of the public was still not aware of the existence of these giants by 1900.

The final stage of discovery of the "whale reptiles" came when museums began to realize that these huge skeletons gathering dust in their basements would make for great publicity and draw huge crowds. By 1905, the American Museum of Natural History in New York, the Carnegie Museum of Natural History in Pittsburgh, and the Yale Peabody Museum of Natural History in New Haven, Connecticut, had mounted skeletons of large sauropods labeled with Marsh's invalid name "Brontosaurus."

Sadly, the name "Brontosaurus," which is so entrenched in our culture, is a junior synonym of *Apatosaurus* and cannot be used. Marsh named *Brontosaurus* in 1890 based on a particularly complete adult skeleton of a large sauropod from Como Bluff. As was the custom in those days, nearly every fossil that was even slightly different from the other known specimens got a new name. This name was then attached to the mounted skeletons at the American Museum and Peabody Museum, where it became an icon and "Brontosaurus" entered every book about dinosaurs.

As it turned out, in 1877 Marsh had given the name *Apatosaurus* to a slightly less complete and juvenile specimen of the same dinosaur. In 1903, Elmer Riggs looked closely at Marsh's specimens and concluded that *Apatosaurus* was the same animal as *Brontosaurus*. By the rules of the International Code of Zoological Nomenclature, the first name given is the correct one, so as far as paleontologists go, the name "Brontosaurus" has been a junior synonym since 1903. But the most influential paleontologist of the time, Henry Fairfield Osborn of the American Museum of Natural History, refused to believe Riggs's analysis, and he helped the incorrect name "Brontosaurus" to survive in the popular imagination long after all other paleontologists had abandoned it.

Unfortunately, the popular literature and media often do not keep up with the science, so the name was still common until the 1980s and 1990s, when museums began to redo their mounts to put them in more realistic poses. Because the specimen from Como Bluff had no skull, the original skull added to the skeletons at the American Museum and Peabody Museum was that of a short-faced brachiosaur. John Ostrom and Jack McIntosh showed that *Apatosaurus* had a long-snouted skull, much like that of *Diplodocus*. Finally, enough paleontologists had complained and children's books and news articles began to reflect the change—90 years after *Apatosaurus* was published.

Osborn also commissioned the legendary artist Charles R. Knight to paint the iconic reconstruction of "Brontosaurus," and soon the public became obsessed with brontosaurs and the imagery of huge long-necked, long-tailed sauropod dinosaurs (figure 17.3). They appeared in the earliest stop-motion animated films, including *The Lost World* (1925), with animated dinosaurs by the legendary Willis O'Brien, based on Knight's artwork. Sauropod dinosaurs were soon everywhere—editorial cartoons, more movies, merchandise, and even the logo of the Sinclair Oil Company—so it's hard to imagine that just 110 years ago, nobody but a few scientists had heard of these creatures.

Figure 17.3 ▲
Charles R. Knight's iconic painting, from 1905, of "Brontosaurus" as a sluggish, tail-dragging swamp dweller, an idea that is now completely obsolete. (Image no. 327524, courtesy American Museum of Natural History Library)

LIFESTYLES OF THE HUGE AND ANCIENT

The study of sauropods has come a long way in the past century. The number of genera recognized is at least 90 and probably more, although deciding which named sauropods are valid taxa is a bit of a problem. Due to their huge size, many of their bones are robust and durable and easily preserved, no matter how much the skeleton has been broken up and washed away. As a result, most named sauropods are known from only a few bones: typically some of the backbone elements, or vertebrae, and occasionally the limbs. There are numerous partial skeletons, but even they have the annoying habit of losing their heads before they are fossilized (skulls tend to be lighter and more fragile than the other bones). Only a few sauropods are known from reasonably complete skeletons, and they are the ones featured in museums over and over again: *Apatosaurus*, *Diplodocus*, *Brachiosaurus*, *Camarasaurus*, *Barosaurus*, *Mamenchisaurus*, and a few others.

The sauropods originated from a group of Triassic dinosaurs called prosauropods, which are classic intermediate forms linking the big Jurassic

Figure 17.4 ▲
Skeleton of the prosauropod *Plateosaurus*, from the Triassic of Germany. (Photograph by the author)

monsters with early dinosaur lineages, some of which were as small as chickens. Prosauropods such as *Plateosaurus* were up to 10 meters (33 feet) in length and weighed up to 4000 kilograms (8800 pounds), but were nowhere near as large as their descendants (figure 17.4). Nevertheless, they had the beginnings of the long neck and tail. Although *Plateosaurus* was almost completely bipedal, the limbs of some prosauropods (such as *Melanorosaurus*) allowed them to walk on either four feet (quadrupedal) or two feet (bipedal), and they had well-developed fingers for grasping, unlike their much heavier descendants, with their elephantine limbs.

By the Middle and Late Jurassic, it was truly a world of giants. These monsters were not the slow, sluggish tail-dragging lizards of the swamps that people imagined in 1905. Early scientists were so impressed by their size that they could not imagine sauropods supporting their weight on land, and hence put them in swamps. In reality, a number of important specimens (including trackways) plus many good biomechanical analyses have radically transformed our view of sauropods and their paleobiology. First of all, trackways show that sauropods walked with their tails held straight, because almost none show tail-drag marks. In addition, analyses of the Morrison Formation and other rock formations full of sauropods demonstrate that they did not live in swamps, but were adapted not only to coastal regions but even to drier habitats. Sauropods were excellent walkers that covered long distances to find forage, and they fed on foliage in trees that they were able to reach with their long necks. Finally, they had so many air sacs in their bodies that they could not have sunk very far into the water, let alone dove beneath the surface.

Their anatomy is quite remarkable. Their head was very small for such large animals, and most had simple peg-like or blade-like teeth. Many scientists have puzzled over how they could feed such enormous bodies with such a limited dental apparatus. (By contrast, both duck-billed dinosaurs and horned dinosaurs evolved dental batteries of hundreds of grinding teeth to process lots of vegetation.) Some paleontologists have speculated that sauropods fed indiscriminately on nearly everything they could eat, from the tops of trees to the dense carpet of ferns. Remember that no flowering plants, especially not grasses, evolved until the Early Cretaceous, long after the heyday of the giant sauropods of the Jurassic.

The individual vertebrae over the entire neck, back, and tail are marvels of engineering, with many bony struts and braces to make them light but very strong and to help them hold together, bound by many powerful tendons. Like those of many dinosaurs and birds, the bones of sauropods (especially along the spinal column) were full of air sacs, which made them relatively light. Some recent research has suggested that sauropods did not hold their head very high for very long (contrary to most reconstructions), because they would have needed an extraordinarily high blood pressure to pump blood to their head. The studies on which this idea is based, however, were conducted with domesticated animals, which have been bred to have unhealthy high blood pressure. Other recent research (not yet published) has argued that sauropods would have had manageable blood pressure and would not have required an extraordinarily large heart to pump blood to their head while raising it. Like giraffes, they probably had special valves in the blood vessels of the neck that prevented a sudden drop in blood pressure and kept them from fainting when they raised their head high.

As the largest land animals ever, sauropods had massive limbs and feet with the toes compacted into short disks or columns of bone, as do elephants. Unlike elephants, however, sauropods walked on the tips of their stumpy toes (digitigrade, as did almost all dinosaurs), rather than on the soles of their feet and toes (plantigrade, as do humans as well as elephants), although their feet were partially digitigrade and partially plantigrade. The huge sauropod leg bones, the spacing of their trackways, and the immense mass they carried rule out the idea that large sauropods were fast moving. Most of the time, they ambled along at a slow but steady pace, although they may have been capable of a bit of running (as elephants can even now). But with their long legs, they could cover extensive amounts of territory without the need to run.

There are several major branches of sauropods, including the very-long-necked, whip-tailed diplodocines (like *Diplodocus* and *Apatosaurus*); the tall brachiosaurs, with their elongate front legs and giraffe-like neck; and the small-headed, stocky titanosaurs (which flourished mainly in Africa and South America, but lived on every continent, including Antarctica), among others. The majority of sauropods reached their heyday in the Late Jurassic, but some groups (such as the titanosaurs) were still flourishing in the Southern Hemisphere in the Cretaceous (even though they had nearly vanished in North America) and may have survived to nearly the end of the Cretaceous.

SIZE MATTERS!

Naturally, for animals that were always the biggest creatures in their own habitat, and the largest land animals that ever lived, size matters. A number of candidates have been championed as the "largest dinosaur," only to be toppled by new discoveries a few years later. Complicating the claim is that the larger the dinosaur, the fewer the bones that have survived. The largest and heaviest dinosaur for which a nearly complete skeleton is known is the famous *Brachiosaurus* (now called *Giraffatitan*) in the Museum für Naturkunde (Humboldt Museum) in Berlin (figure 17.5), which was found in the Tendaguru beds of German East Africa (now Tanzania) in 1909 to 1912. This impressive specimen is a composite of five partial skeletons (mostly juvenile), and it towers several stories (13.5 meters [44 feet]) above the floor of the gallery and reached a length of 22.5 meters (74 feet). Its mass would have been about 30 to 40 metric tons (33 to 44 tons).

Bigger dinosaur bones have been found—for example, a shin bone in the same collection is 13 percent larger than that of the mounted *Giraffatitan*—but accurately estimating the size of an animal based on a few vertebrae or limb bones is fraught with problems (figure 17.6). For example,

DIE WELT IM OBEREN JURA

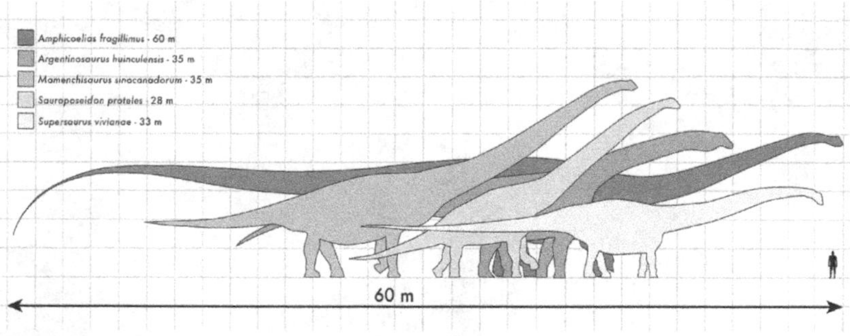

Figure 17.6 ▲

Comparison of the sizes of sauropods, some of which—*Amphicoelias, Sauroposeidon*, and *Supersaurus*—are too incompletely known to accurately calculate their true size. (Drawing by Mary P. Williams)

Mathew Wedel and Richard Cifelli recovered four neck vertebrae from a huge titanosaur from the Early Cretaceous of Oklahoma that they named *Sauroposeidon* (after Poseidon, the Greek god of the sea and earthquakes). The bones are so huge that they were first misidentified as petrified tree trunks until someone cleaned them thoroughly and realized that they are dinosaur bones. Cifelli found them in 1994 and brought them to the Sam Noble Oklahoma Museum of Natural History, but only when his student Wedel looked closer did they realize what they had. *Sauroposeidon* is known from only the four neck bones, but they are truly gigantic. If they can be used to estimate size based on *Giraffatitan*, then *Sauroposeidon* could have reached 17 meters (56 feet) in height with its neck upright, making it the tallest known dinosaur. It was about 34 meters (112 feet) long and weighed around 40 metric tons (44 tons).

If *Sauroposeidon* was the tallest creature ever to live, several other sauropods were longer and heavier. The largest specimen for which enough bones are known to reliably estimate size is *Argentinosaurus*, from the Huincul Formation of (where else?) Argentina (especially Patagonia in southern Argentina), which dates to the Late Cretaceous (figure 17.7). The first bones were found in 1987 by a rancher who, once again, mistook them for

Figure 17.5 ◀

The most complete mounted skeleton of a large sauropod, *Giraffatitan* (= *Brachiosaurus*) *brancai*, from the Tendaguru beds in Africa, displayed at the Museum für Naturkunde (Humboldt Museum) in Berlin. (Photograph by M. Wedel)

Figure 17.7 ▲
Mounted skeleton of *Argentinosaurus*, displayed at the Museo Municipal Carmen Funes in Plaza Huincul, Argentina. (Photograph courtesy R. Coria)

petrified logs. Then the specimens were collected, and the dinosaur was formally named *Argentinosaurus huinculensis* by José Bonaparte and Rodolfo Coria in 1993. *Argentinosaurus* consists of part of the backbone, the hip region, some ribs, thighbones, and a right shin bone. Although few in number, the individual bones are huge. Each vertebra is staggering, over 1.59 meters (5.2 feet) tall (figure 17.8*A*), and the shin bone is 1.55 meters (5 feet) long! The size estimates based on these incomplete fossils range from 30 to 35 meters (98 to 115 feet) in length and 80 to 100 metric tons (88 to 110 tons) in weight, although a more recent calculation suggests that it weighed about 50 metric tons (55 tons). Another estimate based on the smaller but more complete *Saltasaurus* (another titanosaur) places its length at 30 meters (98 feet), with a weight between 60 and 88 metric tons (66 and 97

Figure 17.8 ▶
Sauropod vertebrae: (*A*) gigantic vertebra of *Argentinosaurus*; (*B*) much smaller vertebra of *Giraffatitan* (= *Brachiosaurus*) *brancai* (see figure 17.5), the largest nearly complete dinosaur skeleton ever found or mounted, for comparison. ([*A*] photograph by the author; [*B*] photograph courtesy M. Wedel)

A
B

tons). The mounted skeleton in the Museo Municipal Carmen Funes is 40 meters (130 feet) long and 7.3 meters (24 feet) high, even longer and taller than the original estimates (see figure 17.7). This would make it by far the longest and largest land animal to have ever lived. Thus we will consider it to be the current record holder.

But not so fast! A number of huge sauropods from about 97 to 94 million years ago, the same time as *Argentinosaurus*, are close to the same size, including *Paralititan*, from Egypt, and *Antarctosaurus* (figure 17.9) and *Argyrosaurus*, from South America. Unfortunately, none of them are known from more than a few leg bones, so estimating whether they were bigger than *Argentinosaurus* is uncertain at best. Recently, news releases announced the discovery in Argentina of even larger limb bones. They claimed that it is from the largest dinosaur ever found (the usual hype associated with any specimen like these), but most paleontologists think that it is just a large adult *Argentinosaurus*.

In 2014, another gigantic sauropod was reported from Argentina. The discoverers dubbed it *Dreadnoughtus*, because it reminded them of the huge "Dreadnought" class of battleships during World War I that dreaded no other ship because of their huge size and guns. *Dreadnoughtus* is more complete than most other sauropods and is claimed to be 70 percent complete. It consists of mostly the back end of the animal and its forelimbs, but very little of the head and neck, so its length is purely conjectural. Once again, the discoverers got sucked into the media game of declaring their find the "biggest ever," with a weight estimate of 59 metric tons (65 tons), but many other paleontologists have commented that the specimen is not complete enough to reliably calculate the weight, and certainly not the length. Many paleontologists regard all these titanosaurs of slightly different sizes, excavated from the Cretaceous rocks of Argentina, as one highly variable genus and maybe a few species that have been excessively split into dozens of named genera because of the competition for press attention. Biologists know that animals as large as these tend to have small populations and much variability within species, not dozens of closely related genera sharing a habitat.

Currently, there are suggestions that there were even more massive titanosaurians than *Argentinosaurus*. One of them, *Bruhathkayosaurus* (from the South Indian Sanskrit for "huge heavy body," and the Greek for "lizard"), is from the Late Cretaceous of India. Described by P. Yadagiri and K.

Figure 17.9 ▲

Huge thigh bones of *Antarctosaurus*, the largest known argentinosaur, which are larger than any dinosaur thigh bone yet found, with Francisco Novas for scale. (Photograph courtesy Fernando Novas)

Ayyasami in 1989, it may have weighed 175 to 220 metric tons (190 to 240 tons), but a later estimate knocked that down to 139 metric tons (153 tons). If this is true, it was much larger than any other known sauropod. However, the fossils consisted of only part of the hip bones, part of a thighbone and shin bone, a forearm, and parts of some vertebrae. The shin bone, however, was 2 meters (6.6 feet) long, 29 percent larger than that of *Argentinosaurus*, as was the thighbone. Most paleontologists have reserved judgment on *Bruhathkayosaurus* until more complete material is found (which is not likely). Sadly, the original specimens were lost when monsoonal flooding destroyed their storage area, so all that remains is the original publication with its simple line drawings.

If that is not staggering and frustrating enough, consider the case of *Amphicoelias fragillimus*. It was based on a single vertebra from the backbone, found by pioneering paleontologist Edward Drinker Cope and named in 1877. He published one figure that showed the specimen, and if the measurements he gave are to be believed, it was immense! The single vertebra would have been 2.7 meters (8.8 feet) tall if it were complete! If the size of that vertebra were plugged into the body plan of other sauropods, *Amphicoelias* was 40 to 60 meters (130 to 200 feet) long and weighed up to 122metric tons (135 tons), which would beat any other dinosaur except *Bruhathkayosaurus* (see figure 17.6). Unfortunately, the material of *Amphicoelias* vanished some time after Cope described it. Possibly, it was falling apart in his crowded storage area, since hardeners and preservatives were not in use yet, and was unrecognizably broken by the time people came to move his collection after he died. Thus not only are the two candidates for the largest land animal based on inadequate fossils, but all the fossils are lost! The title is still held by *Argentinosaurus* until better material dethrones it.

A LIVING DINOSAUR IN THE CONGO?

One modern legend concerns an alleged sauropod dinosaur still living in the modern Congo River Basin. Known as Mokele Mbembe, it has been the subject of many books, media reports, television "documentaries," and even the Hollywood movie *Baby: Secret of the Lost Legend* (1985). A number of people have brought back reports about it, so it is almost as famous as the Loch Ness monster and Bigfoot.

If you look closer, however, you will find nothing but smoke and mirrors. As Daniel Loxton and I have carefully documented, there is no good physical evidence whatsoever to support the claim of its existence. Most of the "evidence" consists of eyewitness reports by native peoples that were translated and relayed by American explorers (almost always missionaries or modern-day creationists, not biologists). Such accounts are highly problematic because many native peoples do not distinguish between their mythical creatures and what we consider "real" animals. In addition, many of the accounts are really vague, highly conflicting, and useless scientifically. Some seem to identify a stegosaur or *Triceratops*, not a sauropod. Some even describe a rhinoceros, an animal that is unknown to Congo Basin peoples, because rhinos live in the savannah, not the jungle. Many are suspect because Western explorers often show natives their sketches of the beast and ask them to confirm their depictions, thus "leading the witness." In many cultures, it is normal for native peoples to tell visitors what they want to hear, just as a matter of courtesy to their guests. Most important, recent research by many psychologists has shown that "eyewitness testimony" is virtually worthless as evidence (even in a court of law). Humans are not good video recorders. We are so good at "seeing" what we expect to see, coloring what we originally saw with later expectations, and imagining things that we later believe we really saw that no scientist takes the words of an "eyewitness" as anything more than an individual's experience (and possibly delusion or hallucination).

In addition, there are a huge number of problems with the accounts and the evidence that make the existence of Mokele Mbembe extremely unlikely. All the photographs and video footage are so distant and blurry that that it is impossible to decipher what they show, let alone that they are truly proof of Mokele Mbembe. Of those that can be identified, most turn out to be hippos, people in canoes, or other blurry objects of no diagnostic features. Population ecology tells us that animals as large as sauropods need huge home ranges and would have a significant population, including adults and juveniles—yet all we have are eyewitness accounts and bad videos, with not a bone or a carcass or any other physical evidence.

As time goes on and more and more people look for Mokele Mbembe without finding even one, the case grows even weaker. In fact, the "uncharted jungles" of the Congo are a myth. Real wildlife biologists travel through the Congo Basin all the time, and they never hear reports of or see

Mokele Mbembe. The only ones who believe these reports are credulous missionaries who know nothing about biology. In fact, with Google Earth, anyone can study the region from space and easily see large animals. If you type the coordinates 10.903497 N, 19.93229 E into Google Earth, you can see elephants in great detail from space. Certainly, an animal as big as Mokele Mbembe would have been spotted by now if huge herds of them were roaming the Congo Basin.

The paleontological record for sauropods is excellent, and huge bones like theirs fossilize well. So the fact that *not one* fossil sauropod bone has ever been found in deposits younger than 65 million years is pretty conclusive that sauropods did not survive into the present (although there are many beds of the right environmental setting that do fossilize large mammals).

Finally, there is something about the entire Mokele Mbembe story that just does not ring true. The dinosaur that "eyewitnesses" have described is a version of sauropods that was popular in 1905, when the first skeletons and paintings were in the public eye—but those creatures never actually existed. The slow, sluggish creatures that dragged their tails and hid in swamps have been transformed, as a result of scientific research, into creatures that held their tails straight and lived on land near coasts, not in the water. The accounts of Mokele Mbembe describe it submerging in the Congo River and lingering for hours. In fact, sauropods could not even immerse themselves halfway because they had too many air sacs along their spine. They could not dive, let alone stay underwater for hours.

Instead, the myth of Mokele Mbembe has a strange twist. The only people looking for it are creationist ministers, not wildlife biologists. A few years ago, I was asked to be the "token skeptic" on an episode of *MonsterQuest* that focused on Mokele Mbembe. The entire film shoot was truly bizarre, since the producers spent most of their time trying to get me to say things that could be construed as supporting the existence of Mokele Mbembe. They attempted a "gotcha" moment on camera when they handed me a shapeless lump of plaster and hoped that I would identify it as a "dinosaur track." When I saw the final program, what surprised me the most were the two Mokele Mbembe "hunters" whose search for the dinosaur took up most of the airtime. They revealed themselves as incompetent wildlife biologists, not having a clue as to what they were doing or even how to use their fancy equipment. They made bizarre statements about a tiny hole in the bank of the river, as if a giant sauropod could dig into a low bank, completely hide itself, and leave leaving only a tiny air hole.

Later, I found out that both "explorers" were creationist ministers with no formal training in wildlife biology. One of them, William Gibbon, has made numerous trips to the Congo, wasting lots of money with absolutely no results. Somehow, these people seem to think that the discovery of a living dinosaur would cause the theory of evolution to collapse—never mind the mountains of evidence that support it!

The quest for Mokele Mbembe is no longer just an idle search for a cryptid by amateurs. The "explorers" spending their time looking for it have an anti-science agenda and cannot be trusted with their data or their interpretations. Their search is a part of the global effort by creationists to overthrow the evidence of evolution and undermine the teaching of science by any means possible. As such, it cannot be dismissed or treated lightly, but must bear the full scrutiny of the scientific community as an effort to destroy science.

SEE IT FOR YOURSELF!

Many natural history museums around the world display originals or replicas of sauropod skeletons. In the United States, those with original material include the American Museum of Natural History, New York (*Apatosaurus* and *Barosaurus*); Carnegie Museum of Natural History, Pittsburgh (*Apatosaurus* and original *Diplodocus*); National Museum of Natural History, Smithsonian Institution, Washington, D.C. (*Diplodocus* and *Camarasaurus*); Natural History Museum of Los Angeles County, Los Angeles (*Mamenchisaurus*); and Yale Peabody Museum of Natural History, New Haven, Connecticut (*Apatosaurus*).

The nearly complete skeletons of *Giraffatitan* (= *Brachiosaurus*) *brancai* and *Dicraeosaurus* are exhibited at the Museum für Naturkunde (Humboldt Museum), Berlin; and, in Chicago, replicas of *Giraffatitan* are just outside Field Museum of Natural History and at O'Hare International Airport. The vertebrae of *Sauroposeidon* are displayed at the Sam Noble Oklahoma Museum of Natural History, University of Oklahoma, Norman. In Argentina, reconstructed skeletons of *Argentinosaurus* can be seen at the Museo Municipal Carmen Funes, Plaza Huincul; and Museo Argentino de Ciencias Naturales "Bernardino Rivadavia," Buenos Aires. A replica is displayed at the Fernbank Museum of Natural History, Atlanta.

FOR FURTHER READING

Brett-Surman, M. K., Thomas R. Holtz Jr., and James O. Farlow, eds. *The Complete Dinosaur*. 2nd ed. Bloomington: Indiana University Press, 2012.

Curry Rogers, Kristina, and Jeffrey A. Wilson, eds. *The Sauropods: Evolution and Paleobiology*. Berkeley: University of California Press, 2005.

Fastovsky, David E., and David B. Weishampel. *Dinosaurs: A Concise Natural History*. 2nd ed. Cambridge: Cambridge University Press, 2012.

Holtz, Thomas R., Jr. *Dinosaurs: The Most Complete Up-to-Date Encyclopedia for Dinosaur Lovers of All Ages*. New York: Random House, 2007.

Klein, Nicole, Kristian Remes, Carole T. Gee, and P. Martin Sander, eds. *Biology of the Sauropod Dinosaurs: Understanding the Life of Giants*. Bloomington: Indiana University Press, 2011.

Loxton, Daniel, and Donald R. Prothero. *Abominable Science: The Origin of Yeti, Nessie, and Other Cryptids*. New York: Columbia University Press, 2013.

Paul, Gregory S. *The Princeton Field Guide to Dinosaurs*. Princeton, N.J.: Princeton University Press, 2010.

Sander, P. Martin. "An Evolutionary Cascade Model for Sauropod Dinosaur Gigantism—Overview, Update and Tests." *PLoS ONE* 8 (2013): e78573.

Sander, P. Martin, Andreas Christian, Marcus Clauss, Regina Fechner, Carole T. Gee, Eva-Marie Griebeler, Hanns-Christian Gunga, Jürgen Hummel, Heinrich Mallison, Steven F. Perry, Holger Preuschoft, Oliver W. M. Rauhut, Kristian Remes, Thomas Tütken, Oliver Wings, and Ulrich Witzel. "Biology of the Sauropod Dinosaurs: The Evolution of Gigantism." *Biological Reviews of the Cambridge Philosophical Society* 86 (2011): 117–155.

Tidwell, Virginia, and Kenneth Carpenter, eds. *Thunder-Lizards: The Sauropodomorph Dinosaurs*. Bloomington: Indiana University Press, 2005.

18 | THE FIRST BIRD | *ARCHAEOPTERYX*

A FEATHER IN STONE

And if the whole hindquarters, from the ilium to the toes, of a half-hatched chick could be suddenly enlarged, ossified, and fossilised as they are, they would furnish us with the last step of the transition between Birds and Reptiles; for there would be nothing in their characters to prevent us from referring them to the Dinosauria.

THOMAS HENRY HUXLEY, "FURTHER EVIDENCE OF THE AFFINITY BETWEEN DINOSAURIAN REPTILES AND BIRDS"

NATURAL ART

For more than 300 years, stonemasons had cut slabs of the beautiful, finely layered cream-colored limestones from the Solnhofen quarries near Eichstätt in southern Germany. These incredible rocks were so fine-grained (without any visible fossils, so typical of most limestones) that they were world famous for their use in making lithographic plates by acid etching. There were no flaws or impurities or fossil fragments to ruin the fine detail of the hand-carved plates. Many great works of art had been carved from this rock. It had been used to print some of the first lithographs in the earliest days of printed books, including legendary works by the artist Albrecht Dürer and others. Its completely uniform color and lack of pattern or grain also make it a popular building stone, and it can even be ordered online from a number of commercial operations.

By the mid-nineteenth century, the quarries at Solnhofen were very extensive, with many quarrymen working hard to find good unbroken exposures of limestone from which to cut large flat slabs that could be turned into printing plates or building stones. Occasionally, when they split the slabs along bedding planes, they found art of an entirely different kind: exquisitely preserved fossils of many different creatures, including numerous kinds of bony fish and an occasional crustacean or horseshoe crab or brittle star. But there were also fossils of the chicken-size dinosaur *Compsognathus* and of the first well-preserved specimens of pterodactyls ever found, described by naturalists as early as 1784. The quarrymen were not deliberately looking for these fossils, but when they were exposed by accident, they were nice rewards for all the hours of backbreaking work. Some of them were so beautiful that they were sold to collectors and rich gentlemen who were accumulating these natural objects for pleasure or for scientific reasons.

Then one day in 1860, a quarryman made a surprising discovery in the limestone. It was the distinct impression of a single feather, very much like the asymmetric wing feathers of modern birds. The specimen eventually ended up in the hands of the distinguished paleontologist Christian Erich Hermann von Meyer, who had already described most of the Solnhofen dinosaurs and pterodacyls, as well as the early dinosaur *Plateosaurus* (chapter 17). Based on this one fossil feather, in 1860 von Meyer gave it the formal scientific name *Archaeopteryx lithographica* (ancient wing from the lithographic stones).

DARWIN'S GODSEND

A few months later, a nearly complete skeleton was found in a quarry near Langenaltheim, Germany (figure 18.1), and traded to a local doctor, Karl Häberlein, in exchange for his medical services. This specimen was missing most of the head and neck, and was a jumble of bones, but it clearly showed imprints of feathers around a skeleton that most closely resembled that of a dinosaur. German museums dithered about buying the specimen, so Häberlein took the best offer he could get: £700 (about $72,000 in today's dollars, a fortune in those days!) offered by the British Museum of Natural History. Thus it became known as the "London specimen" from its current place of residence. Once it was in London, the fossil came under the

Figure 18.1 ▶
The "London specimen" of *Archaeopteryx*. (Courtesy Wikimedia Commons)

supervision of the distinguished British anatomist and paleontologist Richard Owen. Already famous for his description of many other fossils and for naming the Dinosauria, Owen soon set to the task and published an extensive description of the specimen in 1863. Even in its incomplete state, Owen could not ignore the fact that its bones were very reptilian, yet it clearly had feathers on its wings.

This discovery was a godsend for another British naturalist, Charles Darwin. His controversial new book, *On the Origin of Species*, had been published in 1859, just two years earlier. Despite the strong case he had built for the reality of evolution, he had to apologize for the absence of good transitional fossils to support his theory. With perfect timing, *Archaeopteryx* offered just such a transitional fossil to bolster his case, and Darwin was ecstatic. He could not have predicted a more perfect example of how it was possible for reptiles to have evolved into birds, a completely different group. In the fourth edition of *Origin*, he bragged that at one time some scientists had argued

> that the whole class of birds came suddenly into existence during the Eocene period [54 to 34 million years ago, as we now date it]; but now we know, on the authority of Professor Owen, that a bird certainly lived during the deposition of the Upper Greensand [Late Early Cretaceous in modern terminology, about 100 million years ago; this specimen was a pterosaur]; and still more recently, that strange bird, the *Archaeopteryx*, with a long lizard-like tail, bearing a pair of feathers on each joint, and with its wings furnished with two free claws, has been discovered in the oolitic slates of Solnhofen. Hardly any recent discovery shows more forcibly than this how little we as yet know of the former inhabitants of the world.

Yet Owen believed in his own form of "trans-mutation," not Darwinian evolution. When he described the fossil in 1863, he studiously avoided or dismissed all the clear connections between birds and reptiles that it suggested. Thomas Henry Huxley, the pugnacious young scientist whose brilliant defense of evolution earned him the nickname "Darwin's bulldog," took Owen to task for his failure to admit the obvious. Huxley argued not only that *Archaeopteryx* perfectly filled the role of "missing link" between reptiles and birds, but, even more important, that it was clearly dinosaurian in most of its bony features. In fact, it turned out that one of the *Archaeopteryx* specimens was originally misidentified as the small Solnhofen dinosaur *Compsognathus*—until a century later, when John Ostrom of Yale University looked closer and saw the feathers.

MORE AND MORE SPECIMENS

The real clincher for the debate came when a local farmer, Jakob Niemayer, found the best of all the known *Archaeopteryx* specimens in 1874 near Blumberg, Germany (figure 18.2). To raise funds to buy a cow, he sold this amazing fossil to the innkeeper Johann Dörr. He, in turn, sold it to Ernst Otto Häberlein, son of the doctor who had sold the first fossil of *Archaeopteryx* to the British Museum about 12 years earlier. This is the most famous and most photographed of all the 12 known specimens, because it is nearly complete and is splayed out on the rock showing all its feathers, with its neck and head pulled backward. This is a typical posture in dying animals as the nuchal ligament that holds up the neck and head contracts.

When Häberlein sought bids for this incredible find in 1877, many institutions wanted to buy it. Not only were the British interested, but Yale paleontologist Othniel C. Marsh also made an offer. But the Germans did not want any foreigners to scoop up their heritage so easily after the first *Archaeopteryx* flew the coop. Financed by Ernst Werner von Siemens (whose famous company is still a giant in many fields), the Museum für Naturkunde (Humboldt Museum) in Berlin bought it for 20,000 Goldmarks (about $21,000 in today's dollars), and it is now known as the "Berlin specimen." It has been studied and restudied many times, and it forms the basis for most of what we know about *Archaeopteryx*. It is an even better example of a "missing link" in evolution than the "London specimen," since it is so complete and displays a mix of dinosaurian and bird-like features with unambiguous clarity.

Even though fossils of *Archaeopteryx* are rare (only 12 specimens found in nearly 500 years), more have turned up since the "Berlin specimen" was officially announced in 1877. One fossil (in the Teylers Museum in Haarlem, Netherlands) was originally misidentified as the wing of a pterosaur after it was found in 1855, before the first specimen identified as *Archaeopteryx* was revealed in the limestone. But in 1970, Ostrom looked a lot closer and realized that it is a wing bone of *Archaeopteryx*, not of a pterosaur; it even has faint feather impressions. Another specimen (in the Jura Museum in Eichstätt) was found in 1951 near Workerszell, Germany, and is one of the

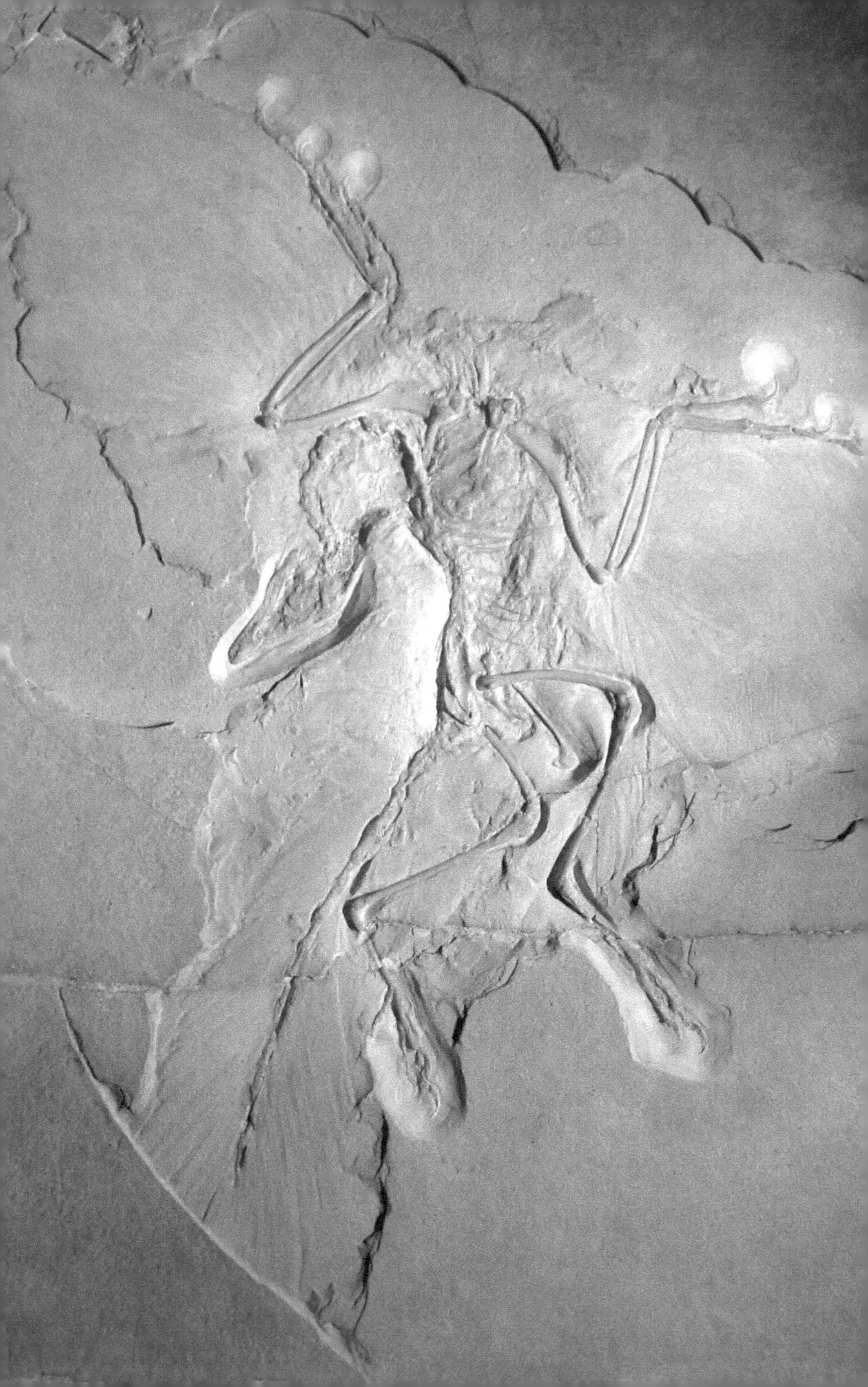

smallest but most complete skeletons known. Yet another fossil, discovered in 1992, was sold in 1999 to the Paläontologisches Museum München for 1.9 million Deutschmarks (about $1.3 million in today's dollars). It is also nearly complete, although it was folded almost in half as it fossilized. The torso of another specimen (no head or tail preserved) was discovered in 1956 near Langenaltheim and was on display for many years at the Maxberg Museum before its owner, Eduard Opitsch, took it back. After he died, it could not be found, so it was either stolen or sold into the black market.

Two other fragmentary fossils are still in private hands. The "Daiting specimen" (from the Daiting beds, slightly younger than Solnhofen) has been displayed only briefly. Another fossil, on temporary loan to the Burgermeister-Müller Museum in Solnhofen is of only a wing. Yet another important specimen was long in private hands before it was donated to the tiny Wyoming Dinosaur Center in the isolated town of Thermopolis. It is one of the more complete fossils, with good feet and a head, but no lower jaw or neck. Finally, the discovery of the twelfth specimen was announced in 2011, but it is privately owned and was just recently described.

BIRD . . . OR DINOSAUR?

As Huxley realized in the 1860s, most of the skeleton of *Archaeopteryx* is so dinosaurian that one specimen was mistaken for the little Solnhofen theropod dinosaur *Compsognathus* (figure 18.3). Like most dinosaurs (but no living birds), *Archaeopteryx* had a long bony tail, a highly perforated skull with teeth, dinosaurian (not bird-like) vertebrae, a strap-like shoulder blade, a hip bone midway between that of typical dinosaurs and of later birds, gastralia (rib bones found in the belly region of dinosaurs), and unique dinosaurian and bird-like specializations in the limbs. The most striking of these are in the wrist. All birds and some predatory dinosaurs, such as the dromaeosaurs (*Deinonychus* and *Velociraptor* and their kin), have a half-moon-shaped wrist bone formed by the fusion of multiple bones; this feature is unique to these animals. This bone serves as the main hinge for the movement of the wrist, allowing dromaeosaurs to extend their wrist and grab prey with a rapid downward flexing motion. It so happens that exactly the same motion is part of the downward flight stroke of birds. *Archaeopteryx*

Figure 18.2 ◀

The "Berlin specimen" of *Archaeopteryx*. (Courtesy Wikimedia Commons)

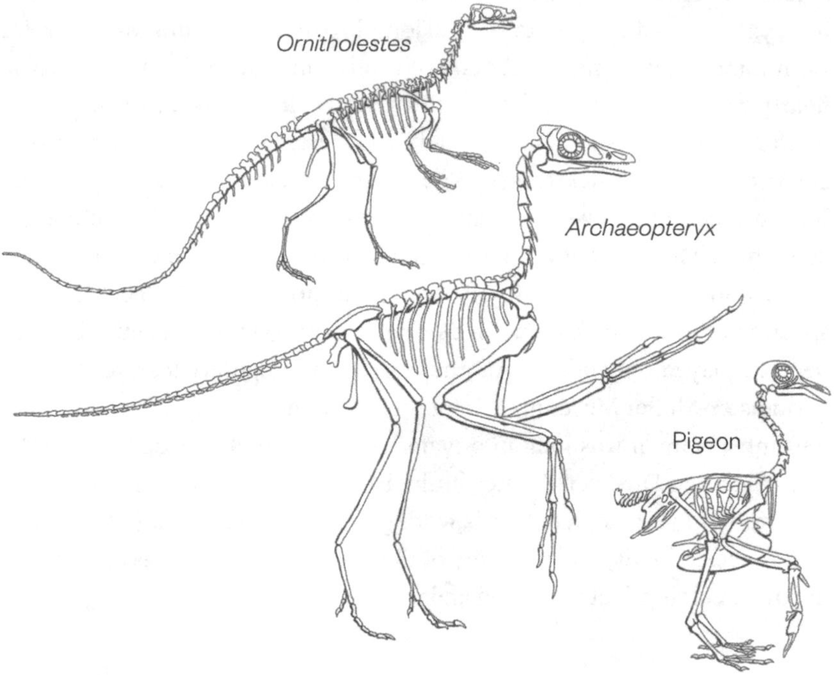

Figure 18.3 ▲

Comparison of the skeletons of the small dinosaur *Ornitholestes*, *Archaeopteryx*, and a pigeon. (Drawing by Carl Buell; from Donald R. Prothero, *Evolution: What the Fossils Say and Why It Matters* [New York: Columbia University Press, 2007], fig. 12.6)

had the same three fingers (thumb, index finger, and middle finger) as most other dinosaurs, and the index finger was by far the longest of the three. In addition, the claws of *Archaeopteryx* were very similar to those of predatory dinosaurs.

The hind limbs of *Archaeopteryx* have many dinosaurian hallmarks as well. The most striking of these is in the ankle. All pterosaurs, dinosaurs, and birds have a unique ankle arrangement known as the mesotarsal joint. Instead of the typical vertebrate ankle, which hinges between the shin bone (tibia) and the first row of ankle bones (as does your ankle), the pterosaurs, dinosaurs, and birds developed a hinge between the first and the second row of ankle bones—that is, within the ankle. The first row of ankle bones thus has little function, and in many birds and dinosaurs, it actually fuses onto the end of the shin bone as a little "cap" of bone. The next time you eat a chicken or turkey drumstick (which is its tibia), notice that the inedible cap of cartilage at the less meaty "handle" end of the drumstick is actually a relict of the dinosaurian ancestry of birds! In addition, part of the front of the first row of ankle bones has a bony spur that runs up the front of the tibia, another feature unique to certain dinosaurs and birds. Finally, the details and structure of the toe bones and the short big toes are unique to predatory dinosaurs and birds as well. *Archaeopteryx* did not have the large bird-like opposable big toe that would have enabled it to grasp or perch on branches well. But recent research has shown that *Archaeopteryx* did have the small "slashing claw" on its hind feet evidenced by the *Velociraptors* in *Jurassic Park*.

With all this evidence that *Archaeopteryx* is basically a feathered dinosaur, why call *Archaeopteryx* a bird? In fact, it has only a few uniquely bird-like features not found in other predatory dinosaurs: its big toe is almost completely reversed; its teeth do not have serrations on their edges like those of a steak knife; and its tail is relatively short compared with that of other dinosaurs, but its arms are long compared with those of most other predatory dinosaurs. All the other features of *Archaeopteryx*, including the feathers and the fused collarbone (wishbone), have now been found in other dinosaurs. Some say that the feathers of *Archaeopteryx* were more advanced than those of predatory dinosaurs and had an asymmetric shape, with the shaft running down one side, that suggests that *Archaeopteryx* could fly, although not as well as most living birds.

BIRDS TAKE OFF

Archaeopteryx was revolutionary as the first transitional fossil found after Darwin's *On the Origin of Species* was published and showed how some dinosaurs evolved into birds. But the fossil record of early birds has grown explosively, especially in the past 30 years, as a huge number of beautifully preserved fossil birds have been found in China. The most earth-shaking discoveries come from the famous Liaoning fossil beds of northeastern China, dating to the Early Cretaceous, which have become one of the world's most important fossil deposits. The delicate lake shales have preserved extraordinary features in fossils—including body outlines, feathers, and fur—as well as complete articulated skeletons with not a single bone missing and sometimes even the feather color and the stomach contents.

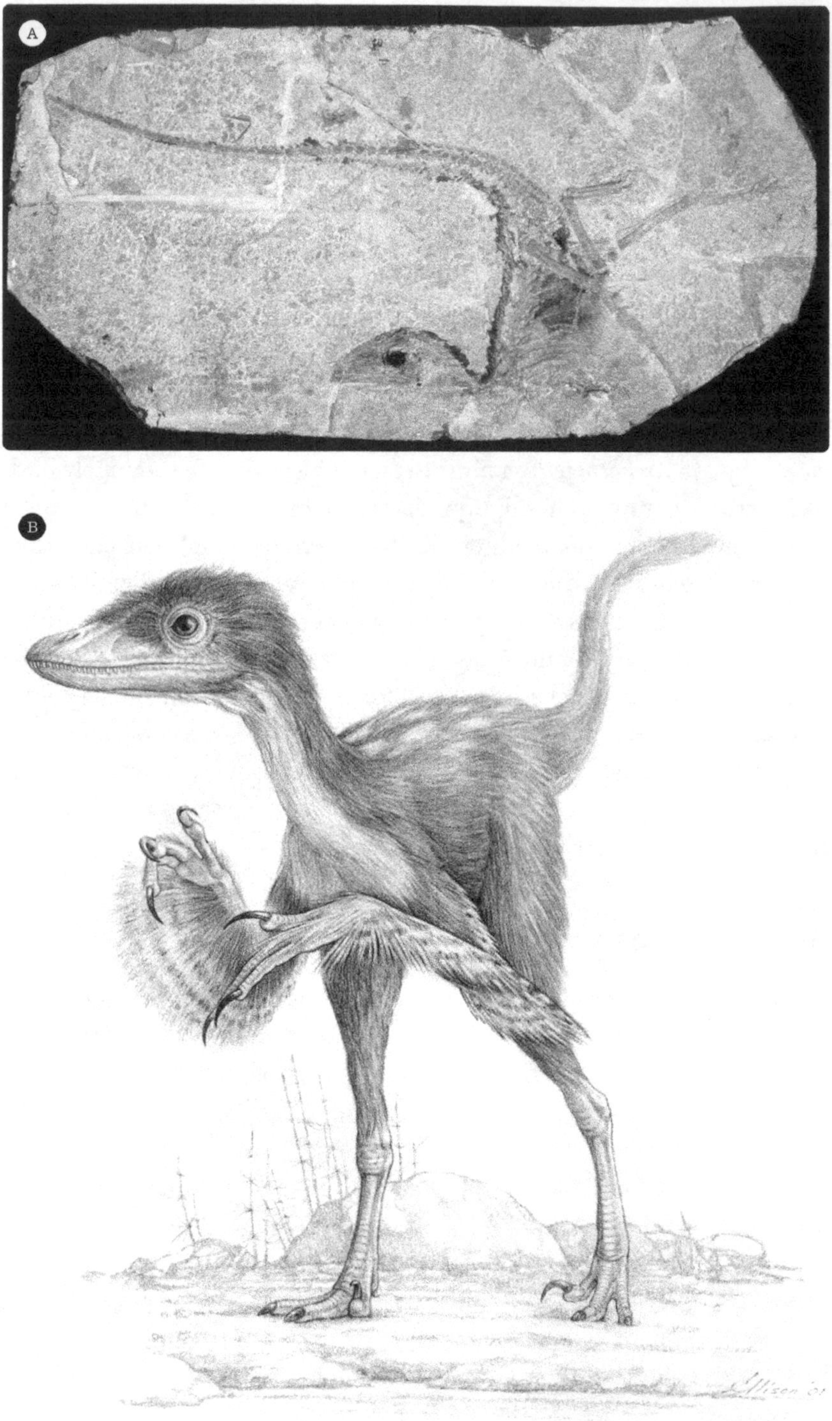
A
B

In the past decade, a major new discovery has been announced from these deposits every few months that renders obsolete almost all previous ideas about birds and dinosaurs. The most amazing fossils of all are those of a number of clearly non-flying, non-bird dinosaurs with well-developed feathers (figure 18.4). They include such incredible complete specimens as *Sinosauropteryx*, *Protarchaeopteryx*, *Sinornithosaurus*, *Caudipteryx*, the large theropod *Beipiaosaurus*, and the tiny *Microraptor*.

Most of these dinosaurs clearly did not have flight feathers or other indications that their feathers were used for flight. Instead, their fossils show that feathers were apparently a widespread feature among predatory dinosaurs (and among most other dinosaurs as well, and maybe even pterosaurs). Feathers, then, did not evolve for flight, but presumably for insulation, and later were modified to become flying structures. In 2003, Richard Prum and Alan Brush published an article that completely rethought the origin of feathers. They showed that feathers are not modified scales (as once believed), but arise from a similar embryonic primordium with different genes controlling development (figure 18.5). Type 1 feathers are simple, hollow pointed shafts, which appeared in the primitive dinosaurs and in the other branch of dinosaurs that includes *Triceratops* and its relatives. Type 2 feathers are down with no vanes (as in the dinosaur *Sinosauropteryx*). Type 3 feathers have vanes and a shaft, but no barbules linking them together like Velcro (as in *Yutyrannus* and, by extension, *Tyrannosaurus rex*). Types 2 and 3 are found in the large dinosaur *Beipiaosaurus*, suggesting that they were present in most advanced predatory dinosaurs, such as dromaeosaurs. Type 4 feathers have barbules that link the vanes into a continuous surface, but the shaft is in the middle of the feather. This kind of feather appeared in *Caudipteryx*, which suggests that it was a feature of more advanced predatory dinosaurs, such as ostrich dinosaurs, oviraptors, and dromaeosaurs. The classic asymmetric flight feather with the shaft near the leading edge of the vane appeared in *Archaeopteryx*, and for this reason many scientists think that *Archaeopteryx* was one of the first transitional dinosaur-birds to modify the long heritage of feathers for true flight.

Figure 18.4 ◀

Sinosauropteryx, a nonflying, nonbird feathered dinosaur from the Liaoning beds of China: (*A*) fossil; (*B*) reconstruction of its appearance in life. (Courtesy M. Ellison and M. Norell, American Museum of Natural History)

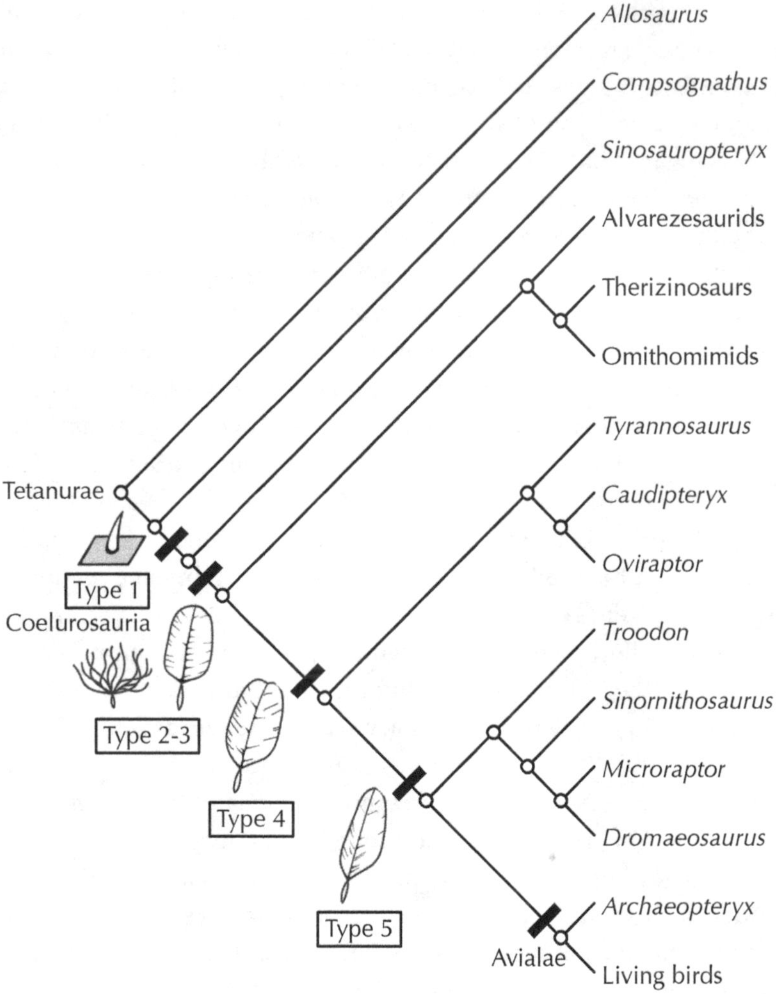

Figure 18.5 ▲

The evolution of feathers in dinosaurs and birds. (Drawing by Carl Buell, modified from Richard O. Prum and Alan H. Brush, "Which Came First, the Feather or the Bird?" *Scientific American*, March 2003; from Donald R. Prothero, *Evolution: What the Fossils Say and Why It Matters* [New York: Columbia University Press, 2007], fig. 12.9)

Moving up from *Archaeopteryx* on the family tree of birds (figure 18.6), we come to *Rahonavis* (figure 18.7A) from the Cretaceous of Madagascar. About the size of a crow, it had a primitive sickle-like claw on its hind feet, a long bony tail, teeth, and many other dinosaurian features, but also such bird-like features as the fusion of its lower back vertebrae with its pelvis (the synsacrum); holes in its vertebrae for all the blood vessels and air sacs found in living birds; fingers with quill knobs (little bumps on the bone where flight feathers attached), suggesting that it was feathered and could fly (no surprise here); and a fibula (the smaller shin bone), which did not reach the ankle. Birds have reduced the fibula to the tiny splint of bone that you bite into when you are eating a chicken or turkey drumstick, but *Archaeopteryx* had a fully developed fibula like that of dinosaurs.

The next step is marked by *Confuciusornis* and its relatives, which had a toothless beak—the first bird to do so—as well as a unique feature found in all higher birds: the pygostyle, formed by the fusion of all the dinosaurian tail vertebrae into a single "parson's nose." These more advanced birds also had an increased number of lower back vertebrae fused to the synsacrum, and longer bones that reinforced the shoulder, which improved flight. Recently, embryological experiments have managed to unlock the bird genes

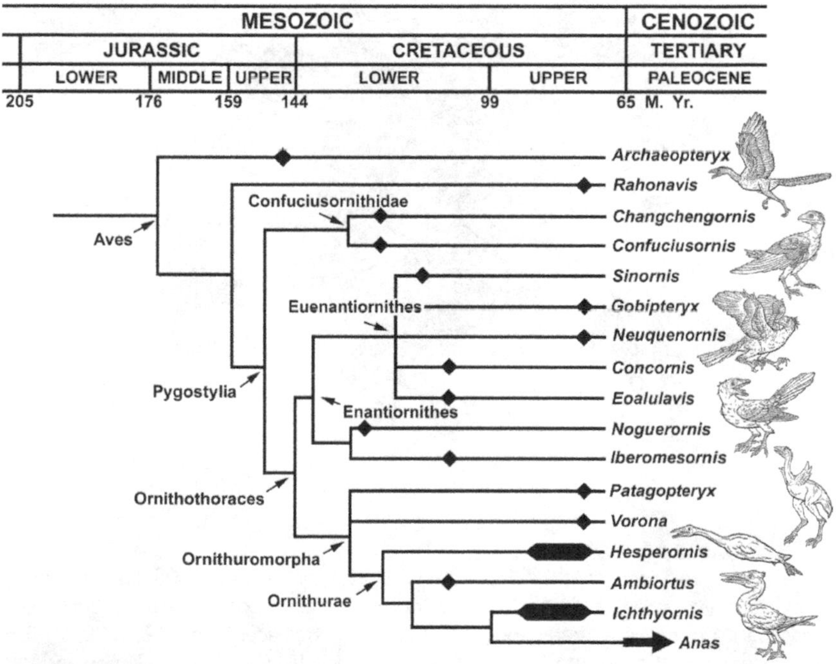

Figure 18.6 ▲

Family tree of birds of the Mesozoic. (Courtesy L. Chiappe, Natural History Museum of Los Angeles County)

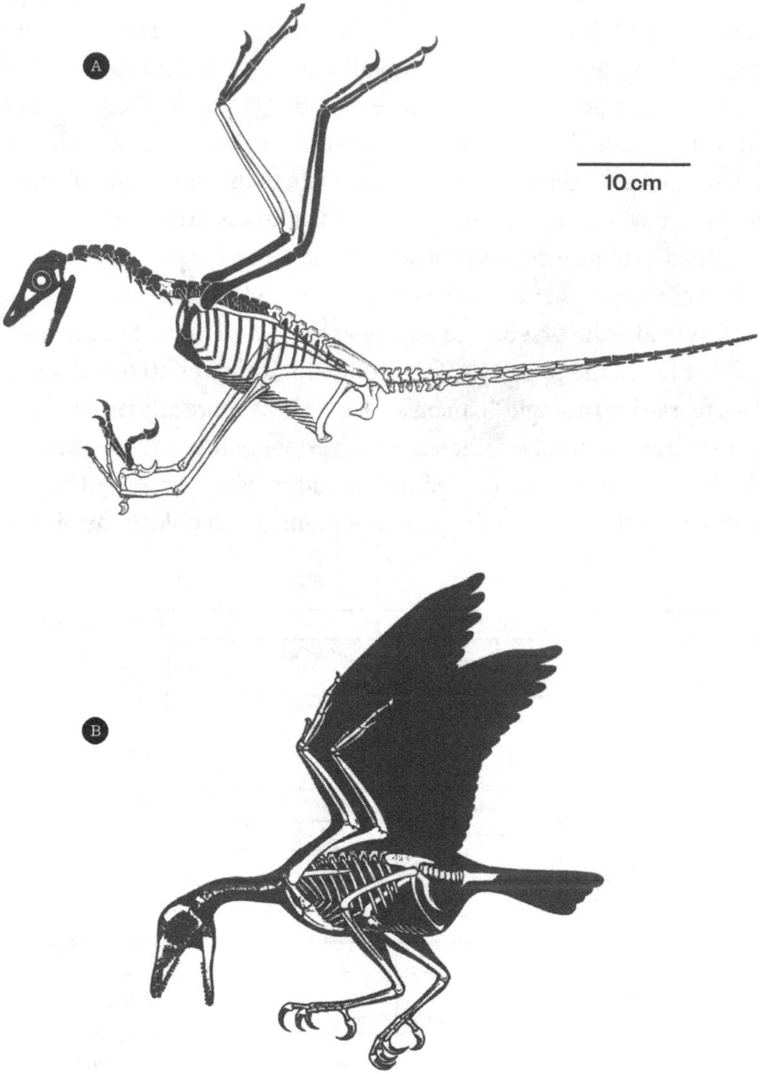

Figure 18.7 ▲

Birds of the Cretaceous: (*A*) *Rahonavis* from Madagascar; (*B*) reconstruction of *Sinornis* from China. ([*A*] after Catherine A. Forster et al., "The Theropod Ancestry of Birds: New Evidence from the Late Cretaceous of Madagascar," *Science*, March 20, 1998; © 1998 American Association for the Advancement of Science; [*B*] from Paul C. Sereno and Rao Chenggang, "Early Evolution of Avian Flight and Perching: New Evidence from the Lower Cretaceous of China," *Science*, February 14, 1992, fig. 2; © 1992 American Association for the Advancement of Science) that suppress tailbone development and keep the tailbones short, and have produced a chick with a long bony tail like that of its dinosaurian ancestors.

Following this transitional form is another branch point, which leads to the extinct Enantiornithes, or "opposite birds" (so named because their leg bones ossified in the reverse direction from that found in modern birds, and because of the odd condition of the shoulder bones) (see figure 18.6). These include *Iberomesornis* from the Las Hoyas locality in Spain, which dates to the Cretaceous; *Sinornis* from China (see figure 18.7*B*); *Gobipteryx* from Mongolia; *Enantiornis* from Argentina; and many others. All these birds were more specialized than *Archaeopteryx* or *Rahonavis* or *Confuciusornis* in that they had a reduced number of trunk vertebrae, a flexible wishbone, a shoulder joint that was better for flying, hand bones that had fused into a bone called the carpometacarpus, and finger bones that mostly had fused into a single element (the meatless bony part of the chicken wing that you never eat).

Continuing up the family tree, we come to several Cretaceous birds, such as *Vorona* from Madagascar, *Patagopteryx* from Argentina, and the well-known aquatic birds *Hesperornis* and *Ichthyornis* from the chalk beds of Kansas. These birds are united by at least 15 well-defined evolutionary specializations, including the loss of the belly ribs (gastralia), reorientation of the pubic bone to the modern bird-like position parallel to the ischium, and reduction in the number of trunk vertebrae, as well as many other features of the hand and shoulder that improved flight performance. *Ichthyornis* is even closer to modern birds in having had a keel on its breastbone on which to attach the flight muscles and a knob-like head on the upper arm bone that made the wing more flexible. Finally, the group that includes all modern members of class Aves is defined by the complete loss of teeth and by a number of other anatomical specializations, such as the fusion of the leg bones to form the tarsometatarsus.

• ◉ •

We have come a long way since the first fossil of *Archaeopteryx* was found. When it was discovered, it played an important role in bolstering the evidence for Darwin's theory of evolution. For decades, it was at the center of every argument about the origin of birds and of flight. Now it is just one among hundreds of amazing specimens of fossil birds from the Age of Dinosaurs that have completely transformed the way we think about dinosaurs—and especially birds. Dinosaurs are not extinct. They are perching in your birdcage or flying around your garden right now. So the next time you see a feathered dinosaur take flight, marvel at how evolution transformed a scary predator like *Velociraptor* into the entire range of amazing birds, from ostriches to hummingbirds. All are living feathered dinosaurs.

SEE IT FOR YOURSELF!

Nearly all the original Solnhofen quarry sites are privately owned, so collecting is not allowed without the owner's permission. Since only 12 specimens of *Archaeopteryx* have been found in nearly 500 years, the odds of finding another are extremely poor.

Most of the original specimens of *Archaeopteryx* are extremely valuable, and some are still privately owned, so they are not on public display. For example, the fossil once on view at the Maxberg Museum is now lost, the "Daiting specimen" is not on display, nor is the fossil that was only recently described, but is privately owned. Accurate replicas are exhibited in many natural history museums and even are available commercially. Many museums, such as the American Museum of Natural History in New York, exhibit replicas of not only the "Berlin" and "London" specimens, but most of the publicly available fossils of *Archaeopteryx*.

The following original fossils of *Archaeopteryx* are on display, as far as I know:

- The "London specimen," at the Natural History Museum, London (see figure 18.1)
- The "Berlin" specimen" (in a secure vault behind glass), at the Museum für Naturkunde (Humboldt Museum), Berlin (see figure 18.2)
- The "Thermopolis specimen," at the Wyoming Dinosaur Center, Thermopolis
- A partial specimen at the Paläontologisches Museum München, Germany
- A nearly complete specimen at the Jura Museum, Eichstätt, Germany
- A wing specimen at the Burgermeister-Müller Museum, Solnhofen, Germany
- A wing specimen at the Teylers Museum, Haarlem, Netherlands

FOR FURTHER READING

Chiappe, Luis M. "The First 85 Million Years of Avian Evolution." *Nature*, November 23, 1995, 349–355.

Chiappe, Luis M., and Gareth J. Dyke. "The Mesozoic Radiation of Birds." *Annual Review of Ecology and Systematics* 33 (2002): 91–124.

Chiappe, Luis M., and Lawrence M. Witmer, eds. *Mesozoic Birds: Above the Heads of Dinosaurs*. Berkeley: University of California Press, 2002.

Currie, Philip J., Eva B. Koppelhus, Martin A. Shugar, and Joanna L. Wright, eds. *Feathered Dragons: Studies on the Transition from Dinosaurs to Birds*. Bloomington: Indiana University Press, 2004.

Gauthier, Jacques, and Lawrence F. Gall, eds. *New Perspectives on the Origin and Early Evolution of Birds*. New Haven, Conn.: Yale University Press, 2001.

Norell, Mark. *Unearthing the Dragon: The Great Feathered Dinosaur Discovery*. New York: Pi Press, 2005.

Ostrom, John H. "*Archaeopteryx* and the Origin of Birds." *Biological Journal of the Linnean Society* 8 (1976): 91–182.

——. "*Archaeopteryx* and the Origin of Flight." *Quarterly Review of Biology* 49 (1974): 27–47.

Padian, Kevin, and Luis M. Chiappe. "The Origin of Birds and Their Flight." *Scientific American*, February 1998, 28–37.

Prum, Richard O., and Alan H. Brush. "Which Came First, the Feather or the Bird?" *Scientific American*, March 2003, 84–93.

Shipman, Pat. *Taking Wing:* Archaeopteryx *and the Evolution of Bird Flight*. New York: Simon & Schuster, 1988.

NOT QUITE A MAMMAL

Of all the great transitions between major structural grades within vertebrates, the transition from basal amniotes to basal mammals is represented by the most complete and continuous fossil record, extending from the Middle Pennsylvanian to the Late Triassic and spanning some 75 to 100 million years.

JAMES HOPSON, "SYNAPSID EVOLUTION AND THE RADIATION OF NON-EUTHERIAN MAMMALS"

PROTO-MAMMALS

One of the most complete and best-documented transitions in the fossil record is the sequence that shows the evolution of mammals from the earliest amniotes (figure 19.1). Literally hundreds of excellent specimens document almost every stage. The proper name of all these fossil "proto-mammals" is the Synapsida, a group that includes not only the ancestors of mammals, but also the mammals themselves. Paleontologists no longer use the obsolete term "mammal-like reptiles" because the mammal lineage (as represented by *Archaeothyris* and *Protoclepsydrops* from the Late Carboniferous) originated at the same time as, and evolved concurrently with, the earliest members of the reptile lineage (defining reptiles as turtles, snakes, lizards,

Figure 19.1 ▶

The evolution of the synapsid skeleton from that of primitive "pelycosaurs" like *Haptodus*, through those of noncynodont therapsids like *Lycaenops* and cynodonts like *Thrinaxodon*, to that of true mammals like *Megazostrodon*. (Drawing by Carl Buell; from Donald R. Prothero, *Evolution: What the Fossils Say and Why It Matters* [New York: Columbia University Press, 2007], fig. 13.4)

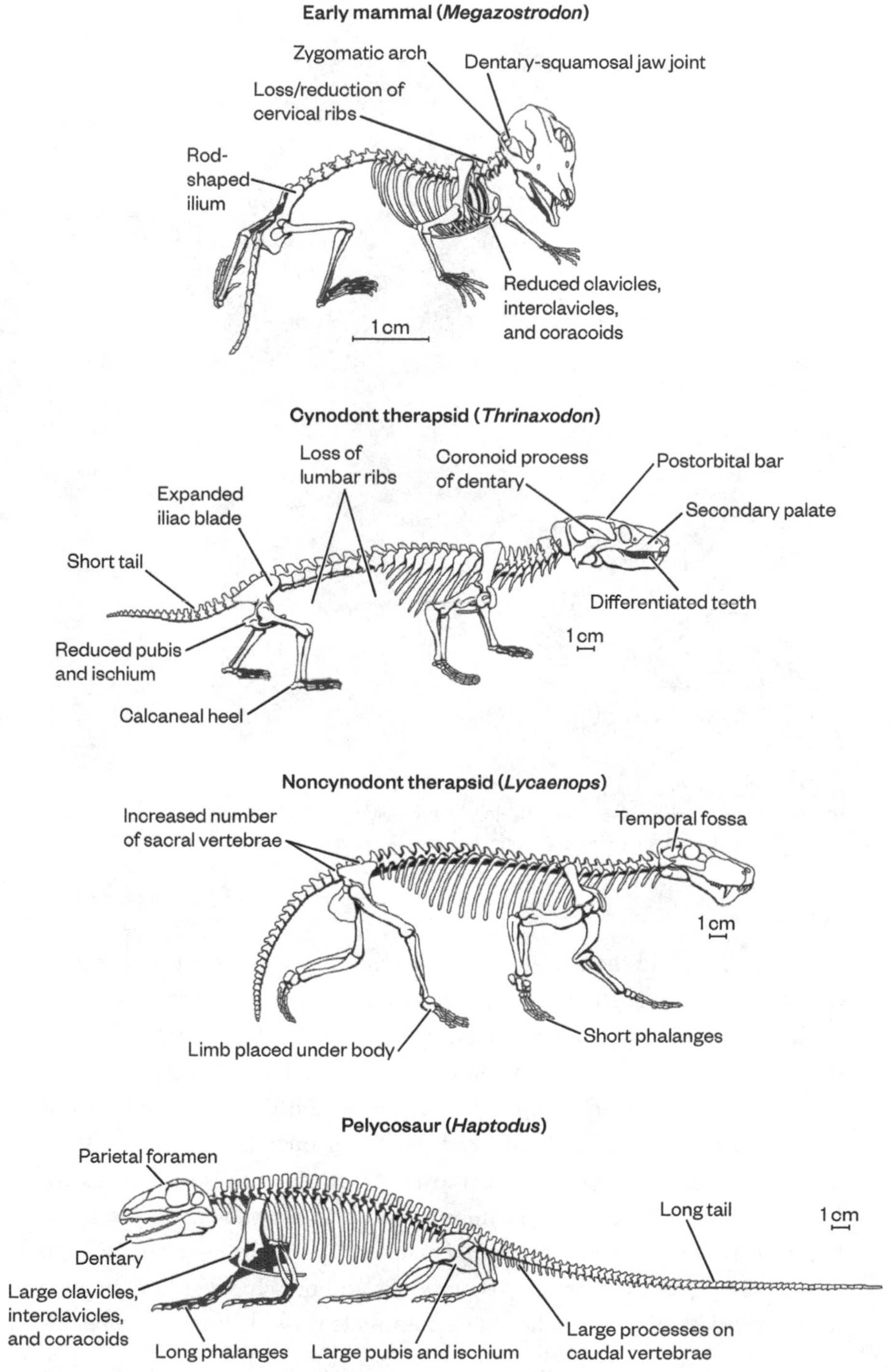

Early mammal (*Megazostrodon*)
Zygomatic arch
Dentary-squamosal jaw joint
Loss/reduction of cervical ribs
Rod-shaped ilium
Reduced clavicles, interclavicles, and coracoids
1 cm
Cynodont therapsid (*Thrinaxodon*)
Loss of lumbar ribs
Coronoid process of dentary
Postorbital bar
Expanded iliac blade
Secondary palate
Short tail
Differentiated teeth
1 cm
Reduced pubis and ischium
Calcaneal heel
Noncynodont therapsid (*Lycaenops*)
Increased number of sacral vertebrae
Temporal fossa
1 cm
Limb placed under body
Short phalanges
Pelycosaur (*Haptodus*)
Parietal foramen
Long tail
1 cm
Dentary
Large clavicles, interclavicles, and coracoids
Long phalanges
Large pubis and ischium
Large processes on caudal vertebrae

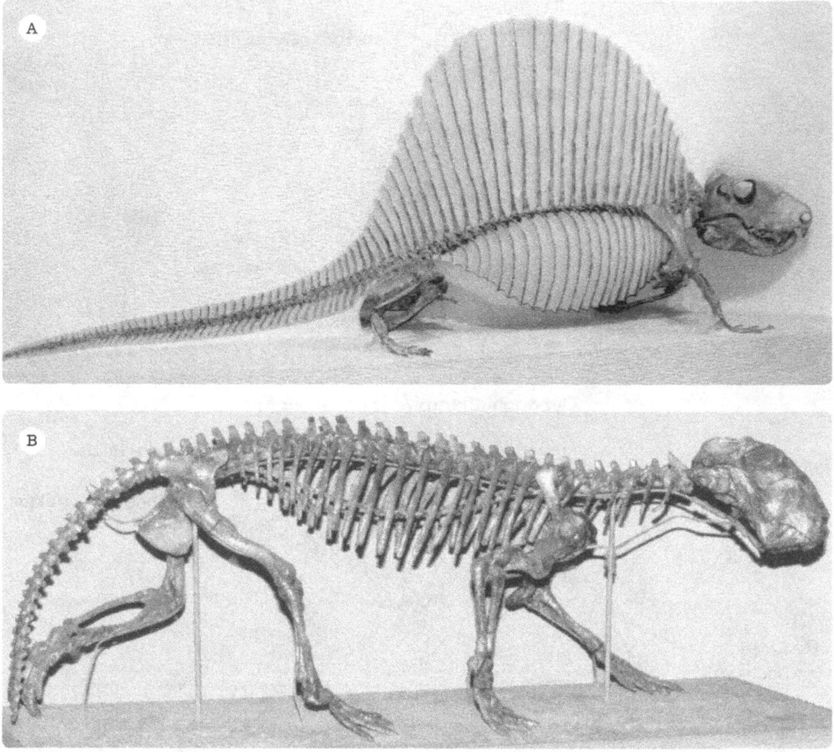

Figure 19.2 ▲

Skeletons of typical synapsids: (*A*) the finbacked "pelycosaur" *Dimetrodon*; (*B*) the wolf-like gorgonopsian *Lycaenops*. (Photographs courtesy R. Rothman)

and crocodiles and their relatives). At no time were the earliest ancestors of mammals part of the Reptilia. Unfortunately, obsolete terms that people learn early in their careers are hard to unlearn, so the mistaken "mammal-like reptiles" still appears widely in books and documentaries.

The first well-known Synapsida are from the Early Permian red beds of northern Texas, site of the discovery of the "Frogamander" and many other important fossils (chapter 11). The most spectacular of these synapsids are fin-backed creatures such as the huge predator *Dimetrodon* (figure 19.2; see figure 19.1) and the herbivore *Edaphosaurus*. Although these animals are often included in children's dinosaur books and merchandise, and in plastic toy sets with dinosaurs, they have nothing to do with dinosaurs whatsoever—they are part of *our* ancestry! (Sadly, much of the public thinks that if an animal is extinct, it was a dinosaur. Most merchandise of prehistoric animals contains a lot of non-dinosaurs labeled as dinosaurs, including mammoths and sabertooths; ichthyosaurs and plesiosaurs; and flying reptiles, or pterosaurs.) Being prehistoric or extinct does not make an animal into a dinosaur. Instead, being a dinosaur has to do with a specific set of unique anatomical features, including a hole through the hip socket, a distinctive hand with only three functional fingers (thumb, index, and middle finger) and reduced ring finger and pinkie; and , the joint in the middle of the ankle; among other characteristics.

Dimetrodon was the top predator in the Early Permian of Texas. It is known from many nearly complete skeletons and dozens of skulls and partial skeletons (see figure 19.2*A*), since it is one of the most abundant fossils in these beds. Large individuals were more than 4.6 meters (15 feet) long, with a sail that reached about 1.7 meters (5 feet) above the ground, and they weighed up to 250 kilograms (550 pounds). *Dimetrodon* had a narrow compressed skull, with strong curved jaws sporting a wicked set of conical stabbing teeth. They varied in size from the big canine-like teeth in the front of its jaw to the more simple conical teeth diminishing in size from front to back along the sides of its mouth. In fact, this feature led Edward Drinker Cope to name the genus *Dimetrodon* (two-size teeth) in 1878. About the only mammalian feature in its skull besides the specialized teeth is the hole (temporal fenestra) low on the side of the head. The lower temporal fenestra is one of the defining features of the Synapsida and appears in modified form in all mammals. It probably served as an attachment point for stronger jaw muscles and allowed the muscles to bulge during chewing, a characteristic that is very important in later synapsids.

The reason for the amazing sail on *Dimetrodon* (and on the herbivore *Edaphosaurus*, which comes from the same beds) has long been controversial. The list of suggested functions is very long, but some paleontologists regard it as a device for warming or cooling its body, since *Dimetrodon* was cold blooded. When the sail was turned perpendicular to the sun, it would absorb heat rapidly; when it was turned parallel to the sun, it would release heat. However, since most other synapsids at that time did not have a sail for thermoregulation, other paleontologists argue that it was used for display—recognizing members of its own species and signaling its size and strength to other animals—just as large horns and antlers serve today in antelopes and deer.

THE GREAT KAROO

In the heart of South Africa is a huge desert region called the Karoo. Like most deserts, it experiences extremes of both heat and cold, and both drought and flood. It receives an average of less than 25 centimeters (10 inches) of rain a year, most of it falling in a few huge flash floods during the limited wet season. For the South African settlers heading north out of Cape Town, it was a great barrier to cross in order to reach the grassy Highveld in the northern part of the country. The vegetation in the Karoo consists largely of succulents, such as the euphorbias, which mimic the appearance of cactuses in the New World, as well as aloes, desert ephemerals, and many other kinds of plants adapted to floods and droughts and extreme temperature change. Animals that can survive these conditions roam the Karoo, including many antelope (especially the springbok, a South African icon), wildebeest, ostriches, rare elephants, rhinos, and hippos, and at one time the half-striped species of zebra known as the quagga (now extinct). Lions, leopards, jackals, hyenas, and other carnivores preyed on them. But the introduction of irrigation has allowed sheep and cattle ranching to take hold on this poor forage, nearly wiping out the limited populations of wild animals.

The Karoo is also important in our study of life's history. The beds of the Karoo Supergroup begin with the Dwyka Group, an Upper Carboniferous (310 million years old) unit with some of the earliest glacial deposits in Gondwana; continue through a thick sequence of Permian (300 to 250 million years old) beds of the Ecca and Beaufort groups that span the world's greatest mass extinction (250 million years ago); and come to the end of the Beaufort Group in the Early Triassic (250 to 200 million years old). These Permian–Triassic red beds are capped by more Triassic rocks of the Stormberg Group and, finally, by Jurassic lava flows of the Drakensburg volcanics (about 180 million years old). The Beaufort Group is so rich in important Late Permian and Triassic fossils that it is the basis for telling time on land during these periods. In particular, the Beaufort has produced crucial fossils of synapsids and other Late Permian creatures that demonstrate the next phase of evolution to mammals. In some places, skulls and bones are weathering out in great abundance across the ground, and paleontologists must be selective and retrieve only the least broken and weathered skulls.

These incredible fossils were originally discovered by a Scotsman, Andrew Geddes Bain, at a road cut near Fort Beaufort in 1838. Some of the early specimens were sent to the British Museum, where pioneering paleontologist Sir Richard Owen described them. By the late nineteenth century, more and more fossils were arriving in Britain, where they caught the attention of another Scotsman, Robert Broom. As early as 1897, he realized that these fossils were not of reptiles, but of synapsids related to mammals.

Trained as a doctor and an anatomist in Glasgow, in 1903 Broom emigrated to South Africa, where he began collecting fossils as a hobby while performing his medical duties. Soon he had collected and described hundreds of specimens of Late Permian synapsids, as well as the bizarre reptiles of the Late Permian and gigantic amphibians. He became curator of vertebrate paleontology at the South African Museum in Cape Town, but the job paid very little and he was struggling to survive. His friend Raymond Dart (chapter 24) wrote to Prime Minister Jan Smuts about this shameful situation. Consequently, Broom was hired in 1934 at the Transvaal Museum in Pretoria. There he shifted his focus to the Ice Age caves of northern South Africa, and he soon became famous for his discoveries of early hominids, including most of the specimens of *Australopithecus africanus* and *Paranthropus robustus*. In 1946, he received the Daniel Giraud Elliot Medal of the National Academy of Sciences, and late in life (he lived to the ripe old age of 84) he was honored for his pioneering contributions to both synapsid paleontology and paleoanthropology.

GORGON FACES, TERRIBLE HEADS, AND DOUBLE DOG TEETH

The Late Permian red beds have yielded an incredible diversity of synapsids and have demonstrated the evolution of this group over about 30 million years. Gone are the archaic fin-backed synapsids like *Dimetrodon*, best known from the Early Permian of Texas (see figure 19.2*A*). Instead, there are many types of more advanced and mammal-like synapsids, which have been lumped into a wastebasket group called Therapsida (figure 19.3). Some were among the first herbivorous land animals. They included the squat creatures with a toothless beak and big canine tusks known as dicynodonts (Greek for "double dog teeth"), which reached 3.5 meters (11 feet) in length and weighed up to 1 metric ton (1.1 tons). The other herbivores were the dinocephalians (terrible heads), which sported an array of warts and bumps and thick bony battering rams on their heavily armored skulls. Some

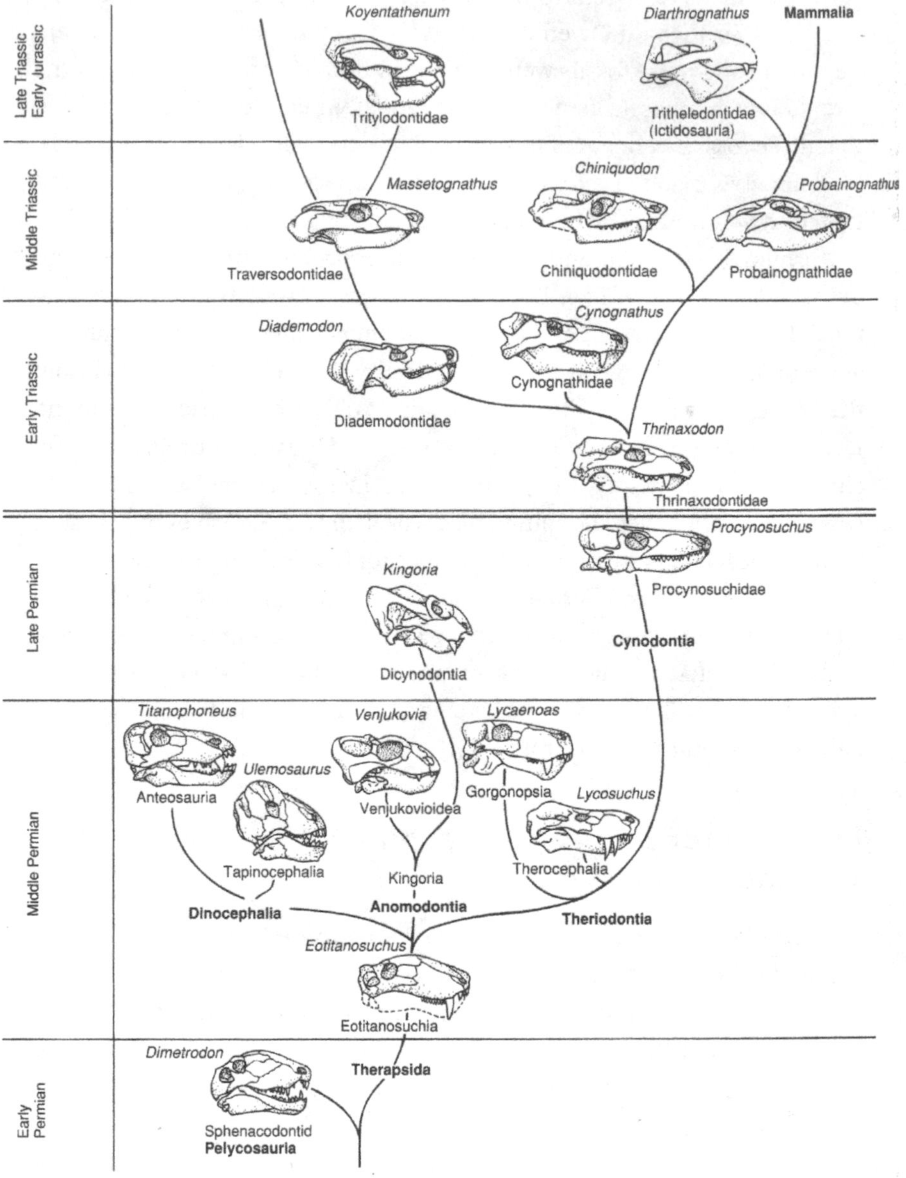

Figure 19.3 ▲

The evolutionary radiation of synapsid skulls from the primitive pelycosaurs, through therapsids and cynodonts, to mammals. (From Kenneth V. Kardong, *Vertebrates: Comparative Anatomy, Function, Evolution* [Dubuque, Iowa: Brown, 1995]; reproduced by permission of the McGraw-Hill Companies) dinocephalians reached up to 4.5 meters (15 feet) in length and weighed up to 2 metric tons (2.2 tons).

Preying on these herbivores was a wide array of ferocious carnivorous therapsids, including the biarmosuchids, the therocephalians, and the bauriamorphs. The most impressive were the terrifying gorgonopsians (Greek for "Gorgon appearance"), which had huge skulls with impressive stabbing canine teeth, strong jaw muscles for chewing, and powerfully built bodies. The largest were bigger than bears, with a skull 45 centimeters (18 inches) long, saber teeth over 12 centimeters (4.7 inches) long, and a long sprawling crocodile-like body up to 3.5 meters (11 feet) long.

Throughout the evolution of these therapsids in the Late Permian, more and more mammal-like features appeared. The small opening on the side of the skull in *Dimetrodon* became a large expanded arch behind the eye into which strong jaw muscles could bulge and allow powerful bite forces and even some chewing. The original reptilian palate began to be covered by a secondary palate, which grew over it and enclosed the nasal passages. (You can feel it if you run your tongue over the roof of your mouth.) The secondary palate allowed therapsids that had it to chew a mouthful of food and breathe at the same time, essential to an animal with a fast metabolism. By contrast, a typical reptile (like a snake or lizard) must hold its breath until its prey is swallowed, but it has a slow metabolism.

Instead of a single ball joint on the back of the skull just below the spinal cord connecting to the neck, therapsids had a double ball joint, allowing for greater strength and flexibility in their neck muscles. Therapsids also showed many modifications of the skeleton (see figure 19.1) that make them more mammalian in appearance than earlier synapsids, including a posture that which no longer sprawled on the belly like a crocodile, but held the body in a semi-sprawling to nearly upright position.

AN EARFUL OF JAWBONES

The most amazing transformation, however, occurred in the jaws and ear region. Primitive synapsids like *Dimetrodon* had a jawbone composed of the primary tooth-bearing bone, the dentary, and a suite of other nondentary bones in the back of the jaw: the angular, surangular, splenials, articular, coronoids, and often more (figure 19.4). The articular bone is particularly important, since it forms the jaw joint against the hinge of the quadrate

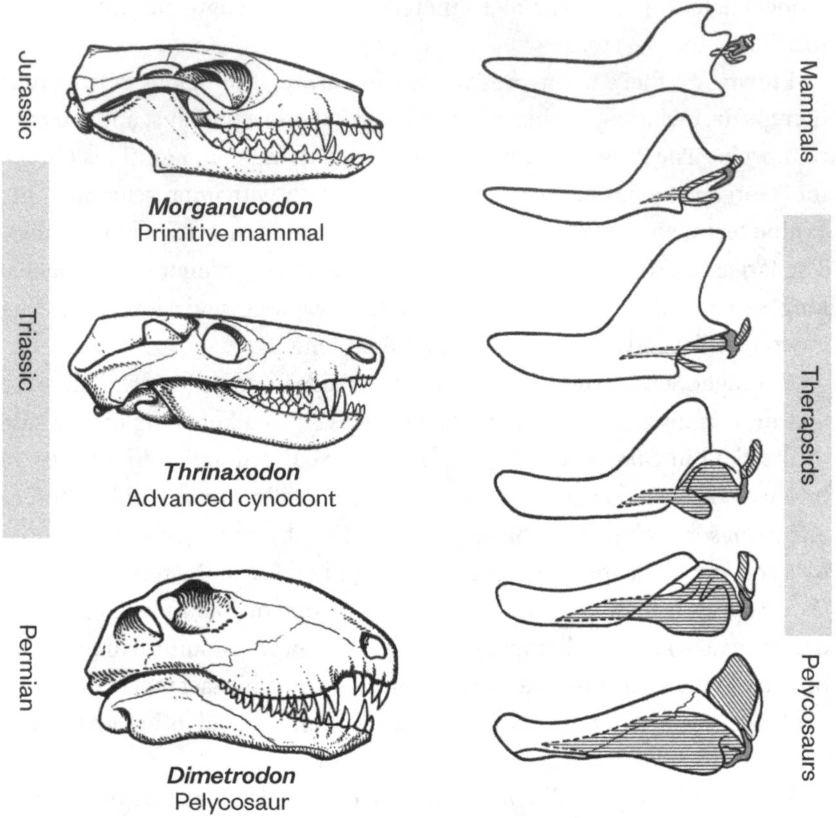

Figure 19.4 ▲

The gradual transformation of the jawbones during synapsid evolution, as the nondentary jaw elements (*shaded*) are reduced, while the dentary bone (*unshaded*) expands backward and crowds them out. All the nondentary jaw elements are lost in mammals except for the articular bone of the jaw, which joins with the quadrate bone of the skull to become the bones of the middle ear. (Drawing by Carl Buell; from Donald R. Prothero, *Evolution: What the Fossils Say and Why It Matters* [New York: Columbia University Press, 2007], fig. 13.5)

bone of the skull. But all these extra bones and their sutures in the back of the jaw made the jaw apparatus complex and weaker than if it were a single bone, a disadvantage when the therapsids evolved complex chewing. Thus as therapsids became more and more specialized for chewing and other complex jaw motions, the dentary bone expanded backward and crowded out the nondentary bones in the back of the jaw. Eventually, these bones became tiny and eventually were lost as their function diminished.

The exception was the articular bone, still attached to the quadrate bone of the skull and serving as the jaw joint. Eventually, the expanded dentary bone made contact with another skull bone, the squamosal, and a new jaw joint was born. In a few synapsids, such as *Diarthrognathus* (Greek for "double jaw joint"), *both* the dentary/squamosal jaw joint and the quadrate/articular jaw joint operated side-by-side, so this animal was literally double jointed on each side of its jaw.

What happened when the dentary/squamosal joint finally took over completely? Did the quadrate/articular joint vanish? No. Instead, in an amazing feat of evolutionary opportunism, it transformed into the bones of the middle ear! The quadrate is the incus, or "anvil," and the articular is the malleus, or "hammer," of the "hammer, anvil, stirrup" that carry vibrations from the eardrum to the inner ear. This may sound incredible, but the fossils prove it. It makes a lot of sense, since many reptiles hear only when their jaw picks up vibrations from the ground, since the quadrate/articular joint has the dual function of both ear bone and jaw joint.

If this still seems incredible, it has happened to you and to every other mammal in your own lifetime. When you were an early embryo, the cartilage predecessors of the quadrate and articular were in your embryonic jaw cartilage. As you developed embryonically, they moved to your middle ear—just as they had over the evolutionary history of synapsids.

THRINAXODON EVOLVING

Then the greatest extinction in Earth history occurred at the end of the Permian (about 250 million years ago), wiping out about 70 percent of the animals on land, including insects, and 95 percent of the animals in the ocean. The causes of the great Permian extinction ("the mother of all mass extinctions" in the words of Douglas Erwin) were complex, but the event was apparently triggered by huge volcanic lava flows pouring across most of northern Siberia. The lava injected large amounts of greenhouse gases (especially carbon dioxide) into the atmosphere and oceans. Earth became a "super-greenhouse" planet, and the oceans then became supersaturated in carbon dioxide, making them extremely hot and acidic and killing nearly everything that lived in them. The atmosphere became too low in oxygen and too loaded with carbon dioxide, so nearly all the terrestrial animals above a certain size vanished, and only a few smaller lineages of synapsids, reptiles, amphibians, and other land creatures made it through the hellish planet of the latest Permian and survived into the aftermath world of the earliest Triassic.

After the Late Permian therapsids nearly vanished in the mass extinction, the synapsids started all over again with a third great evolutionary radiation of much more mammal-like synapsids called cynodonts (Greek for "dog toothed") (see figure 19.3) . They included forms as big as a bear called *Cynognathus* (dog jaw), which was 1 to 2 meters (3.3 to 6.6 feet) long, with a head over 60 centimeters (24 inches) in length, and many smaller species in the size range of raccoons and weasels. Most cynodonts had advanced postures, with their limbs completely under their body for rapid running (see figure 19.1). Their nondentary jawbones were tiny and had been reduced to mere splints in the inside back part of the jaw near the hinge. They had secondary palates going all the way back to the throat, as the palate does in modern mammals, and many other indicators of active living and rapid metabolism. And many had multicusped cheek teeth instead of the simple conical pegs of the primitive synapsids, suggesting that they were capable of complex chewing motions, rather than gulping food whole, as do reptiles.

The transition from primitive amniotes to mammals is demonstrated by such a wealth of transitional fossils within the Synapsida that it is impossible to pick one specific fossil as the most crucial "missing link." If we must pick one, *Thrinaxodon* is as good as any (figure 19.5; see figure 19.1). *Thrinaxodon* represents the start of the cynodont radiation of synapsids after the Early Permian finbacks and the Middle to Late Permian therapsids of the Karoo (see figure 19.4). *Thrinaxodon* was one of the earliest cynodonts, the first fossil to show many of the advanced features of the final phase of the evolution of synapsids into mammals. It was quite common in the Early Triassic (250 to 245 million years ago) of the Beaufort Group in South Africa, so many nearly complete specimens are available, and its anatomy and behavior are better known than are those of most other synapsids.

Figure 19.5 ▶

Thrinaxodon was an Early Triassic weasel-shaped advanced cynodont with many mammal-like features, including hair, a diaphragm, and advanced teeth that enabled chewing: (*A*) skull of a juvenile, showing the distinctive three-cusped molar teeth that gave the animal its name; (*B*) two individuals curled up together and buried in their burrow; (*C*) reconstruction of its appearance in life. ([*A*–*B*] courtesy Roger L. Smith, Iziko South African Museum, Cape Town; [*C*] courtesy Nobumichi Tamura)

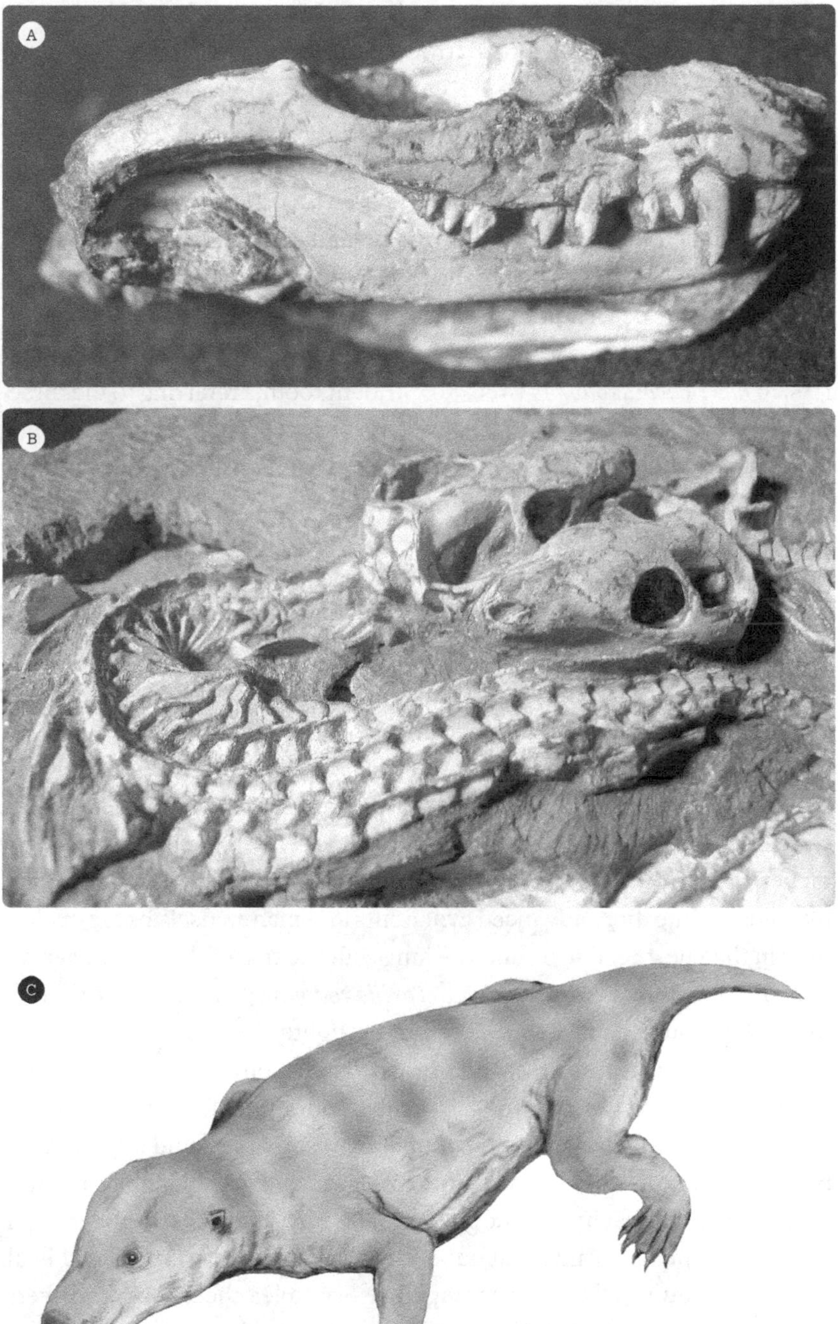
A
B
C

There are two species of *Thrinaxodon*, and both are about the size and shape of a weasel, with a long narrow snout and a long slender low-slung body with short legs. They were typically 30 to 50 centimeters (12 to 24 inches) in length. The dentary bone of *Thrinaxodon* dominates its entire jaw, so the nondentary bones were tiny splints—although it still had the reptilian quadrate/articular jaw joint (see figure 19.4). *Thrinaxodon* had a complete secondary palate, so it could breathe and eat at the same time. It had large eyes (for seeing in the dark or in burrows) and a relatively large head. Like those of its descendants, its cheek teeth were not simple conical pegs, but had complex cusps and could be rightfully called molars and premolars. In fact, *Thrinaxodon* is Greek for "trident tooth," referring to the three-cusped molar teeth in its mouth (see figure 19.5*A*). The temporal opening for the muscles on the side and top of its head was unusually large, allowing for complex chewing motions of the jaw. Yet unlike most mammals, *Thrinaxodon* still had a bony bar that separated the temporal jaw opening from the eye socket.

On each side of its snout were tiny pits in the bone, suggesting that it had whiskers. If *Thrinaxodon* had hair on its snout, it's a good bet that it had hair over its entire body. Hair normally does not fossilize, so this may be the first evidence of hair in the mammalian lineage.

Even though *Thrinaxodon* had short legs, its posture placed its legs beneath its body in a semi-sprawling stance (see figure 19.1). It had advanced shoulder bones and broad hip bones (especially the iliac blade, which attaches the hips to the spinal column and anchors the leg muscles), much like those of the more advanced cynodonts and mammals. Ribs are evident only in the chest region around the lungs; all the ribs of the lower back are lost, as in mammals. This allowed *Thrinaxodon* to bend its back sharply, turn around in a small space, and curl up tightly (see figure 19.5*B*). Even more revealing, *Thrinaxodon* had broad flanges on its thoracic ribs that would have made the rib cage fairly solid and immobile, thus preventing the kind of rib-assisted breathing found in most reptiles (and apparently in primitive synapsids). Instead, *Thrinaxodon* must have had a muscular wall between the lung cavity and the abdominal cavity, known as the diaphragm, which helps pump air into and out of the lungs. This muscle is found in all mammals. Putting all these clues together—complex cheek teeth, whiskers, diaphragm—suggests that *Thrinaxodon* was extremely mammal-like, probably was covered in fur, and had a high metabolic rate and warm-blooded physiology.

In addition, a number of complete articulated *Thrinaxodon* specimens have been found in what appear to be shallow burrows (see figure 19.5*B*). Sometimes two or more individuals were trapped in a den, and fossils of a *Thrinaxodon* and an amphibian, *Broomistega*, were found together in a burrow. Whether the amphibian was prey for the cynodont, or both were seeking shelter and had crawled into the burrow for protection from the flash flood that buried them, or some other cause, it's an odd association.

Thrinaxodon is the perfect transitional fossil between the reptilian features of most primitive synapsids and the more mammalian features of advanced cynodonts. It was extremely mammal-like in its small size, body hair, complex teeth and chewing capability, and high metabolism, yet it still had reptilian jawbones and jaw joint, reptilian bones in its shoulder, and some other primitive features. It lived in burrows as protection from the harsh world of the Triassic aftermath of the great Permian extinction, with its low level of atmospheric oxygen, thin ozone layer, and high level of atmospheric carbon dioxide. Burrows also would have provided protection against the much larger predators of the time and (together with the large eyes) suggest that *Thrinaxodon* emerged mostly at night to hunt. Given its size, it was probably a predator on small reptiles, or especially, insects and other arthropods, which would have been abundant in a world cleared of most of their predators.

Thrinaxodon had vanished by the Middle Triassic, but its more advanced cynodont descendants took over the world. They continued to dominate the Triassic, even as other groups of animals (especially the primitive relatives of crocodiles and the earliest dinosaurs) began to appear. By the latest Triassic, cynodonts were dying out, and the first unquestioned mammals (with a dentary/squamosal joint and complex molar teeth) had emerged (see figures 19.1 and 19.3). They were only shrew-size creatures, but they were living in a world dominated by the rise of the huge dinosaurs. For the next 120 million years (two-thirds of the history of mammals), these Mesozoic mammals remained small (shrew- to rat-size) and evolved complex teeth and other features. They hid from the dinosaurs in the underbrush or came out mostly at night when the dinosaurs were asleep. Then 65 million years ago, the nonavian dinosaurs vanished, and mammals inherited the planet.

SEE IT FOR YOURSELF!

Many large museums display *Dimetrodon* and a number of other synapsids from the Early Permian red beds of Texas. They include the American Museum of Natural History, New York; Denver Museum of Nature and Science; Field Museum of Natural History, Chicago; Museum of Comparative Zoology, Harvard University, Cambridge, Massachusetts; and Sam Noble Oklahoma Museum of Natural History, University of Oklahoma, Norman.

Most of the Late Permian and Early Triassic synapsids are exhibited in museums in South Africa and in Russia, near where they were found, but the American Museum of Natural History does have some of these fossils as well.

FOR FURTHER READING

Chinsamy-Turan, Anusuya, ed. *Forerunners of Mammals: Radiation, Histology, Biology*. Bloomington: Indiana University Press, 2011.

Hopson, James A. "Synapsid Evolution and the Radiation of Non-Eutherian Mammals." In *Major Features of Vertebrate Evolution*, edited by Donald R. Prothero and Robert M. Schoch, 190–219. Knoxville, Tenn.: Paleontological Society, 1994.

Hotton, Nicholas, III, Paul D. MacLean, Jan J. Roth, and E. Carol Roth, eds. *The Ecology and Biology of Mammal-like Reptiles*. Washington, D.C.: Smithsonian Institution Press, 1986.

Kemp, Thomas S. "Interrelationships of the Synapsida." In *The Phylogeny and Classification of the Tetrapods*. Vol. 2. *Mammals*, edited by Michael J. Benton, 1–22. Oxford: Clarendon Press, 1988.

——. *Mammal-Like Reptiles and the Origin of Mammals*. London: Academic Press, 1982.

——. *The Origin and Evolution of Mammals*. Oxford: Oxford University Press, 2005.

Kielan-Jaworowska, Zofia, Richard L. Cifelli, and Xhe-Xi Luo. *Mammals from the Age of Dinosaurs: Origins, Evolution, and Structure*. New York: Columbia University Press, 2004.

King, Gillian. *The Dicynodonts: A Study in Palaeobiology*. London: Chapman & Hall, 1990.

McLoughlin, John C. *Synapsida: A New Look into the Origin of Mammals*. New York: Viking, 1980.

Peters, David. *From the Beginning: The Story of Human Evolution*. New York: Morrow, 1991.

WALKING INTO THE WATER

These dogmatists, who by verbal trickery can make white black, and black white, will never be convinced of anything, but *Ambulocetus* is the very animal that they proclaimed impossible in theory . . . I cannot imagine a better tale for popular presentation of science or a more satisfying, and intellectually based political victory over lingering creationist opposition.

STEPHEN JAY GOULD, "HOOKING LEVIATHAN BY ITS PAST"

WHALE OF A TALE

For thousands of years, people have marveled at some of the most amazing creatures of the sea: the whales and dolphins and their relatives. The ancient cultures of the Mediterranean believed that dolphins swimming beside their ships brought good luck, and the story of Jonah and the whale is a popular one in the Bible. Most of these people regarded whales as just another species of fish, and so the ancients classified whales and dolphins as fish. This is especially true in the biological writings of the Greek philosopher Aristotle, whose ideas became entrenched as part of Church dogma for almost a thousand years. Even today, many people *still* think of whales and dolphins as fish. Members of a number of traditional cultures hunt whales as if they are just another source of food from the ocean, and not mammals—with large brains, complex societies, and a full range of emotions—that are potentially as smart as humans.

The first person to realize that whales are not fish was none other than the inventor of modern biological classification, the Swedish natural historian Carl von Linné. He is better known to us by his Latinized name, Carolus Linnaeus, because he and all scholars of his time wrote in Latin. When he published his classification scheme of animals in the 1750s, he correctly noted that whales breath air through lungs, not gills; are warm-blooded; and have many other anatomical differences that distinguish them from fish.

Even though most people still treated whales as fish, by the nineteenth century, Linnaeus's view was widely accepted by natural historians. As Oliver Goldsmith wrote in *A History of the Earth and Animated Nature* (1825):

> As on land there are some orders of animals that seem formed to command the rest, with greater powers and more various instincts, so in the ocean there are fishes which seem formed upon a nobler plan than others, and that, to their fishy form, join the appetites and the conformation of quadrupeds. These are all of the cetaceous kind; and so much raised above their fellows of the deep, in their appetites and instincts, that almost all our modern naturalists have fairly excluded them from the finny tribes, and will have them called, not fishes, but great beasts of the ocean. With them it would be as improper to say men to go Greenland fishing for a whale, as it would be to say that a sportsman goes to Blackwall a fowling for mackerel.

THE "GREAT SEA SERPENT"

Some of the first good fossils of whales were being discovered in the early nineteenth century, but sadly they were misused by hucksters, not studied by qualified scientists. The most famous of these promoters and con men was "Dr." Albert Koch. A swindler just a few degrees less honest than P. T. Barnum, Koch was always trying to make a buck from outlandish claims about natural history specimens. His prize was a huge skeleton that he called "Hydrarchos," or the "Great Sea Serpent" (figure 20.1). In 1845, it was on display in Philadelphia, where it was the talk of the town. It stretched 35 meters (115 feet) through three rooms, with huge flippers in front. Its skull bore a long snout with huge triangular teeth. It drew throngs of people eager to gawk at it. However good a promoter Koch was, he was no scientist. From some farmers, he had bought the vertebrae of several specimens from the group of primitive whales known as archaeocetes, which are found in rocks of the middle Eocene (50 to 37 million years old) in Alabama, Mississippi,

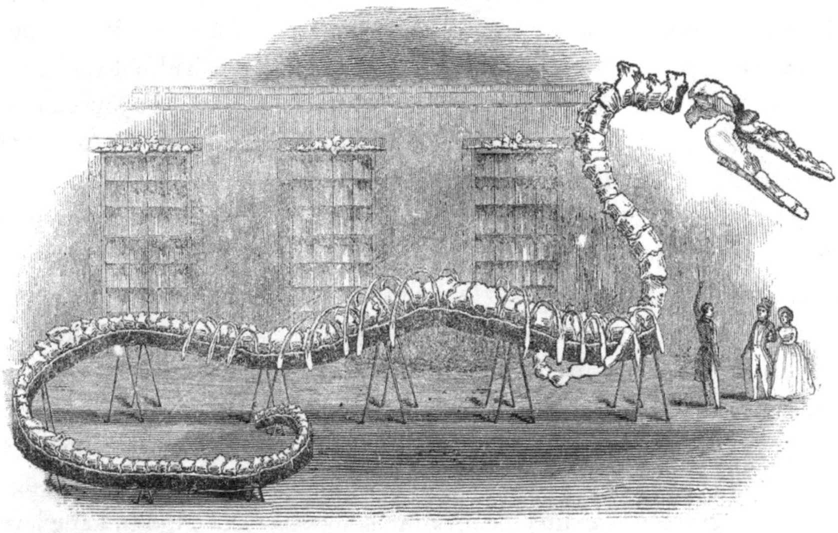

Figure 20.1 ▲
Albert Koch's "Hydrarchos," which toured throughout Europe and North America during the 1840s. It was actually a composite of at least three archaeocete whale skeletons, cobbled together to make the "Great Sea Serpent" seem larger. (From Wikimedia Commons)

and Arkansas. These bones were so abundant that in some places in Alabama farmers built stone walls with them. Koch then cobbled together a composite specimen made of at least three whales to exaggerate its length and size. (This was a favorite strategy of his. Before this incident, he had exaggerated the size of a mastodont skeleton he owned by combining bones from different specimens and calling it the "Great Missourium.")

Koch then took his "sea serpent" on a tour of Europe, where it traveled from city to city, drawing huge crowds to see the "behemoth of the Bible." After leaving London and Berlin because scientists were telling the press that his specimen was a hoax, Koch and "Hydrarchos" visited Dresden, Breslau, Prague, and Vienna. King Frederick William IV of Prussia was so impressed that in 1847 he gave Koch an annual pension of 1000 imperial thalers. Even though his own scientists denounced the skeleton as a fraud, the aging king could not be convinced. Gideon Mantell (who found *Iguanodon*, one of the first named dinosaurs) exposed the hoax and warned people about the damnable swindler. In New York, anatomist Jeffrey Wyman confirmed that the "Great Sea Serpent" was not a reptile, nor were the bones from one animal. As a last resort, Koch was reduced to taking it into rural backwaters, where the words of scientific experts had not yet penetrated. Eventually, he sold his monstrosity to Colonel Wood's Museum in Chicago. There it remained until it was destroyed in the Great Chicago Fire of 1871, allegedly started by Mrs. O'Leary's cow.

Despite Koch's fraudulent skeleton, other whale fossils had reached the hands of legitimate naturalists. In 1834, anatomist Richard Harlan named some huge bones *Basilosaurus* (emperor lizard). Harlan thought that they were the remains of yet another kind of giant reptile we now call a dinosaur, which had just been discovered. In 1839, however, the great British anatomist Sir Richard Owen (who coined the word "Dinosauria" and described some of the first dinosaur fossils ever found) looked at the specimens of *Basilosaurus* and realized that they were not dinosaurs or reptiles, but huge whales. He tried to rename the creature *Zeuglodon* (yoked tooth) to replace the misleading name *Basilosaurus*, but he was too late. By the rules of naming animals, the first name given is the right name, no matter how misleading it might be. This means that the correct name for this whale remains *Basilosaurus*, even though it is a mammal, not a reptile.

As better specimens were found, the archaeocete whales came into focus (figure 20.2). Although not as long as Koch's artificially exaggerated monstrosity, the big archaeocetes were still about 24 meters (80 feet) in length, and weighed about 5400 kilograms (12,000 pounds). They resembled some modern whales in having a long pointed snout with triangular teeth for snagging fish, but they were much more primitive than any living whale. For one thing, they did not have a blowhole near the top of their head (as do all modern whales), but nostrils on the tip of their snout (as do most other mammals). The ears of archaeocete whales were also very primitive, with no specialized ear bones adapted for echolocation in water, like those of modern whales.

The hands and arms of archaeocetes had been modified into paddles, but no hind legs were found on the incomplete fossils excavated in the United States. Then, in 1990, complete articulated skeletons of archaeocete whales were found in Egypt, with their hind limbs still in place. The hind limbs were only about the size of a human arm on a whale more than 24 meters long, so they were no longer functional as hind limbs (although they still anchored the muscles in the back of the body). Since whales no

Figure 20.2 ▲

Mounted skeleton of *Basilosaurus*. (Photograph courtesy Smithsonian Institution, National Museum of Natural History)

longer use them for walking, they are vestiges of the days when whales still walked on four legs. If you see a skeleton of a modern whale on display in a museum, look in the hip region just below the backbone and behind the end of the rib cage. If the specimen is complete and mounted correctly, you will see the tiny nonfunctional remnants of its hip bones and thighbones, buried deep in its body and doing absolutely nothing except proving that whales descended from four-legged land animals. But which ones?

EVOLUTION AND WHALES

When Charles Darwin's *On the Origin of Species* was published in 1859, the fact that whales are mammals took on an even more interesting significance: whales must have descended from land mammals that returned to the water. In the first edition of his book, Darwin speculated about how such a transition may have taken place. He repeated stories about black bears swimming with their mouths open and catching small fish and other aquatic prey. As he wrote: "I can see no difficulty in a race of bears being rendered, by natural selection, more and more aquatic in their structure and habits, with larger and larger mouths, till a creature was produced as monstrous as a whale." Unfortunately, this idea did not go over very well with Darwin's critics, and he dropped this idea from some of the later editions of the book.

The question of the origin of whales remained in limbo for more than a century. Although the fossils of many large archaeocetes resided in many collections, there were almost no decent fossils of even more primitive whales that had been only partially aquatic, or any fossils of mammals that had been fully terrestrial but had whale-like features. In 1966, Leigh Van Valen, a paleontologist at the University of Chicago, reopened the question after decades of neglect. He pointed out that the skulls of archaeocete whales have huge blunt teeth shaped like triangular blades and that these teeth are very similar to those found in a group of large predatory hoofed mammals known as mesonychids (mez-o-NIK-ids). Even though mesonychids had hooves, they were carnivorous or omnivorous and looked like a cross between a wolf and a bear. Many mesonychids had huge, long-snouted skulls that closely resembled those of archaeocetes, and soon other whale-like features began to be noticed as well. The idea that mesonychids were the ancestors of whales became more and more widely accepted over the decades and was still the accepted notion when Robert Schoch and I wrote a book about hoofed mammals.

Meanwhile, the search for more primitive fossil whales began in earnest in the 1970s and the 1980s. At that time, Pakistan owed the United States many millions of dollars for military hardware it had bought from American defense contractors. The Pakistanis were eager to discharge this debt, so through a number of granting foundations, the United States made it relatively easy to obtain grant funds for paleontological research in Pakistan. In addition, paleontologists knew that important early whale fossils (mostly archaeocetes) had been found in the northwestern regions of India (now Pakistan) first by Guy Pilgrim in the 1920s and then by Ashok Sahni and others in the early 1970s. This led a number of paleontologists, especially Philip Gingerich of the University of Michigan and Hans Thewissen of Northeast Ohio Medical University, to explore the rocks in Pakistan that were older than those that had yielded the archaeocete whales and that represented sedimentary environments that were near-shore or shallow marine in origin.

Sure enough, this lucky accident of abundant funding for research in Pakistan led paleontologists to stumble on the time and place where whales actually had evolved from land mammals: in the early Eocene (55 to 48 million years ago), in the shallow tropical seaway known as Tethys. Tethys was a relict of the days of the supercontinent Pangaea and the super-ocean Panthalassa, with a tropical seaway that stretched from the western Mediterranean to Indonesia. The Tethys Seaway was broken up when Africa slid north to close off the Mediterranean, and India collided with the belly of Asia in the middle Eocene to chop the rest of Tethys in half. Before Tethys vanished, however, its shorelines were the home of not only the earliest whales to return to the water, but also the earliest manatee relatives (chapter 21) and many other distinctive mammals (like mastodonts, monkeys and apes, and hyraxes).

The first important transitional whale fossil was *Pakicetus*, reported by Gingerich and his colleagues in 1983 (figure 20.3). Although most of its skeleton is wolf-like, with four long limbs for walking, it had a skull that resembles that of the archaeocete whales, including the large serrated triangular teeth. Its brain was small and primitive, with no special features in the ear for hearing underwater and detecting faint sonar echoes (but it did have dense ear bones and other features that suggest some ability to hear underwater). *Pakicetus* was found in river sediments dating to about 50 million years ago, which indicates that it was primarily a terrestrial animal that spent much time in the water. Although its long legs with short hands and feet were adapted mostly for running and jumping, its limb bones were unusually thick and could have provided ballast in the water, suggesting that it was primarily a wader, not a swimmer.

THE "WALKING SWIMMING WHALE"

The biggest breakthrough, however, came in 1994, when Hans Thewissen reported the discovery of *Ambulocetus natans*, whose name literally means "walking swimming whale" (figure 20.4). Recovered from the Upper Kuldana Formation of Pakistan (a near-shore marine deposit about 47 million years old), it is a nearly complete skeleton of an animal that was truly halfway between a whale and a land mammal. It was about 3 meters (10 feet) long, about the size of a large sea lion. It had a long toothy snout like that of other primitive whales, with the same distinctive triangular teeth. Its

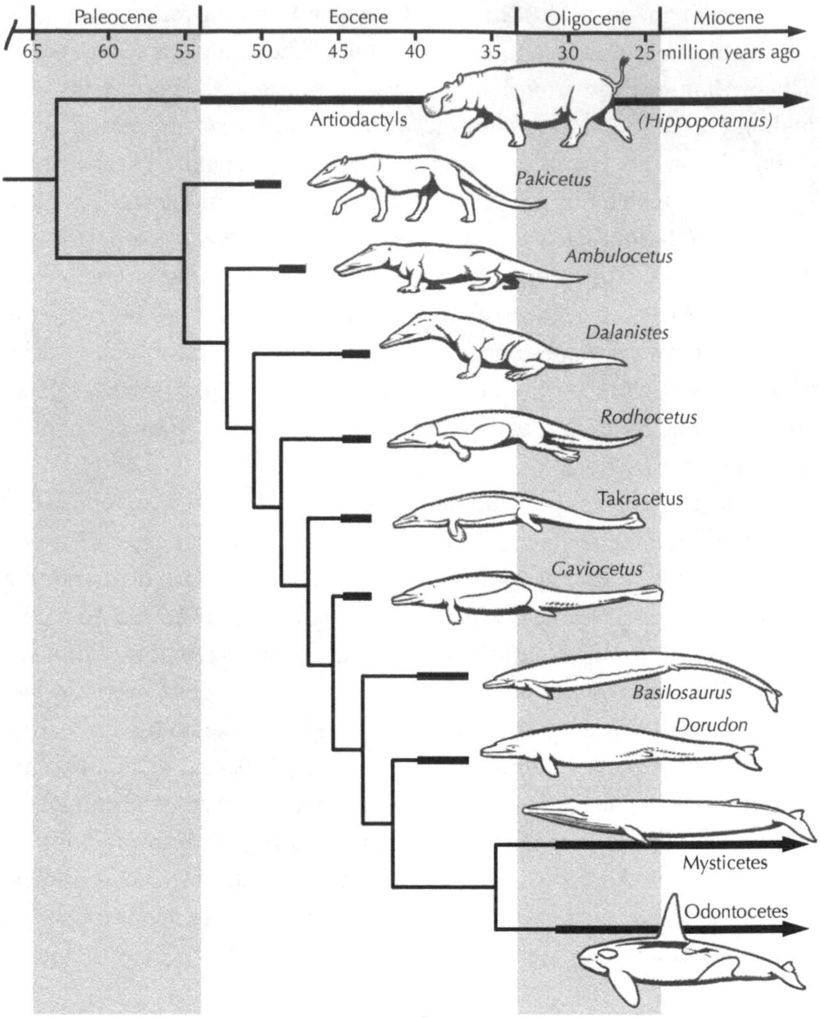

Figure 20.3 ▲

The evolution of whales from land mammals, showing reconstructions of the numerous transitional fossils recovered from beds dating from the Eocene in Africa and Pakistan. (Drawing by Carl Buell; from Donald R. Prothero, *Evolution: What the Fossils Say and Why It Matters* [New York: Columbia University Press, 2007], fig. 14.16)

ears were still not very specialized, nor were they suited for echolocation, but *Ambulocetus* probably used them for hearing vibrations through the ground or water. Its long, strong limbs ended in very long fingers and toes, which probably were webbed. Thus it was a four-legged whale that could both walk and swim, hence its name.

Studies of its spine have shown that it could undulate its back up and down like an otter does, rather than paddling with its feet like a seal or penguin. This kind of up-and-down spinal motion is very similar to that of some whales, although most whales have a rigid torso and use only their tails for propulsion.

Ambulocetus was clearly not a fast swimmer, though. Thewissen suggested that its crocodile-like proportions support the idea that it was an ambush predator, lurking motionless underwater until prey came close and then lunging to catch its food. The location of the specimens in the near-shore marine rocks of the Upper Kuldana Formation suggests that it lived on the margins of lakes and rivers as well as the shores of oceans. Chemical analysis of its teeth further proves that *Ambulocetus* lived both in both fresh- and salt water.

A few years after the discovery of *Ambulocetus*, another nearly complete whale skeleton known as *Dalanistes* was found (see figure 20.3). Like *Ambulocetus*, it had fully functional forelimbs and hind limbs with even longer fingers and toes to support its webbed feet. But its snout was much longer and even more whale-like, as was its robust tail.

In 1994, the year that Thewissen reported *Ambulocetus*, Philip Gingerich and his colleagues found another, more advanced transitional whale from beds about 47 million years old in the southern Baluchistan region of Pakistan (see figure 20.3). Named *Rodhocetus*, it is the best-known representative of a family of primitive dolphin-size whales known as protocetids (although one of them, *Gaviacetus*, was over 5 meters [16 feet] long). The skull of *Rodhocetus* is much larger and more whale-like than that of *Ambulocetus*, with a longer snout and typical archaeocete teeth. The neck vertebrae show that its head and body were merged into a streamlined shape, with no distinct neck that it could turn independently of its torso. Its long limb bones are much shorter than those of *Ambulocetus* and *Dalanistes*, and its fingers and toes are shorter also, suggesting that it had much smaller legs with webbed feet (but not fully developed whale flippers). Yet its hip bones and hip vertebrae were still fused together, suggesting that it was still capable of walking on land. Its skeletal proportions suggest that *Rodhocetus* was a foot-powered swimmer, using alternating strokes of its hind feet to propel it and its tail mostly as a rudder.

Since the discovery of *Rodhocetus*, numerous other transitional whales, such as *Takracetus* and *Gaviocetus*, have been found; they have increasingly specialized hands, developing into whale-like flippers (see figure 20.3),

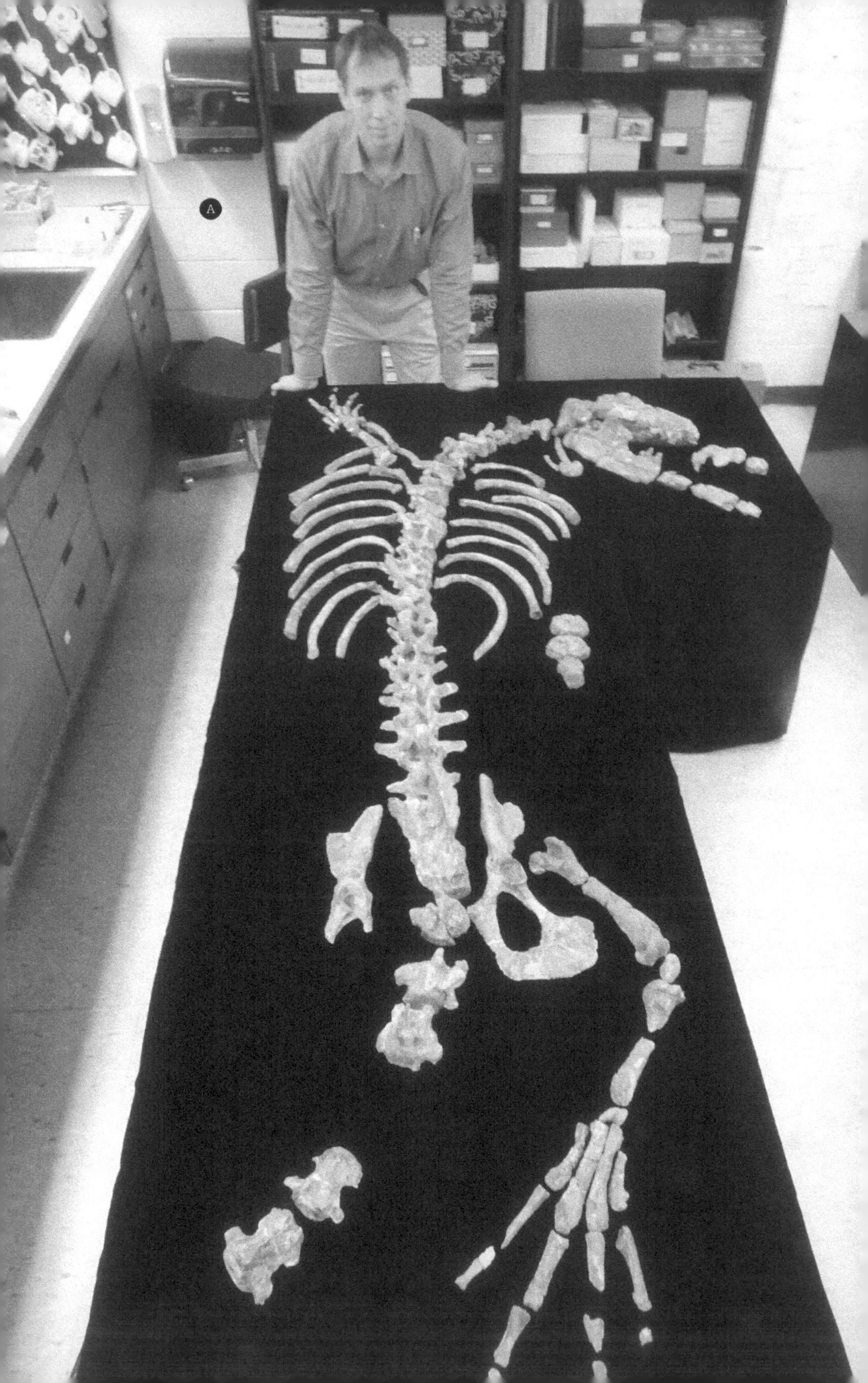
A

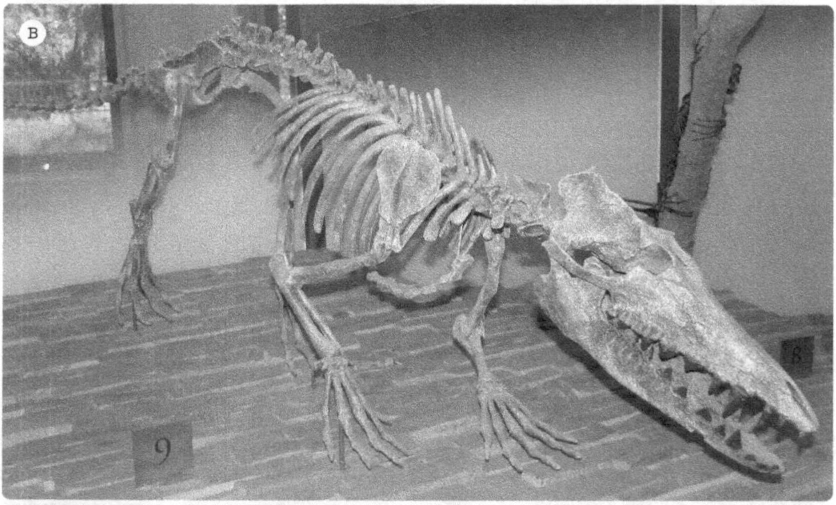

while their hind legs are tiny. Their bodies are also more dolphin-like, with further development of tail propulsion (as in living whales), meaning that they probably had horizontal tail flukes as well. Today, there are so many transitional whale fossils that it is impossible to decide where terrestrial animals end and true whales begin. From a complete mystery in 1980, the origin of whales from land animals now is one of the best evolutionary transitions documented in the fossil record.

Figure 20.4 ◀ ▲

The walking–swimming whale *Ambulocetus*: (*A*) the most complete skeleton, with its discoverer, Hans Thewissen; (*B*) replica of the skeleton, mounted in a walking pose; (*C*) reconstruction of its appearance while swimming. ([*A*] courtesy H. Thewissen, NEOMED; [*B*] photograph by the author; [*C*] courtesy Nobumichi Tamura)

WHIPPING THROUGH THE WHIPPOMORPHA

Since Leigh Van Valen's original suggestion in 1966, most paleontologists have regarded the wolf-like hoofed mammals known as mesonychids as the likely progenitors of whales. The similarities in their teeth and skulls are very striking, and no other group of mammals on the planet had such distinctive teeth. When Earl Manning, Martin Fischer, and I published an analysis of the relationships of the hoofed mammals in 1988, the anatomical characters seemed to strongly support the idea that whales and mesonychids are closely related, and that both are also closely related to the even-toed hoofed mammals known as the Artiodactyla (pigs, hippos, camels, giraffes, deer, antelopes, cattle, and sheep and their kin).

But in the late 1990s, molecular biologists began to analyze the DNA sequences, as well as the sequences of certain proteins that build important molecules, of many mammalian groups. Again and again, the evidence showed that whales not only are closer to the artiodactyls than to any other living mammals, but are *descended* from them. Among the living artiodactyls, whales consistently came out as the nearest relative of the hippopotamus (see figure 20.3). Paleontologists were reluctant to accept the molecular evidence, since the anatomical evidence from fossils of mesonychids seemed much stronger, and the earliest whales look nothing like the earliest hippos. More important, the molecular analyses were based on living animals. We had no DNA or proteins for any of the many fossil forms that suggested the link between whales and mesonychids.

But once again, the fossil record surprised us—and came to the rescue to resolve the problem. In 2001, two independent groups (Thewissen's group and Gingerich's group) found and reported fossils of early whales in Pakistan that have well-preserved ankles. The ankle bones of these whales have the diagnostic "double-pulley" configuration of the astragalus bone (the hinge bone in the mammalian ankle joint), originally known only in the artiodactyls. Unlike any other group of mammals, all artiodactyls have this double-pulley ankle bone, and, indeed, most artiodactyls can be identified as members of that order by this unique bone alone. And, based on the evidence of the fossils from Pakistan, it was now clear that whales have the unique anatomy of the artiodactyl ankle as well.

Resistance to the idea that whales are artiodactyls quickly melted, and paleontologists went back to the drawing board to redo their analyses using even more anatomical features as well as the new molecular evidence. Soon, there was a consensus that whales *are* artiodactyls and should be classified as a group within the branch leading to hippos. Instead of two completely separate orders of Cetacea and Artiodactyla, the new agreement on the evidence requires that we rename the order Cetartiodactyla, and regard Cetacea as a subgroup of one lineage of artiodactyls (the hippos and their relatives, the anthracotheres). However, this ignores the principles of taxonomy. When one group becomes part of another, the name of the larger group usually does not change. Thus Artiodactyla is now understood to include Cetacea and does not have to be renamed Cetartiodactyla, any more than Dinosauria has to be renamed Avedinosauria to include birds. The group of whales plus hippos was named the Whippomorpha (*Wh* for "whale," plus "hippo," plus *morpha* for "shape") by molecular biologists, although most scientists prefer to use the name Cetacodontomorpha for the hippo-whale grouping.

Now, instead of the familiar picture of whales as a group just outside the artiodactyls, they are nested in a group closely related to hippos and many other primitive fossil artiodactyls. The mesonychids are now the odd mammal out, usually regarded as the nearest relative of the whales plus other artiodactyls. This classification requires that their distinctive triangular teeth evolved in parallel with those of archaeocete whales—but this example of convergent evolution is much easier to accept than to dismiss the huge number of molecular similarities between whales and hippos as merely convergent evolution.

But who knows? If mesonychids were alive today and we could sequence their DNA, we might come up with a different answer. They vanished at the end of the Eocene, more than 33 million years ago, so we will never know.

HIPPO-KIN

Imagining whales and hippos as close relatives is not too big a stretch, since both are large-bodied and aquatic. But again, the fossil record helps us out. The fossil record of the modern family of hippopotamids goes back only 8 million years. But the hippos can be linked to an extinct family of artiodactyls known as anthracotheres, which trace back to 50 million years ago. The anthracotheres came in many shapes and adaptations, but many appear to have been partially or completely aquatic.

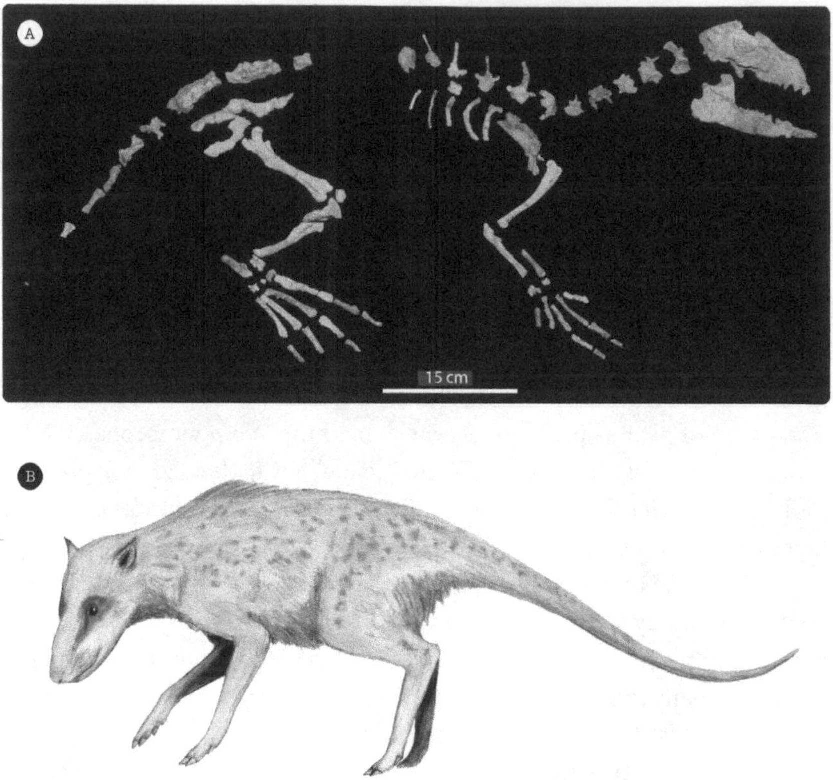

Figure 20.5 ▲

Indohyus, the earliest common ancestor of the whale and hippopotamus lineage: (*A*) the most complete skeleton; (*B*) reconstruction of its appearance in life. ([*A*] courtesy H. Thewissen, NEOMED; [*B*] courtesy Nobumichi Tamura)

Recently, rocks from Kashmir have yielded a spectacular intermediate fossil that links the two groups. Known as *Indohyus* (Indian pig), it was described in 2007 by Hans Thewissen based on fossils collected many years earlier by the Indian geologist A. Ranga Rao (figure 20.5). Even though it was barely larger than a rabbit, with long hind legs for leaping, and had the body of a small deer, its distinctive anatomical features make it the transitional fossil between whales and other artiodactyls. Its ear region shows many features that are found only in whales. Its limbs were made of very dense bone (just like those of whales, hippos, and many other aquatic groups), which provided ballast and helped it wade into or dive under water without floating out of control. Chemical analysis of its bones showed that *Indohyus* was aquatic, but that of its teeth proves that it ate land plants. *Indohyus* provides us with the final link that unites the most primitive whales like *Pakicetus* (also mostly a terrestrial animal) with the anthracotheres, and thus to the anthracothere–hippo lineage.

Thus far from being fish or evolving from swimming bears, whales, according to molecular evidence, descended from a common ancestor with the anthracotheres and hippos. And with fossils such as *Pakicetus*, *Dalanistes*, *Ambulocetus*, *Rodhocetus*, and *Indohyus*, the fossil record also demonstrates how whales evolved from land animals.

SEE IT FOR YOURSELF!

The fossils of *Ambulocetus*, *Dorudon*, and other transitional whales remain in the countries in which they were found. But a number of museums have replicas of those transitional fossils as well as complete skeletons of *Basilosaurus*. In the United States, they include the Alabama Museum of Natural History, Tuscaloosa; American Museum of Natural History, New York; Field Museum of Natural History, Chicago; Museum of Paleontology, University of Michigan, Ann Arbor; National Museum of Natural History, Smithsonian Institution, Washington, D.C.; and Natural History Museum of Los Angeles County, Los Angeles. In Europe, specimens are exhibited at the Naturalis Biodiversity Center, Leiden, Netherlands; and Naturmuseum Senckenberg, Frankfurt, Germany. Farther afield are the Museum of New Zealand / Te Papa Tongarewa, Wellington; and National Museum of Nature and Science, Tokyo, Japan.

FOR FURTHER READING

Berta, Annalisa, and James L. Sumich. *Return to the Sea: The Life and Evolutionary Times of Marine Mammals*. Berkeley: University of California Press, 2012.

Berta, Annalisa, James L. Sumich, and Kit M. Kovacs. *Marine Mammals: Evolutionary Biology*. 2nd ed. San Diego: Academic Press, 2005.

Janis, Christine M., Gregg F. Gunnell, and Mark D. Uhen, eds. *Evolution of Tertiary Mammals of North America*. Vol. 2, *Small Mammals, Xenarthrans, and Marine Mammals*. Cambridge: Cambridge University Press, 2008.

Prothero, Donald R., and Robert M. Schoch. *Horns, Tusks, and Flippers: The Evolution of Hoofed Mammals and Their Relatives*. Baltimore: Johns Hopkins University Press, 2002.

Rose, Kenneth D. *The Beginning of the Age of Mammals*. Baltimore: Johns Hopkins University Press, 2006.

Rose, Kenneth D., and J. David Archibald, eds. *The Rise of Placental Mammals: The Origin and Relationships of the Major Extant Clades*. Baltimore: Johns Hopkins University Press, 2005.

Thewissen, J. G. M., ed. *The Emergence of Whales: Evolutionary Patterns in the Origin of the Cetacea*. Berlin: Springer, 2005.

——. *The Walking Whales: From Land to Water in Eight Million Years*. Berkeley: University of California Press, 2014.

Zimmer, Carl. *At the Water's Edge: Fish with Fingers, Whales with Legs, and How Life Came Ashore but Then Went Back to Sea*. New York: Atria Books, 1999.

WALKING MANATEES

Down to the waist it resembled a man, but below this it was like a fish with a broad, crescent-shaped tail. Its face was round and full, the nose thick and flat; black hair flecked with grey fell over its shoulders and covered its belly. When it rose out of the water it swept the hair out of its face with its hands; and when it dived again, it snuffled like a poodle. One of us threw a fishhook to see if it would bite. Thereupon it dived and disappeared for good.

HERBERT WENDT, *OUT OF NOAH'S ARK*

MERMAIDS!

The legends of mermaids go back millennia in the lore of the sea and are found in many cultures. In the oldest known story, from Assyria around 2300 B.C.E., the goddess Atargatis transforms herself into a mermaid in repentance for having accidentally killed her human lover, a shepherd. In the *Odyssey*, attributed to Homer and possibly dating to the eighth century B.C.E., sirens, mythical women with fish-like bodies, sing so irresistibly that they lure sailors to their deaths on the rocks. Mysterious "sea-girls" are mentioned in some of the tales told by Scheherazade in *One Thousand and One Nights*. Reports of encounters with mermaids and mermen—such as that near Martinique in 1671 by two French sailors, quoted by Herbert Wendt—have been widespread in nearly all western European societies over the past 2000 years. These legends were standardized by such popular stories as Hans Christian Andersen's "The Little Mermaid" (1836), which became a hit Disney movie in 1989. The film *Splash* (1984), which features

Daryl Hannah as a real mermaid, also spread the myth to newer generations. As recently as 2012 and 2013, two "documentaries" on the cable-television network Animal Planet claimed that mermaids are real and had been found, causing a huge number of people to believe this hoaxed "evidence." These pseudo-documentaries were so influential that scientists at the National Oceanographic and Atmospheric Administration had to twice waste their precious time in order to post statements on the agency's Web site stating that the broadcasts were fiction and that mermaids do not exist.

Some of these legends were just products of the fertile human imagination and comparable to the other myths about half-human, half-animal creatures, such as the centaur (horse–human hybrid) and the minotaur (bull–human hybrid). But many scholars believe that some real sightings at sea fed the fancy of sailors and thus were spun into the legendary mermaids. In 1493, on his second voyage near Hispaniola, Columbus reported having seen three "female forms" that "rose high out of the sea, but were not as beautiful as they are represented." The famous English pirate Blackbeard (Edward Teach) claimed to have seen mermaids in the Caribbean and thereafter stayed away from waters where they had been reported. Both sailors and pirates believed that mermaids would enchant them out of their gold and then drag them to the bottom of the sea.

Like reports of sea serpents, there are scattered "sightings" of mermaids by people all over the world, from Canada to Israel to Zimbabwe. In the Indian Ocean, mariners claimed, mermaids appeared in pairs, with one trying to rescue the other if it was harpooned. They were said to cry "tears of secretions," and a mother mermaid supposedly cradled her young in her arms when she was nursing it.

THE SCIENCE OF MERMAIDS

Is there any real basis in truth to all these legends? A number of zoologists have pointed to manatees (found mostly in the shallow tropical waters of the Western Hemisphere) and dugongs (found mostly in the Indian Ocean) and their relatives, an order of marine mammals known as the sea cows, or Sirenia (after the mythical sirens). Both the dugong and the manatee float vertically with their head above water, when they want to observe a ship or another object at the surface (figure 21.1). All sirenians have a pair of breasts on their chests, which might suggest the configuration of human

Figure 21.1 ▲
When a sirenian (like this manatee) floats upright, it is possible to understand how sailors viewing it from a great distance may have mistaken it for a mermaid. (Courtesy Wikimedia Commons)

breasts, and nurse their young in a posture reminiscent of that of human females. But how can an animal that is so plug-ugly be mistaken for a beautiful woman? If a manatee had strands of seaweed across its forehead that resembled hair and was sighted from far enough away (especially in the glare of the open ocean), it is not so hard to imagine it being mistaken for a woman floating out at sea (especially if sailors had been away from land and women for too long). Just a few such "sightings" by Columbus and other early explorers would have confirmed the widespread myth that had been found in nearly every culture for millennia.

When manatees and dugongs were captured and brought to the attention of early naturalists, there was tremendous confusion. Close examination showed that they are nothing like the mermaids of legend. The first naturalists to examine their anatomy classified them as whales, since they are completely aquatic and have fully developed flippers for hands, no hind legs, and a tail fluke. But Carolus Linnaeus spotted many anatomical specializations that allied them with elephants, and he was the first to classify them with the Proboscidea: the order that includes elephants, mammoths, and mastodonts. In 1816, zoologist Henri de Blainville followed Linnaeus's interpretation, although most natural historians were still referring to manatees and dugongs as whales.

But as more anatomical similarities were found, the connection between sirenians and proboscideans became stronger. Both groups have a range of unique specializations. Eventually, zoologists began to give up on the "whale" classification. The argument that Sirenia and Proboscidea are closely related finally reached a critical stage, when in 1975 Malcolm McKenna proposed that sirenians and proboscideans be placed in a group he called Tethytheria. The group was so named because the fossils of both lineages show that they originated around the Tethys Seaway, which ran from the Mediterranean through the Middle East, past India, and on to Australia. A few years later, McKenna, Daryl Domning, and Clayton Ray described a fossil called *Behemotops* from the Oligocene rocks of the northern shore of the Olympic Peninsula of Washington that confirmed the tethythere roots of sirenians. Since then, the idea of the Tethytheria has been supported by numerous molecular analyses that show the close relationship of sirenians and elephants, confirming the many anatomical similarities spotted by Linnaeus.

SEA COWS WALK INTO THE SEA

The anatomical and molecular evidence is overwhelming that sirenians split from the proboscidean ancestral root about 50 million years ago. What does the fossil record show? The earliest fossil sirenian to be studied was also among the most primitive. In 1855, the legendary British anatomist Sir Richard Owen described a strange skull that had been sent to London from a locality called Freeman's Hall in the Chapelton Formation, dated to about 50 to 47 million years ago, on the island of Jamaica. Owen had coined the term "Dinosauria," had described Charles Darwin's *Beagle* fossils from South America; and eventually became Darwin's chief scientific rival. Although the skull is extremely primitive, with parts broken away and teeth worn down to the roots (figure 21.2), Owen correctly realized that it has the slightly downturned snout bones, the nasal opening high on the skull, and many other features of the sirenians. Other skeletal parts of the animal were mere fragments, but they suggested a sheep-size quadruped. The skull and

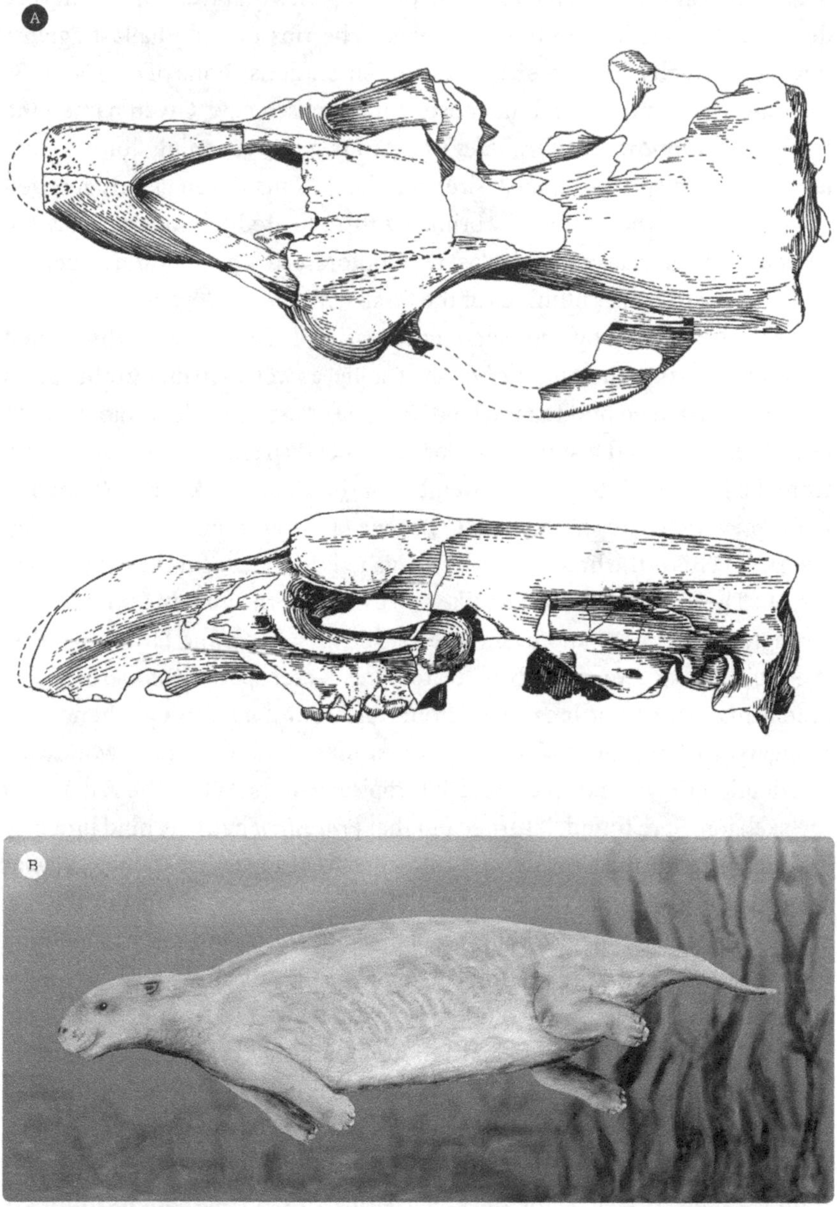

Figure 21.2 ▲

Prorastomus: (*A*) skull, described by Sir Richard Owen; (*B*) reconstruction of its appearance in life. ([*A*] courtesy Daryl Domning; [*B*] courtesy Nobumichi Tamura) bone fragments had been found with pieces of ribs that were thick and very dense, a diagnostic feature of sirenians. The ribs provide ballast against floating too high in the water, and the extremely dense bone of even a single rib fragment is unique and diagnostic for every sea cow. Owen named the fossil skull *Prorastomus sirenoides* (the genus name meaning "broad front jaws," and the species name, "sirenian-like"). Thus Owen clearly realized the fossils were those of a very primitive form related to modern sea cows. Although Owen was one of the last real zoologists to deny natural selection, he could not deny the affinities of the fossil with the modern Sirenia.

As the years went by, more and more fossils of sea cows were discovered along the coasts of the Atlantic and the Pacific, as well as in many other areas where oceans once flooded the land. In 1904, Austrian paleontologist Othenio Abel described a skull of a more advanced sirenian, *Protosiren fraasi*, from the lower Building Stone Member of the Gebel Mokattam Formation in Egypt, which dates to the middle Eocene (47 to 40 million years ago) (figure 21.3). (This is the limestone that furnished the blocks for the pyramids of Egypt.) The skull is much more like those of modern sirenians, with a more strongly downturned snout, the specialized nasal opening farther back on the skull, and other more advanced features. Later specimens were discovered in many far-flung localities—from North Carolina, through France and Hungary, to Pakistan and India—so *Protosiren* had an almost worldwide distribution in warm tropical and subtropical waters. When the remains of the skeleton were found, it turned out that *Protosiren* had tiny hind limbs. In addition, its hips were not strongly attached to its lower backbone, so it was almost completely aquatic and could barely walk on land. Most of the fossils of sirenians younger than *Protosiren* exhibit even more shrunken hind limbs, indicating that the animals from which they came could no longer walk and thus had become completely aquatic. The living manatee and dugong still have tiny remnants of their hips and thigh buried in the muscles around their lower back that no longer have any function except to prove that they evolved from four-legged land animals.

Thus the oldest fossil sirenian (*Prorastomus*) shows the beginning of the skull features, as well as the thick dense ribs, of sea cows, but its limbs are poorly preserved. The next youngest (*Protosiren*) has shortened hind limbs that were weakly connected to the spine, so it was mostly aquatic. What was needed was a fossil that clearly had a sirenian skull and ribs, but four walking legs—final proof that sea cows had evolved from land animals.

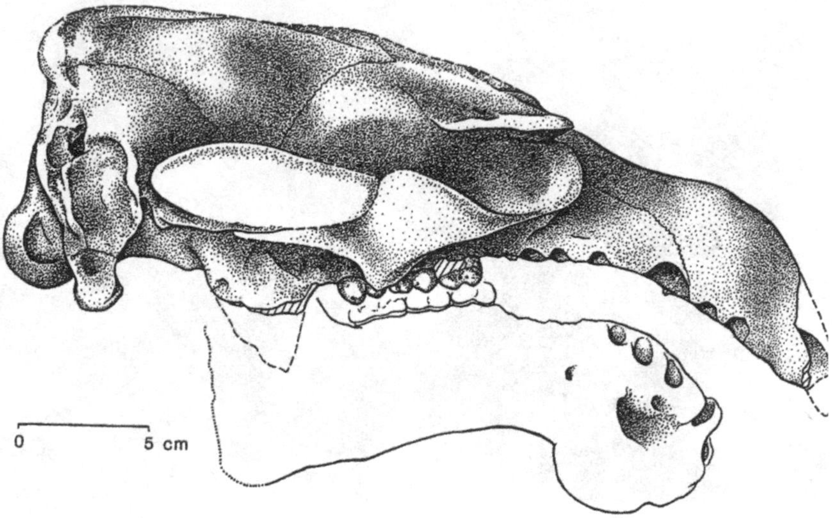

Figure 21.3 ▲
The skull of *Protosiren*, a sirenian more advanced than *Prorastomus*. (Courtesy Daryl Domning)

DOWN IN JAMAICA

The tropical paradise of Jamaica does not resemble the harsh badlands that are so familiar in documentaries about fossil hunters, but Jamaica does have important fossils. About 15 kilometers (9 miles) south of the resort town of Montego Bay are some remarkable bone beds in an area known as Seven Rivers, in the parish of St. James. For years, Roger Portell and other paleontologists collected in the Chapelton Formation, the same lower Eocene unit that had yielded Owen's skull of *Prorastomus*. The mollusc fossils that Portell sought included those of the gigantic marine snail *Campanile* and many other extinct snails and clams. All the bones in these beds were fragmentary because they had been washed into an ancient lagoon or delta and buried alongside the remains of the marine molluscs. Over the years, they came to include bones of an iguana, a primitive rhinoceros, and possibly a lemur-like primate.

By the mid-1990s, the collections of fossils from the Chapelton Formation were growing, and they captured the attention of Daryl Domning of

Figure 21.4 ▲

Pezosiren: (*A*) Daryl Domning with the reconstructed skeleton of *P. portelli*; (*B*) reconstruction of its appearance in life. ([*A*] courtesy Daryl Domning; [*B*] courtesy Nobumichi Tamura)

Howard University, the foremost expert on fossil sea cows. The site promised to yield more specimens of a sirenian like *Prorastomus*, which had been found more than 150 years earlier, so working it was worth the effort. Several major seasons of collecting in Jamaica yielded hundreds of bones. But instead of more fossils of *Prorastomus*, Domning recognized an entirely new genus and species of primitive sea cow—and even better, its skeleton was nearly complete!

In 2001, Domning published his description of the specimen in the world's most prominent scientific journal, *Nature*. He named the creature

Pezosiren portelli (Portell's walking siren) (figure 21.4). *Pezosiren* was about the size of a large pig (about 2.1 meters [6.5 feet] long), with a skull much like that of *Prorastomus*. In some ways (the crest along the top of the skull), it was more primitive than *Prorastomus*, but in most features (the ear region and the downward deflection of the tip of the lower jaw), *Pezosiren* was more advanced. It also had the classic thick, dense rib bones of all sirenians, which were part of a long barrel-shaped trunk, and a short tail. Most important, the fossil of *Pezosiren* has nearly complete hip bones and fore- and hind limbs—and these limbs are short—but perfectly normal hands and feet for walking on land, with no obvious specializations for swimming. Based on the details of the limbs and spine, *Pezosiren* swam by paddling with its feet and propelling itself along the bottom in shallow water (as do hippos), rather than by swimming with an up-and-down motion of its tail (as do otters, sirenians, whales, and seals and sea lions).

Thus the "missing link" between aquatic sea cows and their terrestrial ancestors had been found, a perfect intermediate between the two. *Pezosiren* had the skull and heavy-duty ribs of a sirenian, but retained the fully developed legs and feet of a quadruped. Just like the numerous walking whales and other transitional fossils that have been discovered in recent years, it shows just how yet another group of land animals returned to the sea.

OUT OF AFRICA

Only one piece of the puzzle remained. The closest relatives of the sea cows, the elephants and other tethytheres, emerged in North Africa, Pakistan, and other regions that bordered the Tethys Seaway. Yet the oldest sirenian fossils (*Prorastomus* and *Pezosiren*) came from Jamaica. In 2013, a group of scientists led by Julien Benoit and nine others published newly discovered specimens from the early Eocene (about 50 million years ago) of Tunisia. They included a number of skull bones with ear regions that are distinctly sirenian and some other fragments of the skeleton. The locality is known as Chambi, so the specimens are provisionally known as the Chambi sea cow, since they are too incomplete to merit a formal taxonomic name. Fragmentary though they are, the ear regions of the Chambi sea cow complete the puzzle, showing that the earliest sirenians—like their relatives, the earliest proboscideans, hyraxes, and other tethytheres—first appeared in the Tethys region (primarily Africa). The sea cows were aquatic and soon spread from

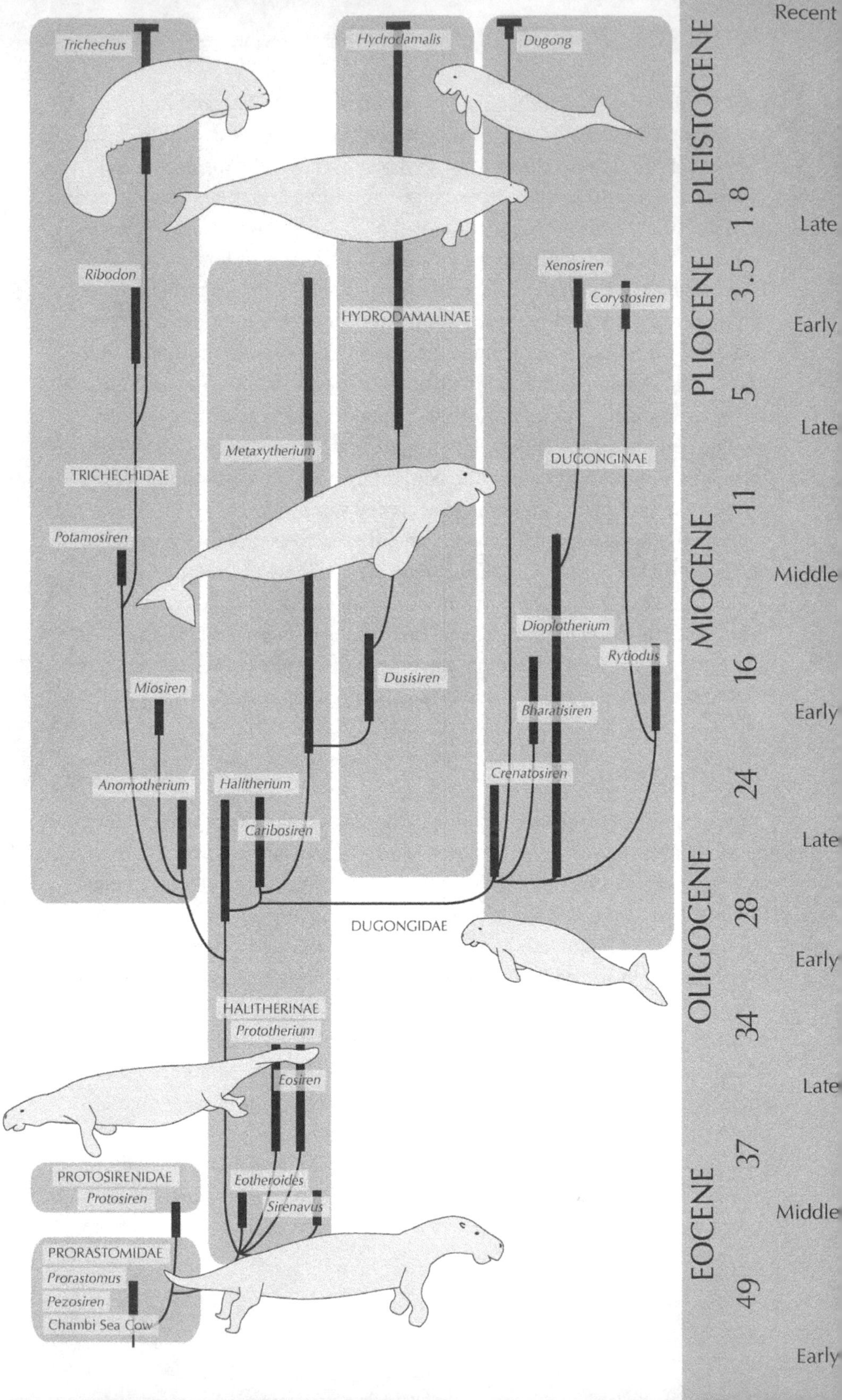

Recent
PLEISTOCENE
1.8
Late
3.5
PLIOCENE
Early
5
Late
11
MIOCENE
Middle
16
Early
24
Late
OLIGOCENE
28
Early
34
Late
37
EOCENE
Middle
49
Early
Trichechus
Hydrodamalis
Dugong
Ribodon
Xenosiren
Corystosiren
HYDRODAMALINAE
Metaxytherium
DUGONGINAE
TRICHECHIDAE
Potamosiren
Dioplotherium
Rytiodus
Dusisiren
Miosiren
Bharatisiren
Anomotherium
Halitherium
Crenatosiren
Caribosiren
DUGONGIDAE
HALITHERINAE
Prototherium
Eosiren
PROTOSIRENIDAE
Protosiren
Eotheroides
Sirenavus
PRORASTOMIDAE
Prorastomus
Pezosiren
Chambi Sea Cow

the Caribbean to India (figure 21.5). Proboscideans, hyraxes, and the rest of the tethytheres, however, remained confined to Africa until about 16 million years ago, when they managed to move out of Africa, by way of the Arabian Peninsula, and migrate around the world. Soon mammoths and mastodonts were found on every northern continent, elephants had arrived in Asia, and hyraxes had spread to many parts of Eurasia. The world was never the same.

STELLER'S MONSTER

In the early eighteenth century, Czar Peter I the Great was attempting to expand the Russian Empire and enlarge his influence over the world. He tried to modernize and civilize the politics and social habits of Russia and, in particular, wanted to emulate European customs and encourage the growth of science and scholarship, as in all the advanced nations like France, England, the Netherlands, and parts of Germany. He sent naval expeditions to the far reaches of his empire in Siberia and, especially, to the Pacific coast, where remote regions like Kamchatka had long been neglected.

In his navy was a Danish sea captain named Vitus Bering, who had enlisted in 1704. By 1725, he was exploring the areas north of the Kamchatka Peninsula, which were virtually unknown. He thought that there might be a sea between Asia and North America, but he had not traveled far enough north and east to before the expedition had to return to Kamchatka. For many years, he sought support and equipment and men to undertake a bigger expedition and discover what was north and east of Kamchatka. Finally, in 1741, he led several ships with large crews to the region northeast of Kamchatka, where they visited many of the Aleutian Islands and reached Kodiak Island and mainland Alaska—the first time Europeans had ever seen these regions. But the weather was harsh and stormy, the ships were separated several times, and one vessel was shipwrecked. In addition, the crew got sick and was dying from scurvy. They were eating nothing but meat and fish from the ocean, but no fruits with vitamin C. By August 1742, the remnant of the crew from one vessel (rebuilt after being wrecked) limped back

Figure 21.5 ◀
The evolutionary history of the Sirenia. (Drawing by Mary P. Williams, modified from Daryl Domning)

to Russia. Bering himself died along the way and was buried on an island near the Kamchatka Peninsula that now bears his name. The Bering Strait, Bering Sea, and Bering Glacier are also named after him.

On the second expedition was a German naturalist, Georg Steller. He had been recruited as the chief naturalist, and, luckily for posterity, he recorded the wildlife in the North Pacific when Europeans first arrived. He convinced Bering to let him roam and collect on land, making him the first European to set foot in Alaska. Steller discovered many species of mammals and birds, many of which are named after him. They include the Steller's jay (a cold-climate jay, with a distinctive black head and crest, found in the mountains of western North America), plus many endangered species, such as the huge Steller's sea lion, Steller's eider duck, and Steller's sea eagle. Two others are already extinct: the spectacled cormorant and the Steller's sea cow.

After the hunters in the crew could no longer find otters for food, they turned to the gentle Steller's sea cow. It was an immense creature, the largest living marine mammal at the time, other than whales (figure 21.6). It grew to a length of 8 to 9 meters (26 to 30 feet) and weighed about 7 to 9 metric tons (8 to 10 tons). Steller's sea cows were completely docile and unafraid of humans, even though the Native hunters of the region had reduced their numbers to just a few thousand in a few remnant populations. As Steller wrote in his report:

> Along the whole shore of the island, especially where streams flow into the sea and all kinds of seaweed are most abundant, the sea cow . . . occurs at all seasons of the year in great numbers and in herds. . . . The largest are four to five fathoms [about 7 to 9 meters (24 to 30 feet)] long and three and a half fathoms [about 2.25 meters (8 feet) diameter] thick about the region of the navel where they are the *thickest*. Down to the navel it is comparable to the land animal; from there to the tail, a fish. The head of the skeleton is not the least distinguishable from the head of a horse, but when it is still covered with skin and flesh, it somewhat resembles the buffalo's head, especially as concerns the lips. The eyes of this animal, without eyelids, are no larger than a sheep's eyes. . . . The belly is plump and very expanded, and at all times so completely stuffed that at the slightest wound the entrails at once protrude with much hissing. Proportionately, it is like the belly of a frog. . . . Like cattle on land, these animals live in herds together in the sea, males and females usually going with one another, pushing the offspring before them all around

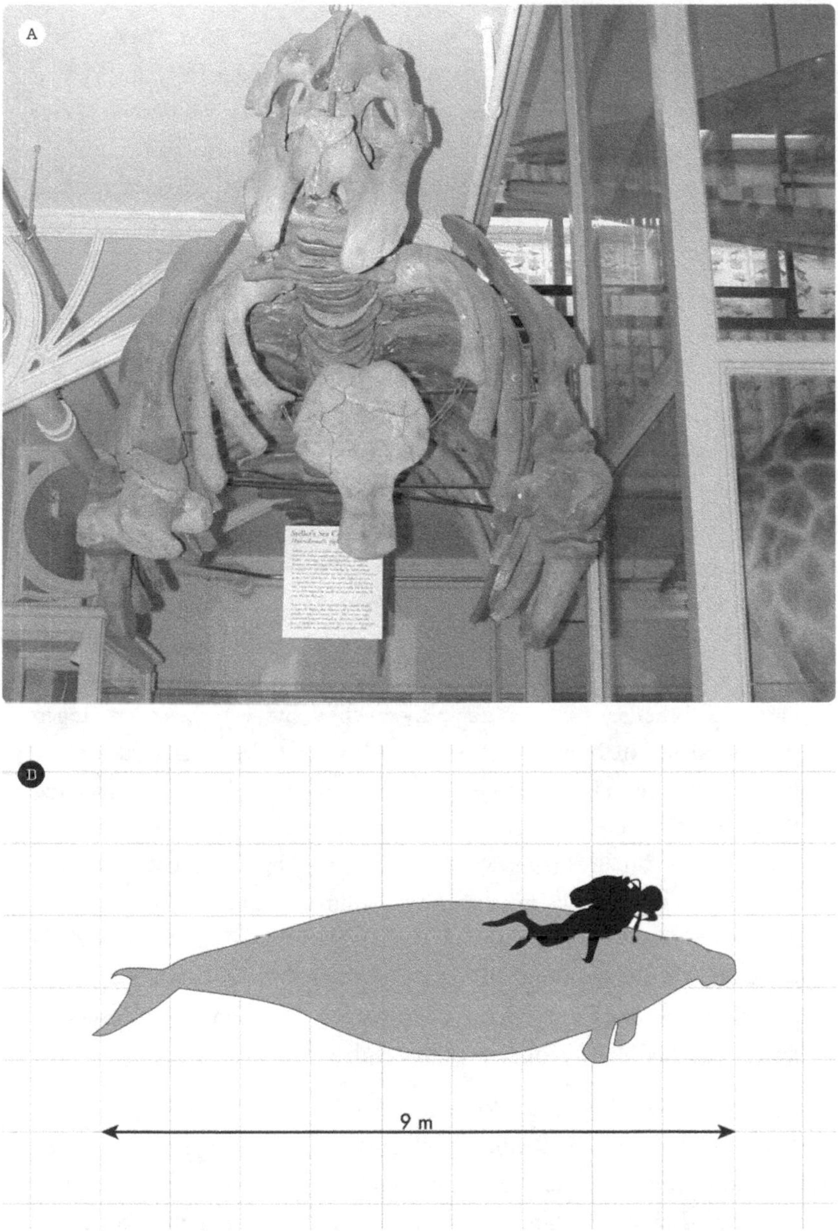

Figure 21.6 ▲

Steller's sea cow: (*A*) skeleton, displayed at the Museum of Comparative Zoology, Harvard University; (*B*) comparison of the sizes of the Steller's sea cow and a human. ([*A*] photograph by the author; [*B*] drawing by Mary P. Williams)

> the shore. These animals are busy with nothing but their food. The back and half of the belly are constantly seen outside the water, and they munch along just like land animals with a slow, steady movement forward. With their feet they scrape seaweed from the rocks, and they masticate incessantly. . . . When the tide recedes, they go from the shore into the sea, but with the rising tide they go back again to the beach, often so close we could reach and hit them with poles. . . . They are not the least bit afraid of human beings. When they want to rest on the water, they lie on their back in a quiet spot near a cove and let themselves float slowly here and there. I could not observe indications of an admirable intellect . . . but they have indeed an extraordinary love for one another, which extends so far that when one of them was cut into, all the others were intent on rescuing it and keeping it from being pulled ashore by closing a circle around it. Others tried to overturn the yawl. Some placed themselves on the rope or tried to draw the harpoon out of its body, which indeed they were successful several times. We also observed that a male two days in a row came to its dead female on the shore and enquired about its condition. Nevertheless, they remained constantly in one spot, no matter how many of them were wounded or killed. The fat of this animal is not oily or flabby but rather hard and glandular, snow-white, and, when it's been lying in the several days in the sun, as pleasantly yellow as the best Dutch butter. The boiled fat itself excels in sweetness and taste the best beef fat, is in colour and fluidity like fresh olive oil, in taste like sweet almond oil, and of exceptionally good smell and nourishment. We drank it by the cupful without feeling the slightest nausea. . . . The meat of the old animals is indistinguishable from beef and differs from the meat of all land and sea animals in the remarkable characteristic that even in the hottest summer months it keeps in the open air without becoming rancid for two whole weeks and even longer, despite its being defiled by blowflies that it is coved with worms everywhere.

As soon as word reached Russia of Steller's and Bering's discoveries, Russian hunters and fur traders followed their tracks from Kamchatka, across the Bering Sea, and to the Aleutian Islands, killing and eating almost any animal they could catch. They were hunting primarily sea otters for their valuable pelts, but they killed seals, sea lions, walruses, whales, and anything else they found. The limited population of a few thousand Steller's sea cows was easily slaughtered for meat or just for sport. By 1768, only 27 years after Steller first saw them, the largest of all sirenians was extinct.

SEE IT FOR YOURSELF!

The original fossils of *Pezosiren* are not on display, but replicas are on exhibit at the Geology Museum, University of the West Indies, Mona, Jamaica; Spanish Bay Conservation and Research Center, Spanish Lookout Caye, Belize; Muséum national d'histoire naturelle, Paris; and National Museum of Nature and Science, Tokyo.

Skeletons of Steller's sea cow are in the collections of 27 institutions around the world and are displayed in a smaller number of them, including the Museum of Comparative Zoology, Harvard University, Cambridge, Massachusetts; National Museum of Natural History, Smithsonian Institution, Washington, D.C.; Natural History Museum, London; National Museum of Scotland, Edinburgh; Muséum national d'histoire naturelle, Paris; Muséum d'histoire naturelle, Lyon; Staatliches Naturhistorisches Museum, Braunschweig, Germany; Naturhistorisches Museum, Vienna; Naturhistoriska Museum, Göteborg, Sweden; Naturhistoriska riksmuseet, Stockholm; Zoologiska museet, Lund, Sweden; Finnish Museum of Natural History, Helsinki; National Museum of Natural History, National Academy of Sciences of Ukraine, Kiev; Museum of the Zoological Institute of the Russian Academy of Sciences, St. Petersburg; and Zoological Museum, Moscow State University.

FOR FURTHER READING

Berta, Annalisa, and James L. Sumich. *Return to the Sea: The Life and Evolutionary Times of Marine Mammals*. Berkeley: University of California Press, 2012.

Berta, Annalisa, James L. Sumich, and Kit M. Kovacs. *Marine Mammals: Evolutionary Biology*. 2nd ed. San Diego: Academic Press, 2005.

Janis, Christine M., Gregg F. Gunnell, and Mark D. Uhen, eds. *Evolution of Tertiary Mammals of North America*. Vol. 2, *Small Mammals, Xenarthrans, and Marine Mammals*. Cambridge: Cambridge University Press, 2008.

Prothero, Donald R., and Robert M. Schoch. *Horns, Tusks, and Flippers: The Evolution of Hoofed Mammals and Their Relatives*. Baltimore: Johns Hopkins University Press, 2002.

Rose, Kenneth D., and J. David Archibald, eds. *The Rise of Placental Mammals: The Origin and Relationships of the Major Extant Clades*. Baltimore: Johns Hopkins University Press, 2005.

22 | THE ORIGIN OF HORSES | *EOHIPPUS*

DAWN HORSES

The geological record of the Ancestry of the Horse is one of the classic examples of evolution.

WILLIAM DILLER MATTHEW, "THE EVOLUTION OF THE HORSE"

HORSING AROUND

When Columbus arrived in the Caribbean in 1492, there were no horses to be found anywhere in the Americas. He brought the first domesticated horses to the Western Hemisphere on his second voyage, in 1493. In 1521, Hernán Cortez conquered the Aztecs. One of the biggest advantages the conquistadors had was not only guns and diseases, but also horses. When the Aztecs first saw the mounted Spanish soldiers, they were terrified and believed that the men and their horses were one creature, something like a centaur.

Horses soon spread throughout the Western Hemisphere. They became the main mode of transport and a primary draft animal, as they had been for millennia in Europe and Asia. They transformed the culture of the Native peoples of the Great Plains, who soon became excellent horsemen, hunting and fighting as they rode. They allowed them to pursue a horse-based nomadic life, following the bison herds. Horses were the foundation of the culture of the Old West, especially as cowboys became essential to the operation of huge cattle ranches. But thanks to the internal-combustion engine and automobiles, horses had become almost obsolete by 1920—

especially as the invention of modern weaponry made cavalry units extremely vulnerable during World War I. Today, horses are primarily a luxury item for the wealthy, although there are still a few places where ranching and the horse culture is still important.

Everyone considered horses a Eurasian native until 1807, when William Clark (of Lewis and Clark fame) found bones of North American horses at Big Bone Lick, Kentucky, which had already produced fossils of extinct mastodonts, mammoths, ground sloths, and other Ice Age creatures. He sent the fossils to his patron, President Thomas Jefferson (who was an avid paleontologist), but Jefferson never wrote anything about significance of this find.

On October 10, 1833, a young Charles Darwin was visiting Argentina on the voyage of the *Beagle*. He was "filled with astonishment" as he found teeth and bones of fossil horses eroding out of a bed that also contained extinct gigantic armadillo-like glytodonts, whose shells were the size and shape of a Volkswagen Beetle. These fossils showed that not only were horses native to the Americas, but they had lived alongside extinct beasts from the late Ice Age. Darwin gave all his fossils to the eminent British paleontologist Sir Richard Owen, who named the horses *Equus curvidens* and commented, "This evidence of the former existence of a genus, which, as regards South America, had become extinct, and has a second time been introduced into that Continent, is not one of the least interesting fruits of Mr. Darwin's palaeontological discoveries." Then in 1848, Joseph Leidy, the founder of American vertebrate paleontology, published on the many different Ice Age horse specimens that he had studied and established that horses had been diverse in North America well before the arrival of Columbus.

Meanwhile, European paleontologists were finding fossil horses as well. There were not only abundant Ice Age horses of the modern genus, *Equus*, in rocks of Pleistocene age, but also more primitive horses from older beds, such as the middle to late Miocene *Hipparion* and the early Miocene *Anchitherium* (and the Eocene *Palaeotherium*, which turned out to be not a true horse or even a member of the horse family, Equidae). By 1872, "Darwin's bulldog," Thomas Henry Huxley, pointed out that the four genera formed a progression showing how horses evolved in Europe. A year later, the Russian paleontologist Vladimir Kowalewsky developed the idea even further.

Yet more and more fossil horses were turning up as American paleontologists such as Leidy, Edward Drinker Cope of the Academy of Natural

Sciences in Philadelphia, and Othniel Charles Marsh of Yale University began to describe the big collections coming from the American West. In 1871 and 1872, Marsh gave the name *Orohippus* to fossil horses from the Rocky Mountains, while Cope called his early Eocene horses *Eohippus*. When Huxley sailed to the United States to give a lecture tour during the 1876 centennial celebrations, he planned to promote Darwin's ideas and talk about the evolution of the horse in Europe. On his tour, he visited Marsh at Yale, spending two whole days in the collection. As his son Leonard Huxley wrote in his biography of his father, "At each inquiry, whether he had a specimen to illustrate such and such a point or to exemplify a transition from earlier and less specialized forms to later and more specialized ones, Professor Marsh would simply turn to his assistant and bid him fetch box number so and so, until Huxley turned upon him and said, 'I believe you are a magician; whatever I want, you just conjure it up.'" Huxley then discarded his original notes and revised his lecture in order to describe the evolution of the horse in North America (figure 22.1).

Soon, it became clear that horses had evolved primarily North America and that European horses like *Anchitherium* and *Hipparion* were immigrants from the North American main stem. By 1926, paleontologists such as William Diller Matthew could draw a highly simplified diagram that showed horse evolution through time (figure 22.2): the tiny horses of the Eocene had three or four toes and low-crowned teeth for eating leaves and fruit; then in the Oligocene came *Mesohippus* and *Miohippus*, which had three toes and much longer legs and toes; they were followed by Miocene horses such as *Merychippus*, which had longer legs and feet, reduced side toes, and higher-crowned teeth for eating gritty grasses; finally, in the Pliocene and Pleistocene, the series culminates with *Equus*, which has very long legs and one toe, side toes completely reduced to tiny splints with no function, and extremely high-crowned teeth.

In the 90 years since Matthew's classic diagram was published, a huge amount has been learned about horse evolution. The simplistic linear sequence has been replaced by a bushy, branching sequence, with multiple lineages of horses living contemporaneously (figure 22.3). For example, Railway Quarry A, in the Valentine Formation of north-central Nebraska, which dates to the Miocene, has yielded 12 species of fossil horses, all of which lived in the same place at the same time. My own research, with Neil Shubin, on *Mesohippus* and *Miohippus* showed that at one point, three

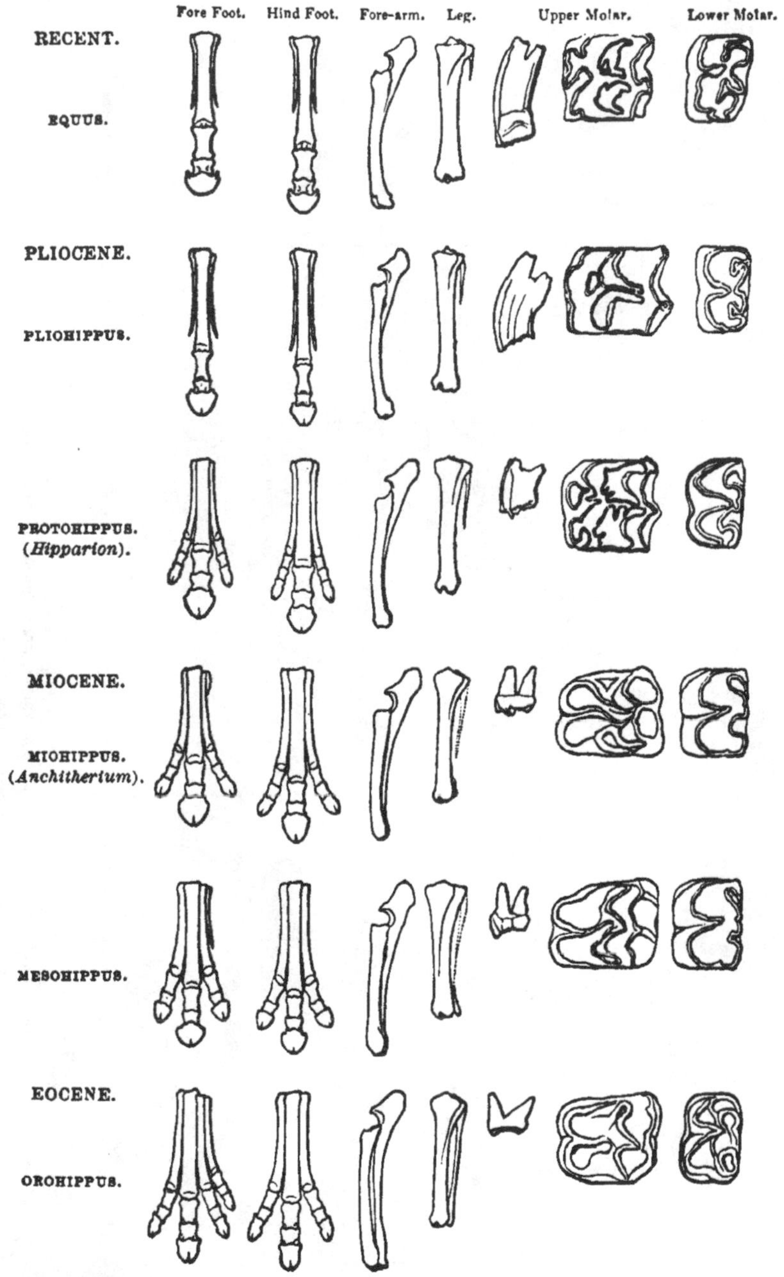

Figure 22.1 ▲

"Genealogy of the Horse": Othniel Charles Marsh's illustration of the transformation of the teeth and limbs of horses, based on fossils from his collection of North American fossil horses. (From O. C. Marsh, "Polydactyl Horses, Recent and Extinct," *American Journal of Science and Arts* 17 [1879])

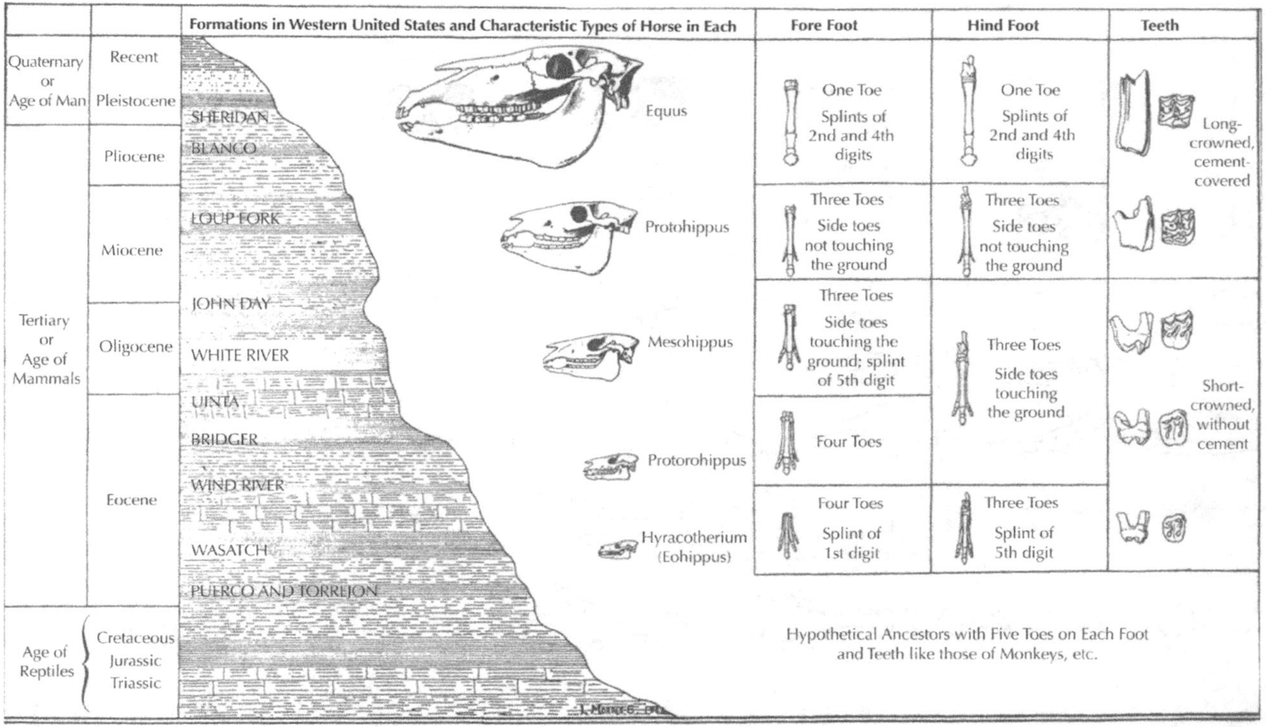

Figure 22.2 ▲

William Matthew's simplified diagram of the evolution of horses, showing the changes in the teeth and skeleton as a simple linear transformation through time. (From William Diller Matthew, "The Evolution of the Horse: A Record and Its Interpretation," *Quarterly Review of Biology* 1 [1926])

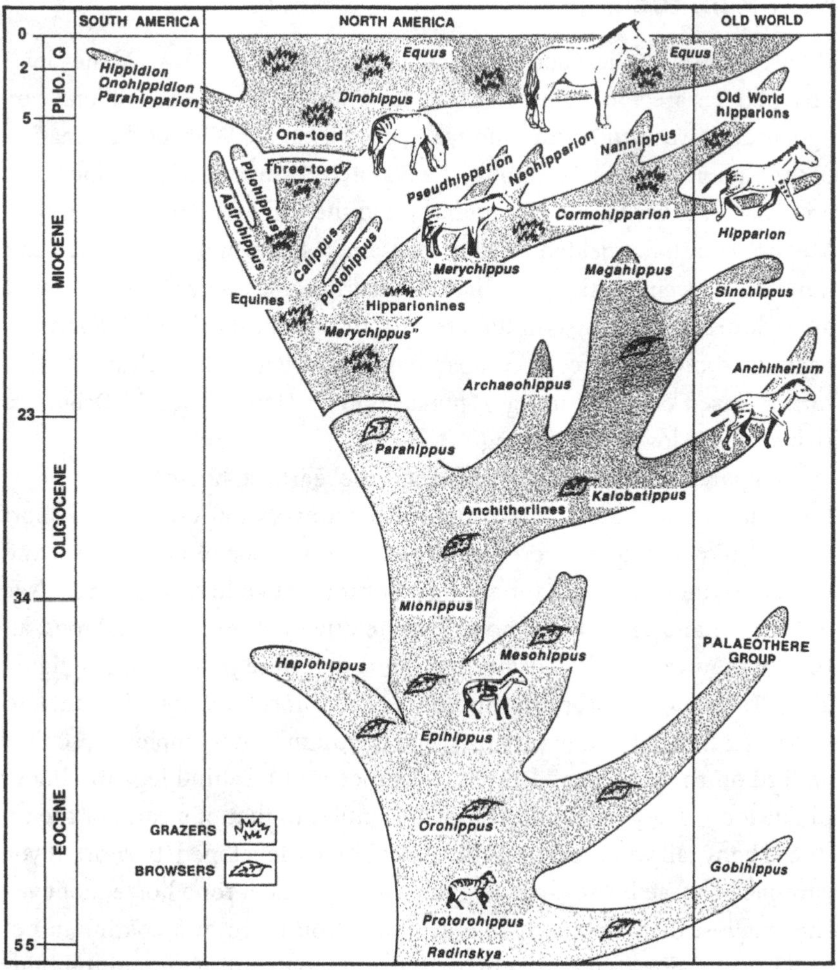

Figure 22.3 ▲

A more modern diagram of the evolution of horses, showing a branchy bushy tree. (Drawing by C. R. Prothero, after Donald R. Prothero, "Mammalian Evolution," in *Major Features of Vertebrate Evolution*, ed. Donald R. Prothero and Robert M. Schoch [Knoxville, Tenn.: Paleontological Society, 1994])

species of *Mesohippus* and two of *Miohippus* were contemporaries, all found at the same level in the same beds in the Big Badlands of South Dakota and equivalent rocks in Wyoming and Nebraska.

"DAWN HORSE"

What about the earliest horses? What did they look like? How did they live? Fossil horses are very common in the lower Eocene (55 to 48 million years ago) beds of western North America, especially in the Willwood Formation of the Bighorn Basin in Wyoming, the Wasatch Formation in the Wind River and Powder River basins in Wyoming, and the San Jose Formation in New Mexico. They have yielded literally thousands of jaws and teeth, as well as a handful of decent partial skeletons. Most early horses were about the size of a beagle or fox (250 to 450 millimeters [10 to 18 inches] in height). For many years, textbooks incorrectly compared their size with the much smaller fox terrier, based on copycatting a publication by Henry Fairfield Osborn, a rich man who loved fox hunting.

Compared with their descendants, the earliest horses had a short head and snout, a small brain, and teeth with very low crowns and short roots. The cheek teeth were composed of a number of cross-crests and low cusps, an adaptation for eating soft browse like leaves and fruit (see figures 22.1 and 22.2). These horses had relatively short limbs and toes, although they ran on the tips of their digits and were good jumpers (figure 22.4). They had the primitive number of four short toes on their front legs (although the pinkie was very tiny, and the thumb lost completely, so they walked on three toes) and three short toes on their hind legs (no big or pinkie toe). They had a long bony tail similar to that of a cat, not the reduced bony tail with long hairs that later horses developed. In short, if you saw one of these horses, you would never mistake it for a horse, not even the smallest dwarf pony. It might remind you more of a coatimundi or another non-horse-like mammal, although no extant mammal remotely resembles it.

The fossil evidence shows that these tiny horses were exquisitely adapted for life in the dense jungles of the super-greenhouse world of the early Eocene. At that time, there was so much carbon dioxide in the atmosphere that the poles were relatively mild and warm enough to support alligators and crocodiles (even though it was dark for six months a year), as well as horses and tapirs. Places such as Montana and Wyoming, whose rocks have yielded these fossils, looked nothing like they do now. Today they are barren steppes with huge snows and long months of subfreezing winter temperatures, but they were tropical forests in the Eocene.

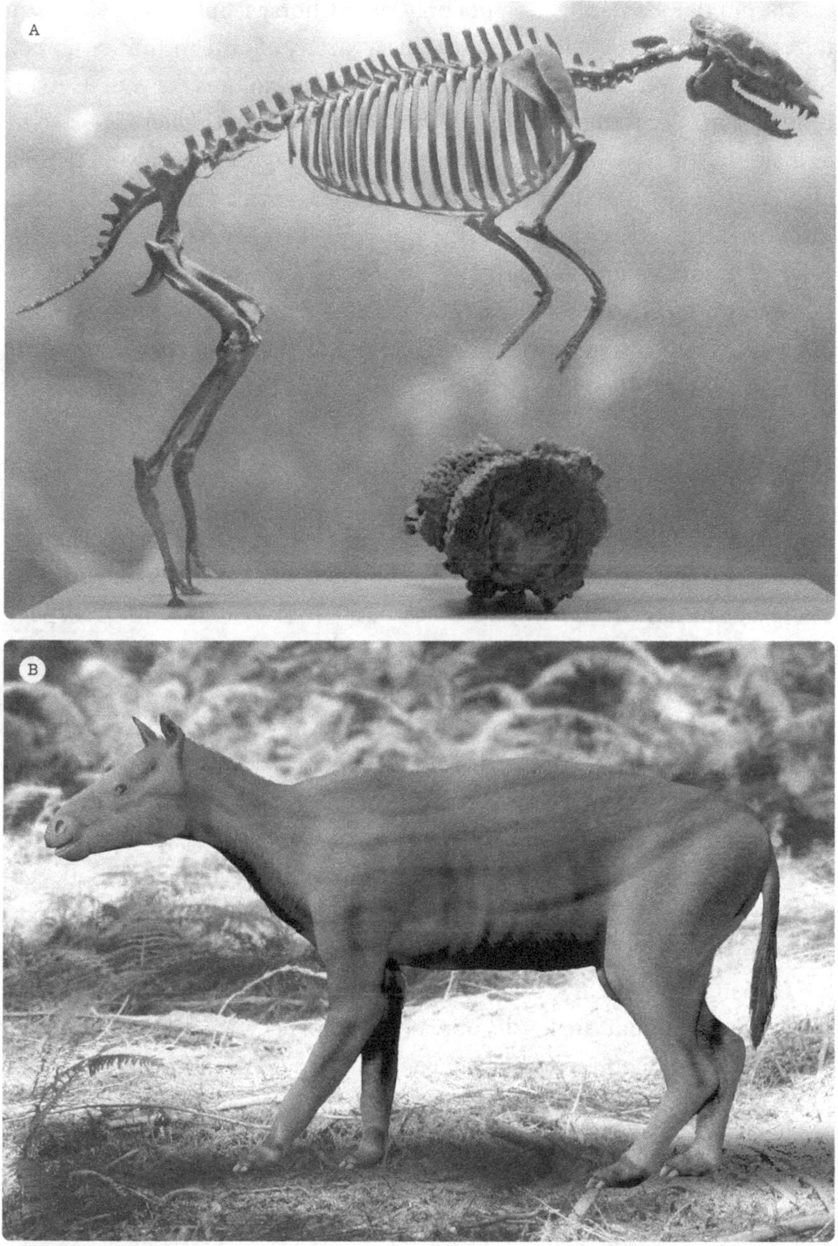

Figure 22.4 ▲

Early Eocene horse of North America: (*A*) mounted almost complete skeleton; (*B*) reconstruction of its appearance in life. ([*A*] Photograph courtesy Smithsonian Institution; [*B*] courtesy Nobumichi Tamura)

The jungles were inhabited not only by tiny horses, but also by abundant tapirs, rhinos that resembled horses, and a variety of other primitive hoofed mammals. The treetops were full of lemur-like primates, as well as many other groups of arboreal mammals that now are extinct. There were some archaic mammalian predators, but they were no bigger than a wolf. In their absence, the top predator of the Eocene jungles was a 2.5-meter (8-foot) tall flightless bird with a huge beak and tiny wings, known as *Diatryma* in the Rocky Mountains and *Gastornis* in Europe. Its prey in Europe were members of a group closely related to horses called palaeotheres, such as *Palaeotherium*. They were not true horses related to the main North American lineage, but they filled the horse role in the early Eocene of Europe.

WHAT'S IN A NAME?

The biggest dilemma about these creatures is what to call them. The first name given to Eocene horses in North America was *Orohippus angustidens*, conferred by Marsh in 1875, based on badly broken tooth and jaw specimens from the middle Eocene of New Mexico. Then in 1876, Marsh named some of his early Eocene fossils *Eohippus* (Greek for "dawn horse"), based on the species *E. validus*, which was represented by a good partial skeleton. Many other good specimens have since been added to the genus *Eohippus*. It soon became evident that the early Eocene *Eohippus* was not the same as the middle Eocene *Orohippus*, so the name *Eohippus* became established for early Eocene horses. It became widely used in the early twentieth century, so the name *Eohippus* appears in nearly all the older diagrams of horse evolution (including those in public use, especially in textbooks).

Then in 1932, Sir Clive Forster Cooper of the British Museum of Natural History noticed that the fossils of American horses were extremely similar to a fossil described by Richard Owen in 1841. Found in from the London Clay, which dates to the early Eocene, the specimen was called *Hyracotherium* (hyrax beast). Because *Hyracotherium* had been named 35 years before *Eohippus*, by the rule of priority in the International Code of Zoological Nomenclature, it became the valid name for this early horse—if, indeed, *Eohippus* and *Hyracotherium* are one and the same. This opinion was enforced by the brilliant paleontologist George Gaylord Simpson in 1951, and so became widely accepted. For most of the rest of the twentieth century, all early Eocene horses from North America and Europe were lumped into *Hy-*

racotherium. That name, also, is still found in many books and other media that have not kept up with the ever-changing science.

But science marches on, new and better specimens are found, and the philosophy of fossils are named changes as well. In the early twentieth century, paleontologists were taxonomic "splitters," conferring a new genus and species name on nearly every fossil they found, no matter how tiny the differences between them. Then in the 1930s and 1940s, paleontologists and biologists began to look at the normal range of variability of natural animal populations in the wild, and soon realized that a lot of characteristics that had been used to justify new species were just normal variants in a single species. That kind of "population thinking" prevailed from the 1940s onward, and most paleontologists still prefer to place many slightly different fossils in the same species, especially if there is no strong evidence from their anatomy, their distribution in space and time, or the statistical measures of normal species variability that justifies their classification as different animals in modern biological terms.

But in recent years, the larger number of specimens, and especially the better specimens with anatomy not previously known, has forced paleontologists to reexamine fossils that had been relegated to "taxonomic wastebaskets." According to the newer thinking in classification (called cladistics), wastebaskets have no evolutionary meaning, nor are they natural groups of organisms, and thus should not be recognized by a formal taxonomic name. For example, some people use the word "fish" for a grouping of all vertebrates that are not four-legged (tetrapods). However, lungfish are much more closely related to tetrapods than they are to bony fish, and bony fish are much closer to humans than they are to jawless fish. Thus modern taxonomy no longer uses a general term like "fish" or "Pisces" because it reflects a common ecology, not a natural group with its own distinct evolutionary history.

Sure enough, the reexamination of earlier fossils and the discovery of many better fossils have blasted apart the idea that all European and North American early Eocene horses should be classified as *Hyracotherium*. First, in 1989 Jeremy Hooker of the Natural History Museum in London looked at all the *Hyracotherium* fossils from the London Clay, did a new analysis, and decided that they were not horses, but European palaeotheres. Thus the name *Hyracotherium* can no longer be used to conveniently lump all the early Eocene horses from North America. (A handful of scientists do not accept this conclusion, but not based on evidence or reasoned analysis. They have thought of North American horses as *Hyracotherium* for so long that they cannot break the habit.) Then in 2002, David Froehlich of the University of Texas did a careful analysis of all the American early Eocene horses. He found that no genus name can be applied to all of them because they had been united by primitive characteristics into one gigantic taxonomic wastebasket. The name *Eohippus* can be revived, but only for Cope's species *angustidens* and Marsh's species *validus*. But most of the specimens of early Eocene horses long called *Hyracotherium* or *Eohippus* cannot be referred to either of these genera, but belong to new genera or to former genera that can be resurrected. For example, Jacob Wortman's genus *Protorohippus*, named in 1894, is the proper one for species of more advanced horses such as *montanum* and *venticolum*. Froehlich established some new horse genera, such as *Sifrhippus* for the dwarfed earliest Eocene horse *sandrae*, *Minippus* for the species *index* and *jicarillai*, and *Arenahippus* for the species *grangeri*, *aemulor*, and *pernix*. And some species, like Cope's *tapirinum*, were not horses, but were related to other perissodactyls, including tapirs, and are now called *Systemodon tapirinum*.

Thus there is no single genus for early Eocene horses that would make it easy to remember their names and to label diagrams. We cannot just call them all *Eohippus* and let it go at that, because that is factually incorrect and grossly oversimplified. Nature is much more complicated and diverse than our simplistic thinking and diagrams, and we must change our views to reflect more recent research—just as we cannot use the long-incorrect name "Brontosaurus" or call Pluto a planet. So every diagram that illustrates the evolution of horses or every textbook section on horses is wrong if it uses only one genus name for early Eocene horses, whether it be *Eohippus* or *Hyracotherium*. A modern diagram should list at least *Protorohippus*, *Sifrhippus*, *Minippus*, and *Arenahippus* if it is to reflect current knowledge.

WHENCE THE HORSE?

For the final project in my undergraduate vertebrate-paleontology class, the professor, Michael Woodburne, gave each member of the class a big mixed sample of bones of the earliest Eocene mammals from the Bighorn Basin near Emblem, Wyoming. Our job was to sort them, identify them using the scientific literature, and create a list of what species we had. It was a difficult task because at that time there was almost no up-to-date classification on the early Eocene mammals. This changed with the flood of papers by Philip Gingerich, Kenneth Rose, David Krause, and Thomas Bown on the mammals from the Bighorn Basin published since the late 1970s. It would have been *so* much easier if these papers had been published by 1975!

What I remember most about the project was that my tray was full of jaws of early perissodactyls, especially of horses (whatever name they would be given today) and the tapir relative *Homogalax*. I found it devilishly hard to tell them apart, even though my youngest son could distinguish a tapir from a horse when he was two years old! Today, their teeth and their entire anatomy are very distinct, but 55 million years ago they were virtually identical in their teeth and in most of their skull and skeleton (figure 22.5). Only a subtle difference or two (particularly on how continuous the cross-crests are in horses versus tapirs) distinguished them, and it took practice and a good eye to see that difference.

Once you look around at the rest of the perissodactyls of the early Eocene, the trend is even more striking. The earliest ancestor of the rhinoceros, known as *Hyrachyus*, is barely distinguishable from the early tapirs and horses, even though rhinos, tapirs, and horses look nothing like one another today. The early members of an extinct rhino-like group of mammals, the brontotheres, are also very similar to the early rhinos, tapirs, and horses. In other words, the hugely diverse modern perissodactyls can be traced back to an early Eocene common ancestor that looked nothing like its modern descendants. Then, through evolutionary divergence, its descendant lineages—once similar in appearance—diverged from one another and became increasingly different over time until they are easy to distinguish. Indeed, by the middle Eocene, the horse, tapir, rhino, and brontothere lineages were distinct, and even a child could have told them apart, even though none of their members looked like any of their modern descendants.

But where did the horses and their perissodactyl kin come from? For the longest time, paleontologists pointed to a group of archaic hoofed mammals that were common in the Paleocene and early Eocene: the phenacodontids. Their teeth were very similar to those of early perissodactyls, and their skulls and skeletons had all the features that could have served to identify them as the common ancestor of the perissodactyls. But in 1989, Malcolm McKenna and three Chinese co-authors described a newly discovered fossil from the late Paleocene of Mongolia, about 57 million years old. They named it *Radinskya* after Leonard Radinsky, one of the giants of early perissodactyl research who had died in 1986 (see figure 22.5). It looks just

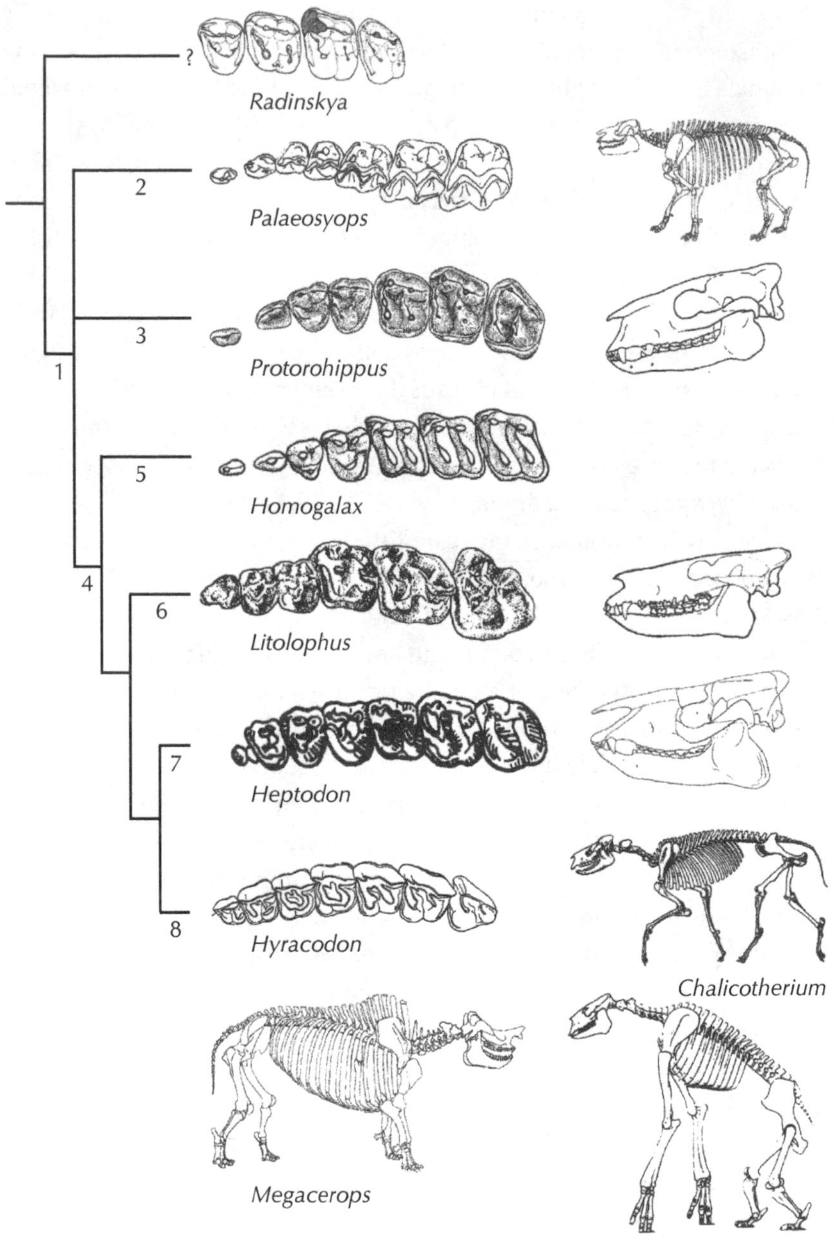

Figure 22.5 ▲

The radiation of primitive perissodactyls. (Modified from Donald R. Prothero, *Evolution: What the Fossils Say and Why It Matters* [New York: Columbia University Press, 2007], fig. 14.5) like a very tiny horse, except that it is even more primitive than the earliest horse. McKenna and his colleagues were stymied by the primitive nature of the fossil and were not sure whether to classify it as a perissodactyl or assign it to another mammal group that is closely related to the perissodactyls. Since then, most scientists have agreed that it is proof that the rapid evolution of perissodactyls in the early Eocene of North America and Europe was not because they evolved from a local phenacodontid. Rather, perissodactyls arrived in North America and Europe from Asia around 55 million years ago, and then quickly began to diversify and drive most of the native archaic hoofed mammals (including their close relatives, the phenacodontids) to extinction by the end of the middle Eocene.

The excellent fossil record of horses demonstrates not only the original similarity and divergent evolution of the perissodactyls, beginning in Asia, but also the evolution of horses in North America and their disappearance from the Western Hemisphere in the Pleistocene—only to return home in the late fifteenth century.

SEE IT FOR YOURSELF!

Many museums in the United States have displays that show the evolution of horses, usually with an early Eocene horse, an Oligocene *Mesohippus* from the White River Badlands of South Dakota, a few Miocene horses, and a Pleistocene *Equus*. They include the American Museum of Natural History, New York; Field Museum of Natural History, Chicago; Florida Museum of Natural History, University of Florida, Gainesville; National Museum of Natural History, Smithsonian Institution, Washington, D.C.; and Natural History Museum of Los Angeles County, Los Angeles.

FOR FURTHER READING

Franzen, Jens Lorenz. *The Rise of Horses: 55 Million Years of Evolution*. Translated by Kirsten M. Brown. Baltimore: Johns Hopkins University Press, 2010.

MacFadden, Bruce J. *Fossil Horses: Systematics, Paleobiology, and Evolution of the Family Equidae*. Cambridge: Cambridge University Press, 1994

Prothero, Donald R., and Robert M. Schoch, eds. *The Evolution of Perissodactyls*. New York: Oxford University Press, 1989.

——. *Horns, Tusks, and Flippers: The Evolution of Hoofed Mammals and Their Relatives*. Baltimore: Johns Hopkins University Press, 2002.

RHINOCEROS GIANTS

All of us had realized that the Beast of Baluchistan was a colossal creature. But the size of the bones left us absolutely astounded. We had brought in only the front of the skull with several teeth. But that was enough for Dr. Granger. "I'm sure," he said, "that the Beast is a giant, hornless rhinoceros. It isn't like any other animal known to science."

ROY CHAPMAN ANDREWS, *ALL ABOUT STRANGE BEASTS OF THE PAST*

QUICKSAND!

In 1922, the famous paleontologist Henry Fairfield Osborn, director of the American Museum of Natural History and a leading figure in science and society at the time, sent an expedition to Mongolia to find fossils of the earliest human ancestors. Osborn (wrongly) thought that humans had evolved in Asia, so he used this pitch to raise money from rich donors and trustees of the museum. The expedition was mounted in grand style, with a caravan of 75 camels (each carrying 180 kilograms [400 pounds] of gasoline or other supplies), three Dodge touring cars, two Fulton trucks, and a large party of scientists and helpers. It was led by the legendary Roy Chapman Andrews, a daring explorer and adventurer whom many people believe was the model for the film character Indiana Jones. Osborn told Andrews, "The fossils are there. I know they are. Go and find them."

The expedition left Beijing, passed through the Great Wall of China, and soon became famous for the amazing fossils of Cretaceous dinosaurs that Andrews and his colleagues found, including the first known nest of dinosaur eggs. But despite their great success at finding fossils, they never discovered evidence of the oldest humans in Asia. This was because Osborn was wrong (and Darwin was right): humans had evolved in Africa. Ironically, the first really ancient fossil human (*Australopithecus africanus* [Taung child; chapter 25]) was found in South Africa in 1924, just when Osborn and Andrews were begging for more money from rich donors to find early humans in Asia—but Osborn, like most scientists of his time, rejected the fossil as just a juvenile ape and of unknown age.

In addition to the spectacular fossils of dinosaurs, museum paleontologist Walter Granger and his Chinese helpers found many very important and impressive fossil mammals. As Andrews wrote in his colorful book (with the very un-politically-correct imperialist title) *The New Conquest of Central Asia* about his third expedition in 1925:

> The credit for the most interesting discovery at Loh belongs to one of our Chinese collectors, Liu Hsi-ku. His sharp eyes caught the glint of a white bone in the red sediment of the steep hillside. He dug a little and then reported to Granger, who completed the excavation. He was amazed to find the foot and lower leg of a *Baluchitherium standing upright*, just as if the animal had carelessly left it behind when he took another stride [figure 23.1]. Fossils are so seldom found in this position that Granger sat down to think out the why and wherefore. There was only one possible solution. Quicksand! It was the right hind limb that Liu had found; therefore, the right front leg must be farther down the slope. He took the direction of the foot, measured off about nine feet, and began to dig. Sure enough, there it was, a huge bone, like the trunk of a fossil tree, also standing erect. It was not difficult to find the two limbs of the other side, for what had happened was obvious. When all four legs were excavated, each one in a separate pit, the effect was extraordinary [see figure 23.1]. I went up with Granger and sat down upon a hilltop to drift in fancy back to those far days when the tragedy had been enacted. To one who could read the language, the story was plainly told by the great stumps. Probably the beast had come to drink from a pool of water covering the treacherous quicksand. Suddenly it began to sink. The position of the leg bones showed that it had settled slightly back upon its haunches, struggling desperately to free itself from the gripping sands. It must have sunk rapidly, struggling to the end, dying only when the choking sediment filled its nose and throat. If it had been partly buried and died of starvation, the body would have fallen on its side. If

Figure 23.1 ▲

The leg bones of *Paraceratherium* were found upright, as they had been buried when the animal was trapped in quicksand. (Negative no. 285735, courtesy American Museum of Natural History Library)

> we could have found the entire skeleton standing erect, there in its tomb, it would have been a specimen for all the world to marvel at.
>
> I said to Granger, "Walter, what do you mean by finding only the legs? Why don't you produce the rest?" "Don't blame me," he answered, "it is all your fault. If you had brought us here thirty-five thousand years earlier, before that hill weathered away, I would have the whole skeleton for you!" True enough, we had missed our opportunity by just about that margin. As the entombing sediment was eroded away, the bones were worn off bit by bit and now lay scattered on the valley floor in a thousand useless fragments. There must have been great numbers of baluchitheres in Mongolia during Oligocene times, for we were finding bones and fragments wherever there were fossiliferous strata of that age.

Andrews's story is colorful, but the details are probably quite different. Unlike movie-style quicksand, which rapidly sucks down victims until they are below the surface, real quicksand is just regular sand saturated with water. When pressure is put on it, quicksand liquefies, so it will flow around victims' legs as they sink. But it's still mostly water, so a person or an animal cannot sink any deeper than either would in a swimming pool when floating. To get out of quicksand, it is necessary to lie flat (as if floating in water) and get traction by grasping a rope or stick held by someone outside the quicksand.

The creature caught in the quicksand probably sank no deeper than its legs and belly. Then the quicksand would have hardened and the creature would have died of thirst. The rest of the body probably was not immersed in the quicksand, but was an easy meal for the scavengers that attacked the dying or dead animal whose legs were trapped.

MONSTERS OF MONGOLIA

The baluchitheres that Andrews and Granger discussed were the gigantic hornless rhinoceroses that now are known as *Paraceratherium*. Some isolated teeth were initially found in 1907, but the first decent specimens were discovered in 1910 by the British paleontologist Clive Forster Cooper in the Baluchistan region of present-day Pakistan. The specimens were just a few broken skulls and jaws and a few bones, though, so the enormity of the creature was not yet appreciated. In 1913, Forster Cooper gave the name *Baluchitherium osborni* to a more complete skull from his collections, and this name remained popular for decades. Four years later, the Russian paleontologist Aleksei Alekseeivich Borissiak named another skeleton—the most complete one yet found—*Indricotherium*, after the Indrik region of the Soviet Union, north of the Aral Sea (northwestern Kazakhstan) (figure 23.2).

The three names were in wide use for decades, even though scientists had long rejected the popular name *Baluchitherium*, which Osborn favored and promoted (because one species in that genus was named after him), since it is clearly another specimen of *Paraceratherium* and was named later, so it is a junior synonym. In 1989, Spencer Lucas and Jay Sobus showed that there was only one, highly variable, population of these animals; therefore, the oldest name, *Paraceratherium*, takes precedence over the newer *Baluchitherium* and *Indricotherium*. Most paleontologists working on these fossils agree that *Paraceratherium* is unlikely to be anything more than one genus with at most three or four species, since animals this large require huge home ranges in order to find enough food to support their enormous

Figure 23.2 ▲

The only relatively complete skeleton of *Paraceratherium*, displayed at the Yuri Orlov Paleontological Museum in Moscow. (Courtesy M. Fortelius)

bodies. It is extremely unlikely on ecological grounds that there were multiple genera and species of such a huge animal covering enormous areas in the same region at the same time.

Whatever the name of these incredible beasts, they roamed widely across Asia during the Oligocene to early Miocene, from about 33 million to about 18 million years ago. Their fossils have been found not only in Mongolia and Pakistan, but also in several places in China, Kazakhstan, and, more recently, Turkey and Bulgaria.

Paraceratherium is the largest land mammal ever found (figure 23.3). It was 4.8 meters (16 feet) tall at the shoulders, 8 meters (26 feet) long, and heavier than any elephant or mastodont. Original estimates put its weight at more than 34 metric tons (37 tons), but more recent methods of calculation place it around 20 metric tons (22 tons), just a bit heavier than the biggest elephant relatives that ever lived: the deinotheres.

Figure 23.3 ▲

Reconstruction of *Paraceratherium* in fiberglass, originally displayed at the University of Nebraska State Museum in Lincoln, and now at the Riverside Discovery Center in Scottsbluff, Nebraska. The modern elephant (*right background*) can be compared for scale, and the reconstruction at its feet (*right foreground*) is of *Hyracodon*, from which *Paraceratherium* evolved. (Courtesy University of Nebraska State Museum)

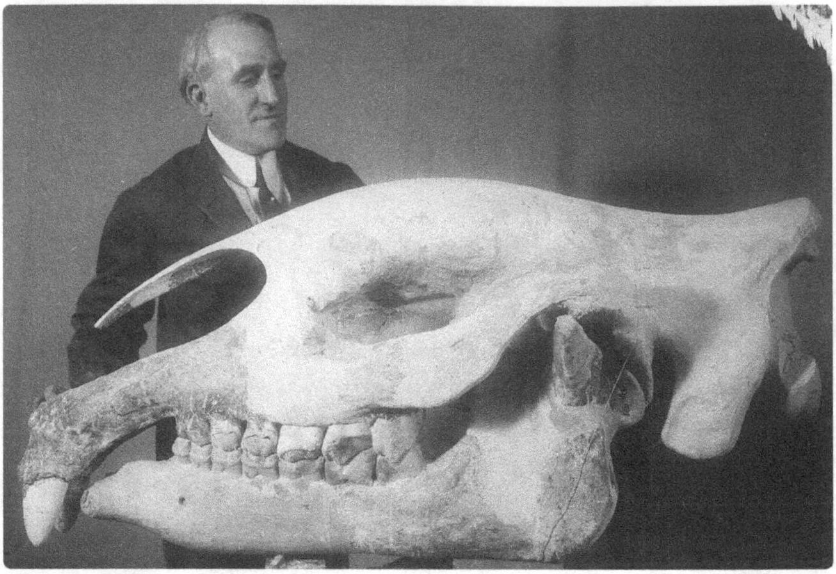

Figure 23.4 ▲

The immense skull of *Paraceratherium*, found in Mongolia by members of the expedition mounted by the American Museum of Natural History in 1922. (Negative no. 310387, courtesy American Museum of Natural History Library)

Paraceratherium had a huge skull, over 2 meters (6 feet) long, with a short proboscis, or trunk (judging from the deep nasal opening on the skull); a prehensile lip for stripping vegetation from branches; and relatively primitive low-crowned teeth that were suitable for only munching leaves, not eating gritty grasses (figure 23.4). There is no indication on the top of the skull that a horn once attached there, so it was hornless (as were most extinct rhinos). Most reconstructions make it look just like a rhinoceros scaled up, with simple rhino ears, but I have argued that its huge body mass would have hindered its ability to lose body heat. It would have needed a large heat radiator (such as the ears of elephants) in order to cool down.

Even though *Paraceratherium* was larger than any elephant, it had relatively long wrist and ankle bones because it had evolved from a group of rhinos (the hyracodonts) that had long limbs and toes, adapted for running (figure 23.5; see figure 23.3). Despite its great size and weight, it never adopted the limb proportions typical of sauropod dinosaurs and elephants, whose toe bones are very short and mostly squashed flat by the huge weight

Figure 23.5 ▲
Reconstruction of the Great Dane–size rhino *Hyracodon*, the ancestor from which *Paraceratherium* evolved. (Drawing by R. Bruce Horsfall)

they bear. In short, *Paraceratherium* was a rhinoceros trying to take over the giraffe niche of treetop-leaf browsing, but doing so by enlarging everything, not just the limbs and neck, as did giraffes.

RHINO ROOTS

Paraceratherium was part of an enormous evolution of rhinoceroses during the Cenozoic. The oldest rhinos come from lower Eocene beds (about 52 million years old), and they are extremely similar to the earliest tapirs and horses (see figure 22.5). But by the late middle Eocene (40 million years ago), rhinoceroses had diverged into three families. One extinct family, the Amynodontidae, was typically adapted for a semi-aquatic life, like hippos, complete with the huge hippo-like skull and jaws and the fat body with short limbs.

The second extinct family is the Hyracodontidae, or "running rhinos." Members of this family were common in the late Eocene of North America and Asia. The best known is the White River Badlands species, *Hyracodon nebraskensis* (see figure 23.5). It was about the size of a Great Dane, but had elongated limbs and hand and foot bones. These bones indicate that it was a faster runner than any rhino, before or since. The hyracodontids continued to thrive in Asia, where they became bigger and more advanced. A slightly larger hyracodontid was the Chinese genus *Juxia*, which was the size of a large horse, with a long neck and long legs as well. These rhinos culminated with the elephant-size *Urtinotherium* and, finally, with *Paraceratherium.*

The third family, the Rhinocerotidae, is that of the extant rhinoceroses of Africa and Asia. In the past, it included dozens of genera and species, evolving rapidly in both Eurasia and North America (and eventually reaching Africa). The rhinoceroses in North America died out about 5 million years ago, at the end of the Miocene, but rhinos carried on in Eurasia until most went extinct at the end of the last Ice Age, about 20,000 years ago. Today, there are only five species in four genera: the two African rhinos, the Indian rhino, and the nearly extinct Javan and Sumatran rhinos. All five species are extremely endangered because of the heavy poaching for their horn, which is used in traditional Chinese medicine. (It has no medicinal value at all, since rhino horns are made of compacted hairs glued together. The chemistry of rhino horn is about the same as that of human hair and nails.) The poaching is so severe that rhinoceroses will be extinct in the wild in another few years despite the best efforts to protect them. This is because powdered rhino horn is more valuable once-per-ounce than gold or cocaine.

BIOLOGY OF THE MONSTER RHINOS

Because *Paraceratherium* was slightly larger than elephants, we can infer a lot about its lifestyle and biology by using the elephant as an analogue. Small herds probably roamed huge areas to find enough food to sustain their gigantic body, and they likely stripped bare the tops of trees as they fed. Based on studies of elephants and biomechanical constraints, they walked slowly, never moving faster than about 30 kilometers (18 miles) per hour, and more often at 10 to 19 kilometers (6 to 12 miles) per hour. Their long legs, however, allowed them to cover a lot of ground even at that leisurely pace. Their height and size indicate that they had a slow pulse rate (elephant hearts beat only 30 times per minute). Large body size also predicts greater longevity, so they would have had a life spans on the order of that of modern elephants (50 to 70 years might be typical).

Paraceratherium populations would have remained quite small, and females would have borne a calf every other year or so. The calf may have taken a decade to mature to an adult. *Paraceratherium* probably spent much of the hot daylight hours resting in the shade or bathing in water holes to manage body heat, and feeding almost nonstop during the cooler hours of evening, night, and early morning. Like modern horses, rhinos, and elephants, it had relatively inefficient hindgut fermentation of food. It did not have the more efficient four-chambered ruminating stomach of cattle, sheep, goats, antelopes, giraffes, deer, and their kin. Consequently, like horses and elephants, *Paraceratherium* would have eaten a huge amount of forage each day, but digested very little of it compared with the ruminating mammals.

Its large body size may have created some problems (especially with managing heat), but conveyed advantages as well. Like modern elephants, adult *Paraceratherium* would have had no fear of predators, because they were too large to be successfully attacked. Most of the predators known from the Oligocene beds of Asia were smaller than wolves, so none of them could have tackled an adult. Calves, however, were vulnerable. Like elephants, *Paraceratherium* probably lived in small female-dominated herds composed of the matriarch, plus sisters, daughters, and nieces. They all shared in keeping the calves and young safe until they were large enough to no longer be targets of predation.

LAND OF THE GIANTS

The habitats of Mongolia and China during the Oligocene were a very peculiar ecosystem in many ways. Most of the localities that yield fossils of *Paraceratherium* are dominated by those of rodents and rabbits, suggesting an environment with few resources for medium-size grazers, but abundant resources for small burrowers. It was apparently mostly an arid semi-desert scrubland, so there were few large areas of grass and thus very few mammals that lived on grasses. Instead, the gigantic *Paraceratherium* browsed on treetops, and only a few medium-size antelope-like species fed on brush (in contrast to the great abundance of smaller herbivores in modern savanna-grassland habitats). Since the predators were relatively small and no match for an adult *Paraceratherium*, they certainly scavenged any carcasses of *Paraceratherium* or other animals they found.

The cause of the extinction of *Paraceratherium* is widely debated, but two events likely contributed. For most of the Oligocene and early Miocene, *Paraceratherium* was unchallenged in its habitat, with no large predators or competing large herbivores. Then, about 20 to 19 million years ago, the first mastodonts left their African homeland and migrated across Eurasia. They almost immediately reached North America as well. Modern elephants (and their prehistoric relatives) have a huge effect on their environment. On the modern Africa savanna, they topple trees and open up dense stands of forest to foster a much greater variety of vegetation. Without them, the trees would grow unimpeded. It is likely that the arrival of mastodonts in Eurasia led to the widespread disruption of forest vegetation and may have destroyed much of the mature forest that *Paraceratherium* required. In addition, the predators of mastodonts followed their prey to Eurasia. They included the bear-size amphicyonids ("bear dogs," an extinct family not related to either bears or dogs) and the huge *Hyaenailouros*. Such big predators were able to take down large mastodonts and probably were too much for *Paraceratherium*, long free from large predators, to handle. Whatever the reasons, the paraceratheres vanished soon after the mastodonts and their large predators arrived in Eurasia.

INDRICOTHERES IN THE MEDIA

As the largest land mammal ever found, *Paraceratherium* has been a popular subject in many media. It was the featured animal in an entire episode of *Walking with Prehistoric Beasts* that focused on the Oligocene of eastern Asia. The animators did an excellent job of inferring what they could about the behavior of these creatures. However, we have only their bones and some knowledge of the basic constraints on their biology from using elephants as analogues. In fact, we have no detailed evidence for what color they were, what they sounded like, or how they behaved.

Apparently, the slow steady motion of these towering beasts had another effect as well. Animator Phil Tippett, who designed most of the sets and props for *The Empire Strikes Back*, the second film in the *Star Wars* saga, was apparently inspired by *Paraceratherium* when he created the models of the towering AT-AT walkers, which lumber along as they prepare to attack the rebels on the ice planet Hoth.

SEE IT FOR YOURSELF!

The only nearly complete skeletons of *Paraceratherium* are displayed at the Yuri Orlov Paleontological Museum, in Moscow, and at several large museums in China. The best skull of *Paraceratherium* is at the American Museum of Natural History, in New York, while others are exhibited at the Natural History Museum, in London, and the Sedgwick Museum of Earth Sciences, Cambridge University.

For many years, a fiberglass reconstruction of *Paraceratherium* was on display at the University of Nebraska State Museum in Lincoln (see figure 23.3). It has found a new home at the Riverside Discovery Center in Scottsbluff, Nebraska.

FOR FURTHER READING

Prothero, Donald R. *Rhinoceros Giants: The Paleobiology of Indricotheres*. Bloomington: Indiana University Press, 2013.

THE APE'S REFLECTION?

The next time you visit a zoo, make a point of walking by the ape cages. Imagine that the apes had lost most of their hair, and imagine a cage nearby holding some unfortunate people who had no clothes and couldn't speak but were otherwise normal. Now try guessing how similar those apes are to us in their genes. For instance, would you guess that a chimpanzee shares 10 percent, 50 percent, or 99 percent of its genetic program with humans?

JARED DIAMOND, *THE THIRD CHIMPANZEE*

THE APE'S REFLECTION?

The subject of human evolution along with the rest of the animal kingdom has always been contentious and emotional. For religious reasons, even today a significant minority of Americans reject the idea that humans are related to the rest of the animal kingdom or that they are just another animal species—even though this fact is not controversial in almost any other modern developed nation in the world. Yet some polls show that a high percentage of Americans accept the idea of evolution in plants and other animals—just not humans.

Ironically, so much research and interest have been focused on the evolution of humans that it is one of the best-supported examples of evolution of all. An entire branch of anthropology (physical anthropology and human paleontology) is devoted to the fossil record of our nearest relatives. Thousands of scientists worldwide are working on an array of research topics in this field—far more than study dinosaurs or any other prehistoric creatures.

Literally hundreds of thousands of specimens of fossil hominins (members of the human subfamily, or the subfamily Homininae) are stored in museums all over the world. The number of specimens is so overwhelming, and the wealth of detail about human evolution is so impressive, that if we were talking about any other species on the planet, it would be a slam-dunk case of evolution, as well documented as that of any family of animals. But so many people hold nonscientific objections to the idea that it receives unfair scrutiny, is distorted, and is denied outright. If the same volume of overwhelming evidence were brought to bear on any other issue, there would be no controversy at all.

But even if we did not have the incredible fossil record of humans, the evidence is *still* overwhelming. All we have to do is look in a mirror. As early as 1735, the founder of modern classification, Carolus Linnaeus, gave humans the scientific name *Homo sapiens* (thinking man) and diagnosed our species with the Greek phrase "Know thyself." In 1766, Georges-Louis Leclerc, Comte de Buffon, wrote in volume 14 of *Histoire naturelle* that an ape "is only an animal, but a very singular animal, which a man cannot view without returning to himself." Other French naturalists like Georges Cuvier and Étienne Geoffroy Saint-Hilaire commented on the extreme anatomical similarity of apes and humans, although they refused to actually say that humans are a kind of ape. The pioneering French biologist Jean-Baptiste Lamarck explicitly argued in *Philosophie zoologique* in 1809:

> Certainly, if some race of apes, especially the most perfect among them, lost, by necessity of circumstances, or some other cause, the habit of climbing trees and grasping branches with the feet, . . . , and if the individuals of that race, over generations, were forced to use their feet only for walking and ceased to use their hands as feet, doubtless ... these apes would be transformed into two-handed beings and . . . their feet would no longer serve any purpose other than to walk.

The issue clearly was a critical one when Charles Darwin published *On the Origin of Species* in 1859. His book was already controversial, so he tried his best to downplay the issue of human evolution. In the entire book, he wrote only one phrase: "Light will be shed on the origin of man, and his history." Although Darwin was reluctant to say more at that time, his supporter Thomas Henry Huxley jumped into the breach and in 1863 published *Evidence as to Man's Place in Nature*. In this book, Huxley described and

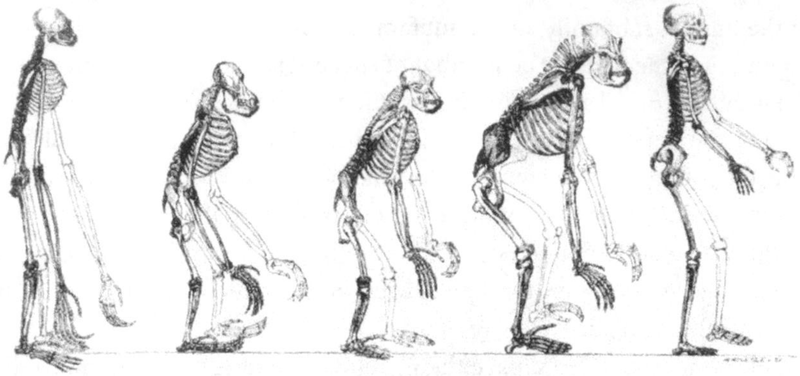

Figure 24.1 ▲

Benjamin Waterhouse Hawkins's illustration of the extreme bone-by-bone similarity of the skeletons of apes and humans. (From Thomas Henry Huxley, *Evidence as to Man's Place in Nature* [London: Williams & Norgate, 1863])

illustrated the detailed anatomical similarity of every bone and muscle and organ in the great apes and in humans (figure 24.1). Finally, in 1871, Darwin published his own thoughts in *The Descent of Man*, although he focused mostly on topics such as sexual selection, without even mentioning fossils. At that time, there was still no fossil evidence for human evolution (other than Neanderthals, who had been misinterpreted).

Jump forward in time 90 years. Unbeknownst to Darwin or any other biologist before the 1960s, another source of data clearly shows our relationships to apes and the rest of the animal kingdom: DNA. Some of the very first molecular techniques demonstrated that human DNA and chimp and gorilla DNA are extremely similar. When the serum of antibodies of humans and of apes is put in the same solution, the immune reactions are much stronger than those with humans and any other animal, suggesting that the immunity genes of humans and of apes are most similar.

Then in the late 1960s, a technique called DNA-DNA hybridization was developed. A solution of DNA of an ape and a human is heated until the two strands of the double helix unzip. Then the mixture is cooled, and each strand binds to the nearest strand, creating some DNA with one strand from the human and the other from the ape. (Some of the strands of the ape's DNA bind to other ape strands, and some of the strands of the human's DNA bind to other human strands, but of greatest interest are the double helices of hybrid DNA.) When the solution with the hybrid DNA is reheated, the more tightly bonded the hybrid strands (which reflects how similar they are), the higher the temperature needed to unzip them. Doing this with the DNA of chimps, gorillas, other apes, plus monkeys, lemurs, and nonprimate animals gives a rough measure of how similar each is to humans—and, once again, chimp DNA is virtually identical to human DNA.

Then, in the past 20 years, technological leaps like the polymerase chain reaction (PCR) have made it possible to directly sequence the DNA not only of humans, but of many other animals and plants. The entire genome of humans was sequenced in 2001, and that of chimps in 2005. When they were compared, the result was exactly the same as that obtained from DNA-DNA hybridization: humans and chimps share 98 to 99 percent of their DNA. Less than 1 to 2 percent of our DNA differentiates us from chimps and from gorillas as well. This is because about 60 to 80 percent of our DNA is "junk" that is never read or used, but is carried around passively generation after generation. Some of this junk is endogenous retroviruses (ERVs), which are remnants of viral DNA inserted into our genes when some distant ancestor was infected, and still carried around even though it no longer codes for anything. A smaller percentage is structural genes that code for every protein and structure in our body, including genes we no longer use. The 1 to 2 percent that distinguishes us from chimps are regulatory genes, the "on-off switches" that tell the rest of the genome when to be expressed and when not to be. They are the reason that humans look so different from other apes, even though our genes are nearly identical.

For example, all apes and humans have the structural genes for a long tail, but do not express those genes, except in rare cases where the regulatory genes fail. When such an error occurs, humans grow a long bony tail. Birds also have the genes for a long bony, dinosaurian tail, inherited from their raptor ancestors, not the stubby "parson's nose" fused tailbones found in modern birds. Once in a while, the regulatory genes fail and birds hatch with dinosaur tails. Likewise, living birds have toothless beaks, and no longer the teeth of their dinosaur ancestors (chapter 18), but they still have the genes to make teeth. Experimentally grafting the mouth epithelial tissue of a mouse into a chick embryo produced a bird with teeth. But the teeth that grew were not mouse teeth, but dinosaur teeth! Thus all animals have many

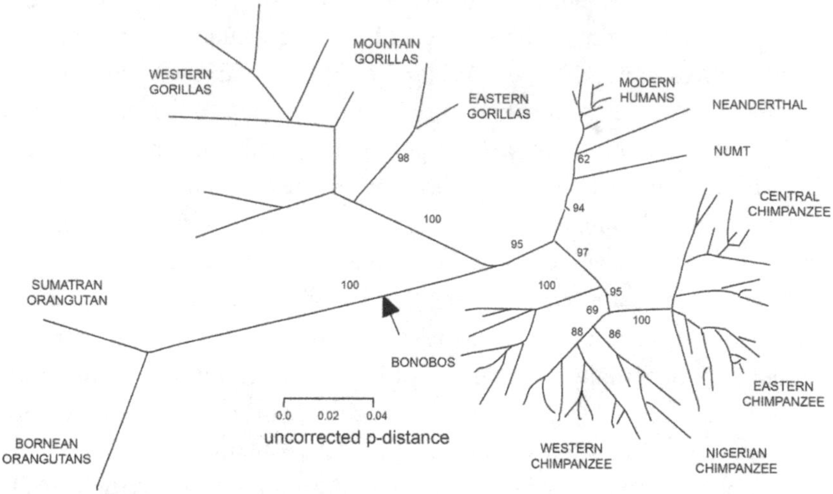

Figure 24.2 ▲

Molecular phylogeny of apes and humans, showing their genetic distance from one another based on mitochondrial DNA. All human "races" are much more similar to one another than two populations of gorillas or chimpanzees are to each other. (Modified from Pascal Gagneux et al., "Mitochondrial Sequences Show Diverse Evolutionary Histories of African Hominoids," *Proceedings of the National Academy of Sciences USA* 93 [1999], fig. 1*B*; © 1999, National Academy of Sciences USA)

ancient genes in their DNA that are no longer expressed, but it takes only some sort of modification of gene regulation to resurrect primitive features.

The extreme similarity of the genes of humans to those of the two species of chimpanzee (common chimp [*Pan troglodytes*] and pygmy chimp, or bonobo [*P. paniscus*]) should, all by itself, be overwhelming and convincing proof of our close relationship. Despite some people's gut reactions and religious ideas, humans are indeed the ape's reflection. Biologist Jared Diamond puts it this way: imagine that some alien biologists came to Earth, and the only biological samples they could obtain were DNA. They sequenced many different animals, including humans and the two chimps. Based on these data alone, they would conclude that humans are just a third species of chimpanzee. Our DNA is more similar to that of the two species of chimp than the DNA of any two species of frog are similar to each other, and even more similar than the DNA of lions and tigers are to each other. Indeed, the differences among the DNA of all the human "races" are smaller than are the differences between the DNA of different populations of chimpanzees from various regions of Africa (figure 24.2)! This suggests two things. First, the genetic differences among the human "races" are tiny and trivial, and are much less significant than many people realize. And second, the big differences between the appearances of chimps and humans are caused by tiny changes in the regulatory genes, which have huge results.

Case closed: humans are slightly modified apes. The evidence from genes, as well as from anatomy, is overwhelming. The DNA in every cell in your body is a testament and witness to your close relationship to chimps, no matter how much this fact makes some people uncomfortable or upset. We know this without a single fossil human showing the transition from apes. But how long ago did humans and apes diverge?

CLOCKS IN ROCKS

Scientists have approached the question of when the ape and human lineages split from each other in two ways. One is to search for fossils that are progressively more ape-like than human-like. This strategy is being tried all the time as exploration continues, although its success depends on the luck of finding the right rocks of the right age and hoping that a primitive hominin fossil might be preserved in them. Human bones tend to be very rarely fossilized, so even in beds with humans fossils, there may be only a few scraps of hominin teeth or jaws compared with the hundreds of specimens of other mammals, such as pigs or antelopes or mastodonts. Nonetheless, as we shall see in chapter 25, paleoanthropologists have spent decades in the field trying to find these elusive hominin fossils, since an important discovery will make a career and burnish a reputation.

Once hominin fossils are found, the next trick is to obtain a reliable date for them. Many hominin fossils are discovered in caves or other places where there is no material that can give a useful date. If the specimen is younger than about 60,000 years (latest Ice Age to Holocene in age), the organic material in the fossil can be dated directly using carbon-14 dating (or radiocarbon dating). This technique is widely employed by archeologists to date human artifacts (most of which are younger than 60,000 years old) and by paleontologists to date late Ice Age fossils. For example, the fossils found in the La Brea Tar Pits in Los Angeles are no older than about 37,000 years, so they have been dated repeatedly using the radiocarbon technique.

For older fossils, however, dating is much more complicated. Radiocarbon dating no longer works on material older than 60,000 years (although the best labs today can sometimes push it out to 80,000 years). The best method to use on older fossils is potassium-argon (K-Ar) dating (or its newer version, argon-argon [$^{40}Ar/^{39}Ar$] dating). With this technique, a fossil cannot be dated directly, by analyzing material either from the specimen or from the sedimentary layers in which it was found. Instead, what is dated are the crystals that formed when they cooled out of a volcano, either a lava flow or a volcanic ash fall. Once the volcanic crystals cool, they lock the unstable parent isotope, potassium-40, into their lattices. As the crystals age, the unstable potassium atoms spontaneously decay, or break down, to form a daughter isotope, argon-40. The rate of decay is very well known, so by measuring the ratio of parent atoms to daughter atoms, geologists can calculate the age of the crystals.

As with any other technique in science, there are limitations and pitfalls that have to be avoided. Because dating is a measure of the time since a crystal cooled and locked in the radioactive parent atoms, potassium-argon dating works only with rocks that cool down from a molten state, or igneous rocks (such as granites or volcanic rocks). A good geologist will tell you that the crystals in a sandstone or any other sedimentary rock cannot be directly dated. Those crystals were recycled from older rocks and have no bearing on the age of the sediment. But geologists long ago circumvented this problem by finding hundreds of places all over Earth where datable volcanic lava flows or ash falls are interbedded with fossiliferous sediment, or where intruding magma bodies cut across the sedimentary rocks and provide a minimum age. From settings such as these, the numerical ages of the geological time scale are derived, and their precision is so well resolved that we know of the age of most events that are millions of years old to the nearest 100,000 years.

If the crystal structure has somehow leaked some of its parent or daughter atoms, or allowed atoms to enter the lattice and contaminate the crystal, the parent/daughter ratio is disturbed and the date is meaningless. But geologists are always on the lookout for this problem, running dozens of samples to determine whether the age is reliable and cross-checking their dates against other sources of determining age. The newest techniques and machinery are so precise that a skilled geologist can spot an error in almost any date and quickly reject dates that don't meet very high standards.

By these methods, most of the fossils found in Africa have been dated very precisely, establishing their ages over the past 5 million years (chapter 25). Anthropologists have frequently collaborated with geochronologists to find fresh ash layers with many unweathered crystals of the appropriate minerals (typically potassium feldspars, but also micas like muscovite and biotite). There have been a few missteps along the way, but generally the age framework of most hominin fossils is well established. In addition, if volcanic ash layers are not present in a given area, then paleontologists can use the differences in fossil assemblages through time to obtain a rough sense of the age of a locality, since the same fossil assemblage occur elsewhere associated with a volcanic ash date.

But what about the fossil record? The story starts with important fossils that were found in the Siwalik Hills of Pakistan. This amazing sequence of rocks spans much of the Oligocene, Miocene, and Pliocene epochs of geologic history and is incredibly fossiliferous. These deposits represent the flood of river sediments that were shed across South Asia as the Himalayas slowly rose high in the sky and that eroded to form the Siwaliks. They have been studied by paleontologists and geologists since 1902, when British geologist Guy Pilgrim did pioneering research throughout South Asia, which was a British colony.

Over the past century, the Siwaliks have yielded huge collections of fossil mammals that offer a very detailed picture of evolution in South Asia during the Miocene. Thanks to the tense nature of Indian and Pakistani politics and American policy toward both countries, Pakistan owed the United States millions of dollars for all the military hardware it had bought. As a result, from the 1970s through the 1990s, there was a lot of grant money (especially from the Fulbright Foundation) for American scholars to go to Pakistan and undertake important research. Lots of paleontologists jumped on the Fulbright opportunity, and there was a flood of studies on the fossils and geology of the Siwalik Hills and nearby areas. Thanks to an abundance of volcanic ash and a technique called paleomagnetic stratigraphy, the Siwalik fossils are extremely well dated. Today, of course, the political situation is so dangerous that few Americans can travel there, and even researchers from other countries who have no ties to the United States are threatened by the pro–Al Qaeda and pro-Taliban tribes in many regions.

But in 1932, paleontologist G. Edward Lewis of the Smithsonian Institution was working in the Tinau River valley in the Nepalese Siwaliks and recovered a jaw that looked very much like that of a primitive hominin. It had relatively small canines, and its shape was more like a broad semicircle in top view (typical of human jaws) than like the *U*-shaped jaw of apes, with its huge canines on a flat lower chin and long parallel back parts. In the 1960s, anthropologist David Pilbeam of Harvard and primatologist Elwyn Simons of Yale and then Duke and others began to champion the view that this jaw (named *Ramapithecus* by Lewis) was the oldest known hominin fossil. (Rama is one of the Hindu gods, and *pithecus* is Greek for "ape"; there are also primates named after the Hindu gods Shiva and Brahma.) Since some of the specimens dated back to 14 million years ago in the well-calibrated Siwalik sequence, this placed the split between apes and hominins at least 14 million years ago. Through the 1960s and 1970s, every student of anthropology, primate evolution, and human paleontology learned that *Ramapithecus* was the "first hominin."

CLOCKS IN MOLECULES

There is an approach other than radiocarbon and potassium-argon techniques to dating the time of divergence between two groups of animals: the molecular clock. As early as 1962, the legendary molecular biologists Linus Pauling (winner of two Nobel Prizes) and Emile Zuckerkandl were among the first to use molecular methods to draw a tree of evolutionary relationships among organisms, the first evidence of evolution to emerge from our own cells and DNA. Pauling and Zuckerkandl noticed not only that the number of amino-acid differences in hemoglobin molecules matched the branching sequence of the animals in their study, but that the number of changes was proportional to how long ago these creatures had diverged from one another over time. A year later, another pioneer in molecular biology, Emanuel Margoliash, noted:

> It appears that the number of residue differences between cytochrome c of any two species is mostly conditioned by the time elapsed since the lines of evolution leading to these two species originally diverged. If this is correct, the cytochrome c of all mammals should be equally different from the cytochrome c of all birds. Since fish diverges from the main stem of vertebrate evolution earlier than either birds or mammals, the cytochrome c of both mammals and birds should be equally different from the cytochrome c of fish.

> Similarly, all vertebrate cytochrome c should be equally different from the yeast protein.

All these data suggested that molecular changes have accumulated through time as different groups of animals branched apart, and that the rate of change of molecules is proportional to the time the lineages split or diverged.

Meanwhile, the evidence that most of the DNA of any animal is "junk" or at least nonfunctional began to emerge. So much of the genome is simply never read when the genes are expressed and thus is invisible to natural selection, or adaptively neutral. Pioneering work by Japanese biochemist Motoo Kimura, in particular, established that most of the molecules in DNA are unaffected by what happens to the organism. These adaptively invisible molecules can spontaneously mutate, and there is no selection to weed them out or favor one version over another. Over time, these mutations continue to accumulate at a regular rate, ticking like a clock. As long as natural selection cannot "see" these changes, the ticking of the "molecular clock" is a good method of estimating divergence time in the geologic past between any two lineages. The only thing needed is calibration by using well-established divergence times of key evolutionary splits, as established in the fossil record.

Soon many molecular biologists were working hard on molecular clock estimates of the branching history and timing of divergence of many groups of animals. Again and again, work by the late Vincent Sarich and Allan Wilson at Berkeley showed that the molecular clock estimate for the divergence between humans and chimps is only 7 to 5 million years ago and no earlier than 8 million years ago, not the 14 million years ago that *Ramapithecus* suggested. Yet the paleontologists stuck by their guns. They distrusted the molecular clock method as unproven and unreliable because it did indeed give some very strange and ridiculous results every once in a while. (This still happens, and we do not always know why.)

As the controversy got more and more heated during the 1970s and 1980s, the major players got into shouting matches at meetings and contentious debates in journals. Sarich and Wilson were convinced that their data were reliable and something must be wrong about *Ramapithecus* or its age. Sarich was a burly, towering, impressive figure with a natty beard, a loud voice, and strong opinions who did not mind ruffling feathers and offending people if necessary. In 1971, he said, "One no longer has the option of considering a fossil older than about eight million years as a hominid no matter what it looks like." This, of course, upset researchers like Simons and Pilbeam, who kept insisting that *Ramapithecus* proved that the molecular biologists were wrong.

The impasse was finally broken by another discovery in the Siwaliks. In 1982, Pilbeam reported on newly discovered specimens that included not only a more complete lower jaw of *Ramapithecus*, but also a partial skull. With the addition of the skull, the specimen now looked much more like a fossil orangutan that had been named *Sivapithecus* by Guy Pilgrim in 1910 when the Siwaliks were first explored. The lower jaw of *Ramapithecus* was just the jaw of a fossil relative of the orangutan that happened to look like a hominin. Soon, the anthropologists were forced to retreat and acknowledge their error, which ceded the victory to Sarich and Wilson and molecular biology. Now that paleontologists knew that there were no hominin fossils as old as 14 million years, the questions then became: What *is* the oldest hominin fossil? And would it indeed fit the prediction from Sarich and Wilson that it is no older than 8 million years old?

"TOUMAI"

Through the past 25 years, paleoanthropologists have been working hard all over the world to push back the fossil record of hominins into older and older beds. As discussed in chapter 25, humans evolved in Africa, and the oldest fossils are found there. Although the early work focused on South Africa, and then on Kenya and Tanzania, since the 1970s the effort has concentrated on even older beds in places like Ethiopia.

Since the discovery of "Lucy" (*Australopithecus afarensis*) in 1974 (chapter 25), there has been a major discovery of even older specimens every few years. In 1984, fossils were found in Kenya of a poorly known species called *Australopithecus anamensis*. This material is much more primitive than "Lucy" and dates to 5.25 million years ago. Then in 1994, an even more primitive species was found in Ethiopia. Named *Ardipithecus ramidus*, it was based on a few scrappy fossils until 2009, when Tim White and his co-workers announced a partial skeleton and many more fossils. Now *Ardipithecus* consists of a number of limb elements and even a partial skull. Recent discoveries of an even older species, *Ardipithecus kaddaba*, push the genus back to 5.6 million years ago.

Meanwhile, a French-British-Kenyan team led by Martin Pickford was working in the Tugen Hills, an area of Kenya that is much older than the classical deposits at Olduvai Gorge and Lake Turkana. In 2000, they announced the discovery of an even older hominin called *Orrorin tugenensis*. Much better fossils were reported in 2007. *Orrorin* is known from only about 20 specimens (the back of the jaw, the front of the jaw, isolated teeth, fragments of the upper arm bone and thighbone, and finger bones). The teeth (as far as they are known) are very ape-like, but the hip region of the thighbone clearly shows that *Orrorin* was bipedal. Like other Kenyan deposits, the Tugen Hills contains dated volcanic ashes, which place the age of the *Orrorin* fossils at between 6.1 and 5.7 million years old.

Thus the hominin fossil record now extends back to at least 6 million years ago, within the window predicted by molecular clocks at about 7 to 5 million years for the split between hominins and apes. But where would one find slightly older beds that might preserve fossil hominins? By 1995, French paleontologist Michel Brunet had spent many years working on Miocene mammals around the world. He specialized in working on some of the most dangerous and remote fossil sites. Brunet had been strafed by fighter jets in Afghanistan, arrested in Iraq, lost a collaborator to malaria in Cameroon, and been held at gunpoint in Chad. By the mid-1990s, he had been digging in the Miocene beds of Chad (once a French colony) for many years.

The conditions in the Djurab Desert in Chad are not exactly easy to tolerate. Brunet was approaching 60, and working in the desert would have been a challenge for a much younger man. Even though the temperatures can reach 43 to 49°C (110 to 120°F), Brunet had to wrap his head in cloth and wear a ski mask and goggles to protect himself from the sand that blows into eyes, ears, nose, and mouth. The temperatures in the shade can be so hot that water bottles can spontaneously explode. As Brunet and his colleagues swept the desert floor looking for fragments of bones and teeth, they had to be careful not only because of the killer heat and the howling winds and sand, but also because of the buried land mines left by combatants in one of the many tribal wars. On January 23, 1995, he found a jawbone of a primitive hominin that was 3.5 million years old, the first such find outside South or East Africa. It was later named *Australopithecus bahrelghazali*.

The following July, he met in Addis Ababa with Tim White at the National Museum of Ethiopia to compare the hominin fossils he had found in

Chad with those that White had unearthed in Ethiopia. Brunet told White that he knew of an older formation below the one that had yielded the jaw he brought to show him. The older formation contained the fossils of extinct gerbils and other mammals that placed it between 7 and 6 million years in age. White was doubtful because gerbils indicate dry climates, and he thought that hominins would not be found there. While they were in the museum, Brunet bet White that he would find older hominins, since he was working in older sediments. "I will win," he said.

Fast-forward to 2001. For six more years, Brunet worked on the older beds, which are late Miocene in age and contain fossils that suggest they are 7 to 6 million years old. Brunet and his colleagues formed the Mission Paléoanthropologique Franco-Tchadienne (MPFT), a collaboration between the University of Poitiers and the University of N'Djamena in Chad. Brunet and three Chadian crew members were working in the broiling heat at a locality called Toros-Menalla. Suddenly, Ahounta Djimdoumalbaye bent down and looked closer at an object protruding from the ground. He called to Brunet and the rest of the crew, and they soon saw that Djimdoumalbaye had found a very important specimen. It looked somewhat like an ape skull, but it also had hominin features (figure 24.3). They quickly recovered it, saturated it with hardeners, and carried it back to camp.

Even though Brunet had not finished analyzing the specimen since bringing it to the University of Poitiers, the rumor mills were buzzing. Everyone was speculating about what had been found, based on a few leaks from people who saw pictures of or heard about the skull. Brunet had no choice but to publish a preliminary analysis before false information was spread. On July 11, 2002, his article appeared as the leading paper in the world's preeminent scientific journal, *Nature*. Brunet named the specimen *Sahelanthropus tchadensis*, after the Sahel region of Chad, where it had been found, and the French spelling "Tchad," But he and his collaborators nicknamed it "Toumai," which means "hope of life" in the Dazanga language of Chad.

Sahelanthropus consists of only a skull, with no jaw or any other part of the skeleton. It was also badly crushed and sheared diagonally, so it looks very odd and asymmetric in its original form. Technicians and computer experts have used morphing software to retrodeform the skull and show its true shape before it was smashed and buried. The fossil is about the size of chimp skull, so *Sahelanthropus* would have been chimp-size in life. The skull encloses a brain cavity of about 320to 380 cubic centimeters (cc) in volume

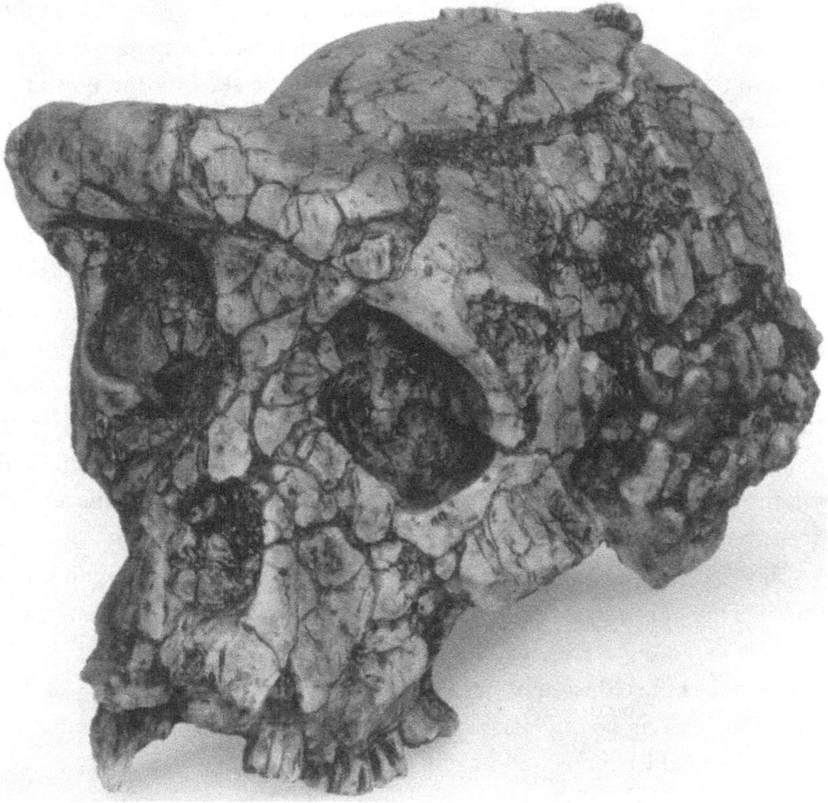

Figure 24.3 ▲
"Toumai," the skull of *Sahelanthropus tchadensis*. (From Michel Brunet et al., "A New Hominid from the Upper Miocene of Chad, Central Africa," *Nature*, July 11, 2002; courtesy Nature Publishing Group)

(compared with modern humans, with over 1350 cc in brain volume). It still has big brow ridges, like apes and many primitive hominins. There are a number of other ape-like features as well, including the relatively primitive cheek teeth.

Yet as Brunet and his colleagues pointed out, *Sahelanthropus* has some features that definitely put it closer to hominins than to chimps or other apes. Its flat face has almost no snout, unlike the face of any ape. It has small canines, unlike the big fangs of apes (even though it appears to be the skull of a male, and most male apes have large canines), and thus its teeth are arranged around the palate in a *C* shape, rather than the elongate *U* shape characteristic of most apes. Most important, the position of the hole in the bottom of the skull (foramen magnum), through which the spinal cord connects to the brain, is directly below the base of the skull, not tilted to the back of the braincase. This indicates that the skull sat upright over the spinal column, rather than hanging forward from the spine, as in chimps and other apes.

This last point is crucial. As we shall see in chapter 25, the biases of anthropologists for most of the twentieth century was that brain size was the most important factor influencing human evolution and that features like bipedal erect posture came later. Yet most of the hominins whose fossils have been found in the past 30 years, from "Lucy" to *Ardipithecus* to *Ororrin*, were clearly fully bipedal, but had small brains. Now *Sahelanthropus*, the oldest hominin fossil yet discovered, also shows evidence that its skull sat directly above its spine. Bipedalism is one of the first adaptations that occurred in human evolution, long before our brains got big.

This realization—combined with the flat face, small canines, and hominin-like upper jaw shape—put *Sahelanthropus* closer to humans than to any ape. Although there are always new discoveries, for now "Toumai" holds the record as the oldest member of the hominin family. And its age, at 7 to 6 million years, is exactly where molecular biologists have been predicting the timing of the chimp–human split for the past 40 years.

SEE IT FOR YOURSELF!

The original fossils of *Sahelanthropus, Ororrin, Ardipithecus, Australopithecus,* and other earliest hominins are kept in special protected storage in the museums of the countries from which they came (mainly, Ethiopia, Kenya, Tanzania, and Chad). Only qualified researchers are allowed to view these collections or to touch these rarest of treasures.

Many museums have exhibition halls devoted to human evolution, featuring high-quality replicas of the most important fossils. In the United States, they include the American Museum of Natural History, New York; Field Museum of Natural History, Chicago; National Museum of Natural History, Smithsonian Institution, Washington, D.C.; Natural History Museum of Los Angeles County, Los Angeles; San Diego Museum of Man; and Yale Peabody Museum of Natural History, Yale University, New Haven, Connecticut. In Europe, they include the Natural History Museum, London; and Museum of Human Evolution, Burgos, Spain. Farther afield is the Australian Museum, Sydney.

FOR FURTHER READING

Diamond, Jared M. *The Third Chimpanzee: The Evolution and Future of the Human Animal.* New York: HarperCollins, 1992.

Gibbons, Ann. *The First Human: The Race to Discover Our Earliest Ancestors.* New York: Anchor, 2007.

Huxley, Thomas H. *Evidence as to Man's Place in Nature.* London: Williams & Norgate, 1863.

Klein, Richard G. *The Human Career: Human Cultural and Biological Origins.* 3rd ed. Chicago: University of Chicago Press, 2009.

Marks, Jonathan. *What It Means to Be 98% Chimpanzee: Apes, People, and Their Genes.* Berkeley: University of California Press, 2003.

Sponheimer, Matt, Julia A. Lee-Thorp, Kaye E. Reed, and Peter S. Ungar, eds. *Early Hominin Paleoecology.* Boulder: University of Colorado Press, 2013.

Tattersall, Ian. *The Fossil Trail: How We Know What We Think We Know About Human Evolution.* New York: Oxford University Press, 2008.

——. *Masters of the Planet: The Search for Our Human Origins.* New York: Palgrave Macmillan, 2013.

Wade, Nicholas. *Before the Dawn: Recovering the Lost History of Our Ancestors.* New York: Penguin, 2007.

LUCY IN THE SKY WITH DIAMONDS

We must, however, acknowledge, as it seems to me, that man with all his noble qualities, still bears in his bodily frame the indelible stamp of his lowly origin.

CHARLES DARWIN, *THE DESCENT OF MAN*

THE DESCENT OF MAN

In *On the Origin of Species*, published in 1859, Charles Darwin never discusses the fossil record of human evolution. Even in *The Descent of Man*, which appeared in 1871, human fossils are never mentioned. There was a good reason for this silence: in the mid-nineteenth century, only a few artifacts suggested prehistoric peoples. The first well-described Neanderthals were found in 1856 in a limestone quarry in the Neander Valley near Düsseldorf, Germany, only three years before *On the Origin of Species* was published. The fossils consisted of only a skullcap and a few limb bones, and they were originally mistaken for those of a cave bear. Later, the fossils were widely misinterpreted as the remains of a diseased Cossack cavalryman or were given other bizarre mistaken identifications. Nobody considered them anything more than the bones of an unusual modern human. The earliest complete Neanderthal skeleton, from La-Chapelle-aux-Saints in France, happened to be that an old diseased individual with rickets, so the early reconstructions falsely showed Neanderthals as stooped and brutish, rather than upright and powerfully built, as we have learned from many better skeletons found since then.

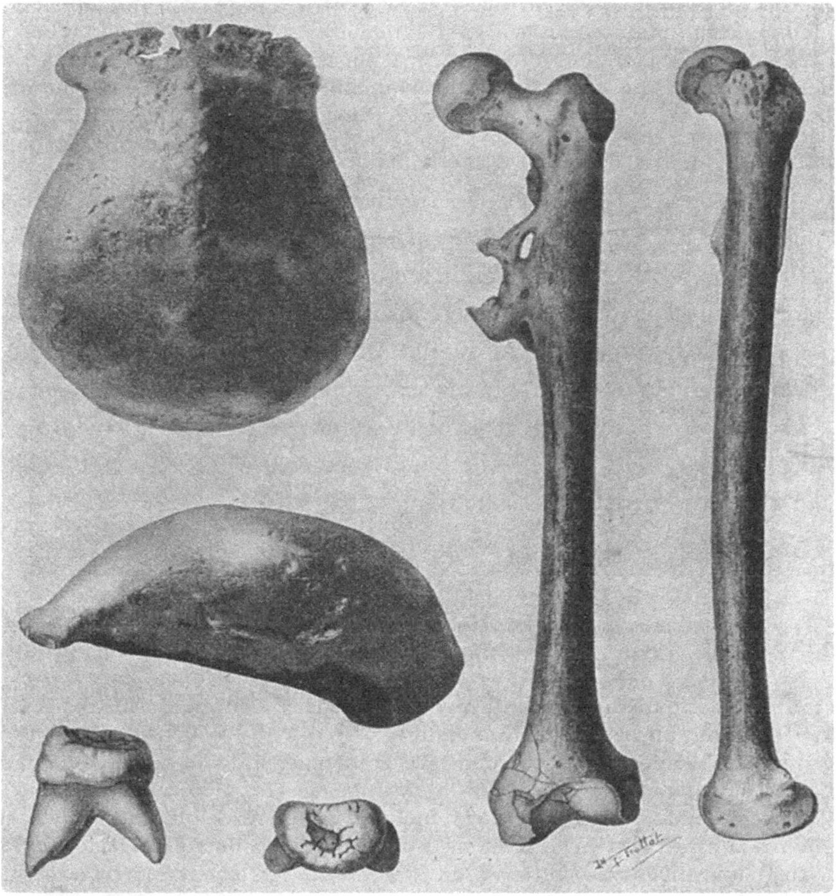

Figure 25.1 ▲
Three fossils of "Java man," as originally drawn by Eugène Dubois: the top of the skull, a molar, and a thigh bone, each in two views. (Courtesy Wikimedia Commons)

The nineteenth century was nearly over before specimens of hominins more primitive than Neanderthals were discovered. The Dutch doctor and anatomist Eugène Dubois was fascinated with Darwin's ideas and convinced that humans had evolved in eastern Asia, so in 1887 he volunteered for the Dutch army as a surgeon, to be posted the Dutch East Indies (present-day Indonesia). Sure enough, he was extraordinarily lucky. After a few excavations, he hit the jackpot. It turned out that there were fossil hominins in the region, and between 1891 and 1895, he and his Javanese crews found a series of specimens, including a skullcap, a thighbone, and a few teeth (figure 25.1). He called them *Pithecanthropus erectus* (Greek for "upright ape-man"), but they came to be known as "Java man" after the island on which they had been found. Although the specimens were incomplete, it was clear from the thighbone that the creatures had walked upright. The skullcap was very primitive, with prominent brow ridges, yet the cranial capacity was about half that of modern humans.

Dubois returned to Holland and received a professorship in 1899. Unfortunately, he did not handle the normal harsh criticism from the scientific community very well. Many anthropologists were not convinced of Dubois's claims from such incomplete material, and thought that the fossils were from a deformed ape. As a result, Dubois withdrew from the debate, an angry and bitter man. He hid his specimens away and refused to show them to anyone or to get involved in the scientific discussion. By the 1920s, opinion was turning in his favor, but he remained withdrawn and embittered until his death, at age 82, in 1940.

OUT OF EURASIA?

In 1871, Darwin argued that humans must have evolved from roots in Africa. His reasoning was simple: all our closest ape relatives (chimpanzees and gorillas) live there, so it makes sense that the common ancestor of apes and humans originated in Africa. But most later anthropologists rejected Darwin's suggestion, insisting that humans had appeared in Eurasia. A number of reasons were given, including Dubois's discoveries in Java, but underlying their view was a deeply held racism that regarded African peoples as sub-human and not even members of our species. The idea that all humans had descended from black Africans was abhorrent to many white scholars in the early twentieth century.

Nearly all the anthropologists and paleontologists in the early twentieth century thought that the homeland of humanity was Eurasia. The prominent paleontologist Henry Fairfield Osborn, director of the American Museum of Natural History, organized and funded the legendary Central Asiatic Expeditions to Mongolia in the 1920s, under the leadership of Roy Chapman Andrews, on the premise that they would find the oldest human ancestors (chapter 23). They didn't, but they did discover very important dinosaur fossils (including the first dinosaur eggs and nests), as well as really interesting and unusual fossil mammals. Eugène Dubois's discoveries of fossil hominins in Java helped confirm the "out-of-Asia" notion.

In 1921, while the American Museum expedition began to explore Mongolia, the Swedish paleontologist Johann Gunnar Andersson found a cave called Choukoutien (Pinyin, Zhoukoudian) near Beijing. The Austrian paleontologist Otto Zdansky took over the excavations, which yielded an excellent fauna of Ice Age mammals, including a giant hyena and two teeth of a hominin. Zdansky gave the specimens to Canadian anatomist Davidson Black (then working at Peking Union Medical College), who published them in 1927 and called them *Sinanthropus pekingensis* (Chinese human from Peking), popularly referred to as "Peking man."

The excavations continued after funding was obtained, with only a few more teeth to show for many years of work. Finally, in 1928, the workers found a lower jaw, skull fragments, and more teeth, and the primitive nature of the species was confirmed. This brought new funding, which prompted a much larger excavation with mostly Chinese workers and scientists. Soon they had unearthed more than 200 human fossils, including six nearly complete skulls (figure 25.2). Black died of heart failure in 1934, and a year later, the German anatomist Franz Weidenreich took over the study and description of the fossils. Although Black had published many preliminary descriptions of the fossils as they were found, it was Weidenreich's detailed monographs that gave complete documentation of them. With this material, it soon became apparent that "Peking man" was very similar to "Java man," and most anthropologists consider them to be the same species: *Homo erectus*.

As the excavations at Zhoukoudian continued, war clouds were gathering on the horizon. The Japanese Empire was expanding, and Japan began to attack and annex parts of China, piece by piece. In 1931, Japan invaded Manchuria, the northeastern part of China, and turned it into a Japanese province, Manchukuo. Japan set up a puppet government headed by Puyi, the last emperor of China. In 1937, the second Japanese invasion of China began; and the Japanese annexed another large chunk of China as they fought the Nationalists under Chiang Kai-shek, and the Communists under Mao Zhedong.

Then in 1941, just before the attack on Pearl Harbor, the alarmed scientists in Beijing could sense that war was coming. The crews at Zhoukoudian were afraid that the fossils would fall into Japanese hands and become war souvenirs, rather than specimens preserved for scientific study. They packed all the specimens at Peking Union Medical College into two large

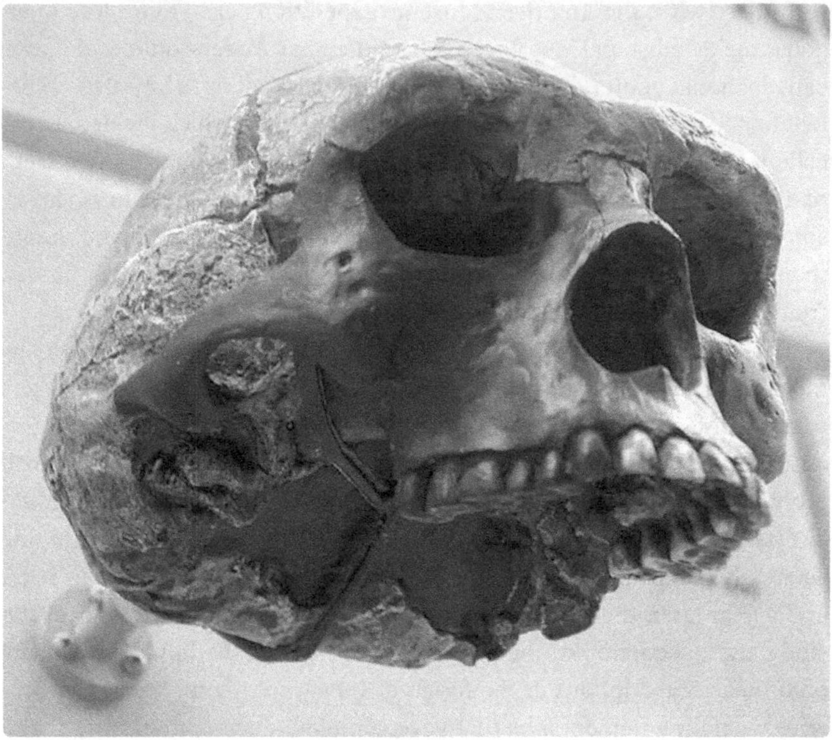

Figure 25.2 ▲
One of the more complete skulls of "Peking man," from Zhoukoudian, China. (Courtesy Wikimedia Commons)

crates, loaded them onto a U.S. Marine Corps truck, and tried to smuggle them out of the country through the port of Qinhuangdao. Somewhere in the secretive scramble to avoid the Japanese invaders, the crates were lost and have never been found. Some say that they were loaded onto a ship that was sunk by the Japanese. Others suggest that they were secretly buried to avoid discovery, and no one knows where they are. Still others think that Chinese merchants who routinely destroy fossils ("dragon bones") and grind them up for traditional "medicine" found them. Fortunately, nearly all the original material was molded, cast, and made into accurate replicas that are housed in many museums, so we know what they look like in detail. In addition, more recent excavations have yielded much more material, so the loss was not irreparable.

OUT OF BRITAIN?

The idea that Asia is the original homeland of humanity goes all the way back to the famous German embryologist and biologist Ernst Haeckel, who forcefully argued the point (long before any fossils could test it). Even though Haeckel was Darwin's greatest protégé in Germany, he disagreed with Darwin's contention that humans had emerged and evolved in Africa. Haeckel directly inspired Dubois, who appeared to have offered evidence to support the view when he found fossil hominins in Java. The other pioneering anthropologists and paleontologists who agreed with Haeckel included not only those working in Zhoukoudian—Andersson, Zdansky, Black, and Weidenreich—but also Osborn and his colleagues at the American Museum of Natural History: paleontologists Walter Granger (who was the chief scientist of the Central Asiatic Expeditions), William Diller Matthew (who argued that most mammalian groups arose in Eurasia and then spread from that center of origin), and William King Gregory.

At that time, the fossil record seemed to support the notion of the Eurasian origin of humans, first with Neanderthals and then with "Java man" and "Peking man." And, surprisingly, a discovery in England seemed to confirm that Eurasia had been the primary center of human evolution. At a meeting of the Geological Society of London in 1912, an amateur collector named Charles Dawson claimed that four years earlier a worker in a gravel pit near Piltdown had given him a skull fragment. The worker thought that the skullcap was a fossil coconut and tried to break it, but Dawson returned to Piltdown again and again and found more pieces. Then he showed them to Arthur Smith Woodward of the British Museum of Natural History, who accompanied Dawson to the Piltdown site. Dawson supposedly came upon more pieces of the skullcap and a partial jaw, although Woodward found nothing.

Woodward soon produced a reconstruction of the skull and jaw, based on the few pieces that were available (figure 5.3). The specimens were very curious. The skull seemed to be very much like that of a modern human, with a bulging cranium, a large braincase, and small brow ridges. However, the jaw was extremely ape-like. Crucially, the hinge of the jaw was broken and missing, as was the face and many parts of the skull, so there was no way to tell if the jaw fit properly with the skull. In August 1913, Woodward, Dawson, and French priest and paleontologist Pierre Teilhard de Chardin

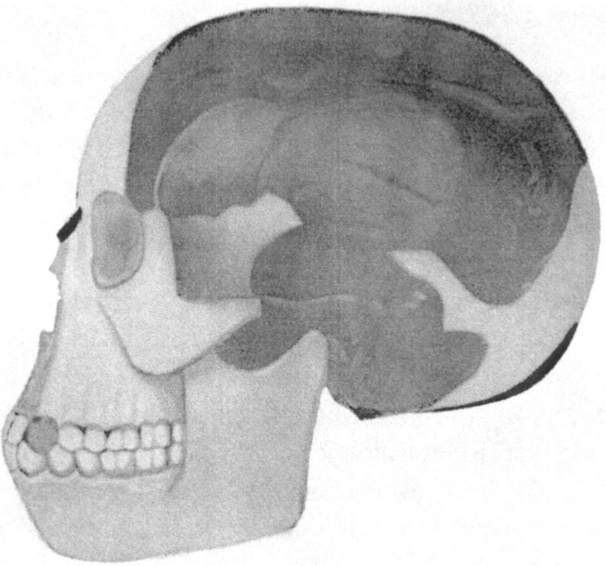

Figure 25.3 ▲

The skull of "Piltdown man," as reconstructed by Arthur Smith Woodward. (Courtesy Wikimedia Commons)

went back to the Piltdown spoil piles, where Teilhard found a canine tooth that fit into the gap between the broken parts of the jaw. The canine was small and human-like, not the large fang-like canines typical of most apes.

Dawson's discovery and Woodward's reconstruction were not unchallenged, however (figure 25.4). Anatomist Sir Arthur Keith disputed the reconstruction and made one that was much more human-like. Anatomist David Waterston of King's College London decided that the two specimens could not belong together and that "Piltdown man" was just an ape jaw attached to a human skull. This was also suggested by French paleontologist Marcellin Boule (who had described the Neanderthals from La-Chapelle-aux-Saints) and American zoologist Gerrit Smith Miller. In 1923, Franz Weidenreich (who had described "Peking man") argued strongly that "Piltdown man" was a modern human skull and an ape jaw, with the teeth filed off so their ape-like appearance was masked.

Although there were always critics of and doubters about "Piltdown man" in the 1920s and 1930s, the pillars of British paleontology (especially Woodward, Keith, and Grafton Eliot Smith) were firm believers (see figure 25.4). Despite its problems, the "fossil" fit all their prejudices. First, it seemed to suggest that human evolution had been driven by the enlargement of the brain, long before our ancestors lost their ape-like teeth and jaws, or began to walk on two legs. This was the accepted dogma of paleoanthropology at the time: the large human brain and intelligence came first, and intelligence drove human evolution. The second factor was simple chauvinism. The British were proud that the "missing link" had been found on their soil, the "first Briton" being even more primitive than "Java man" and "Peking man." Thus it appeared that Europe (especially the British Isles) had been the center of human evolution. The "fossil" fit so perfectly with the biases and myths of anthropology at the time that the questioning soon died down, and "Piltdown man" remained an iconic specimen for 41 years.

It was not until the late 1940s and early 1950s that people began to revive the questions about the specimen, because by then it no longer fit into the

Figure 25.4 ▲

John Cooke, *A Discussion of the Piltdown Skull* (1915): this famous painting shows members of the British anthropological community studying the specimen of "Piltdown man": (*front row, center*) Sir Arthur Keith (in white coat), its chief advocate; (*back row, left to right*) F. O. Barlow, Grafton Elliot Smith, Charles Dawson (who planted the forgery), and Arthur Smith Woodward (curator of geology at the Natural History Museum, who formally described it); (*front row, left*) A. S. Underwood; (*front row, right*) W. P. Pycraft and the famous anatomist Ray Lankester. (Courtesy Wikimedia Commons) fossil record, which was becoming better and better known, especially in Africa. For decades, the Piltdown specimens were kept under lock and key, and only a set of replicas was made available for study, so few people saw the originals close up. Then in 1953, chemist Kenneth Oakley, anthropologist Wilfred E. Le Gros Clark, and Joseph Weiner examined the original "fossils." They confirmed that "Piltdown man" was a hoax: the skull was from a modern human, excavated from a medieval grave; the jaw was from a Sarawak orangutan; and some of the teeth were from chimpanzees. All the specimens had been stained with a solution of iron and chromic acid to make them look old, and the teeth had been deliberately filed to make them look less ape-like—as Weidenreich had surmised.

The identity of all of those involved in the conspiracy is still debated. Charles Dawson, of course, "found" all the "fossils," and further investigation into his past showed that he had had a long history of forging artifacts and human fossils, so he could have been the sole culprit. Some have argued that he needed expert guidance from an anatomist or anthropologist to make such a successful fraud, strategically breaking off all the parts that would demonstrate that the jaw and the skull did not belong together. At various times, scholars have suggested that Pierre Teilhard de Chardin, Arthur Keith, the zoologist Martin A. C. Hinton, the prankster and poet Horace de Vere Cole, or even Sir Arthur Conan Doyle of Sherlock Holmes fame was also behind it. More than a century has now passed, and all the evidence so far has been inconclusive. All we know is that Dawson was the primary (and maybe only) hoaxer and had a long track record of frauds. Whether someone helped him may never be revealed.

THE TAUNG CHILD AND "MRS. PLES"

While research on the Eurasian roots of humans was being undertaken in the museums and universities of Europe, Africa was a scholarly backwater. Most of its cities were sleepy colonial outposts, without major universities or museums. European (especially British, German, French, Portuguese, and Dutch) scientists visited African colonies to collect and remove important specimens for their museums, but left nothing for the host population, who were considered just crude colonials or ignorant natives.

One of the few countries that was not a primitive outpost for science was South Africa. Thanks to its critical position for all shipping passing around the southern tip of Africa and its enormous wealth in gold, diamonds, and precious metals, it had been settled and Europeanized centuries long before the rest of Africa. As a result, there had been a much greater effort in developing a modern European-style state, by both the British masters and the Dutch settlers, who became the Afrikaners. Cape Town, Johannesburg, Durban, Pretoria, and other cities were large and sophisticated. They boasted their own universities and museums, among the few in all of Africa. In addition, South Africa was the second colony in Africa to become independent of its European masters, in 1910, long before most other African colonies became independent in the late 1950s and the 1960s.

Among the European-trained scholars in South Africa was a young Australian, Raymond Dart. He had earned a medical degree from University College London and then emigrated to South Africa, where he took a post in the newly established Department of Anatomy at the University of Witswatersrand in Johannesburg. Upon arriving, he was dismayed to find that the department had no comparative collections of human and ape skulls and skeletons, so essential to teaching anatomy. He announced to his students that there would be a competition to see who could bring in the most interesting bones. One student, the only woman in the class, said that she knew of the skull of a fossil baboon on the mantel of a friend's house. Although Dart doubted that it was really a baboon (since almost no fossil primates were known in sub-Saharan Africa), when he saw the skull in, he knew that his student was right. The skull had come from the house of the director of the Northern Lime Company, which produced cement from limestone dug from a quarry called Taung. Dart then asked him to send over any other fossils the workers found when blasting in the limestone caves.

One summer morning in 1924, Dart was struggling with his stiff winged collar as he was preparing to be best man and host a friend's wedding at his house. He heard the sound of two wooden crates dumped on his doorstep and went out to investigate. The first contained nothing of interest, but when he pried open the second, there was a beautiful braincase with a natural endocranial cast of the brain, right on top! Excitedly, he rummaged around until he found the face that had been attached to the braincase. He could already tell that the brain was much larger than a chimpanzee's, even though the skull was the same size. His friend, the groom, urged him to finish dressing. During the entire ceremony, Dart could not wait to get back to his "treasures." Over the next few months, he cleaned and prepared the

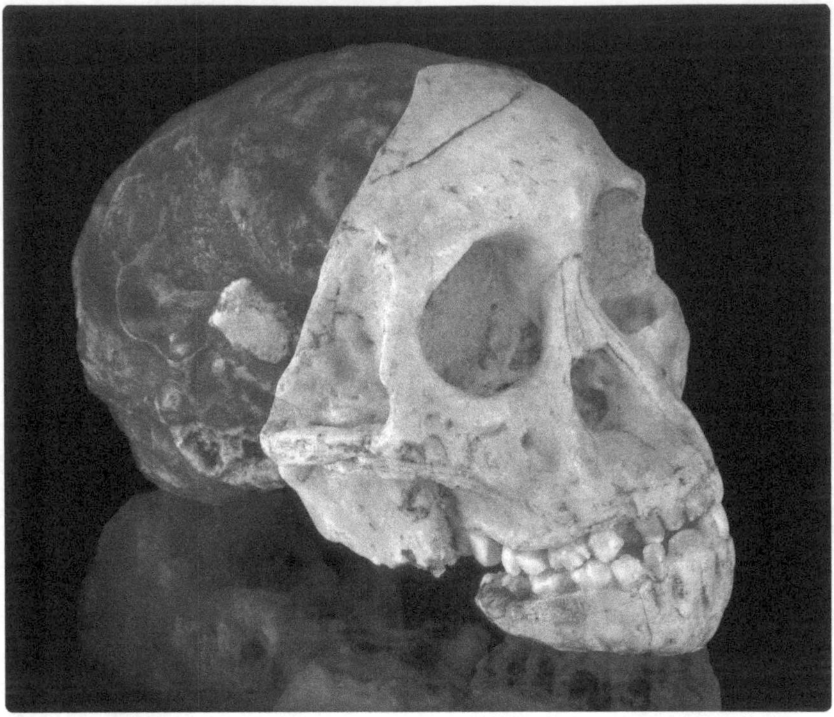

Figure 25.5 ▲

Side view of the skull of *Australopithecus africanus* known as the Taung child. (Courtesy Wikimedia Commons)

specimen and, with one delicate stroke using a hammer and knitting needles, split off the matrix from the front of the face.

The face that emerged was that of a child of about four years, with all its baby teeth still in place (figure 25.5). Although the skull is about the size of that of a modern chimpanzee, it has a number of hominin-like features, including an unusually large brain, a flat face with no snout and small brow ridges, and reduced canine and human-like teeth arranged in a semicircle in palatal view. Most important, the hole in the base of the skull for the spinal column (foramen magnum) is directly below the brain, proving that this creature held its head up and probably walked upright.

Dart wrote his description and analysis and published it in *Nature*, the preeminent scientific journal in the world, in 1925. He called the specimen *Australopithecus africanus* (Greek for "African southern ape"), and it clearly shows that early hominins lived in Africa, especially since it is far more primitive and ape-like than any fossil that had been found so far. Since Dart had studied brain endocranial casts in medical school, he was particularly interested in the natural endocast on this specimen. *Australopithecus* had not only a brain far larger than any ape brain for a skull of that size, but a noticeably more advanced forebrain, like that of humans but not apes.

Dart thought that his evidence was conclusive, and he expected to receive accolades from the scientific community, Instead, he was disappointed to see all the great anthropologists in Europe dismiss his specimen as a "juvenile ape." Part of the problem is that juvenile apes do look a lot more like modern humans than do adult apes. Still, the upright posture, the hominin-like cheek teeth, the reduced canines and semicircular arrangement of teeth around the palate, and the enlarged forebrain were not artifacts of the youth of the specimen.

But the Taung child was running up against a wall of false notions and prejudice. As we have seen, British and other European anthropologists were convinced that a large brain evolved first, followed by smaller teeth and upright posture in hominins. And they had the skull of "Piltdown man," with a human-like brain but ape-like teeth, to prove it. But the Taung child showed just the opposite: a relatively small brain, but upright posture and hominin-like teeth with reduced canines. Thus it could not be accepted. Sir Arthur Keith, one of Piltdown's biggest backers, wrote: "[Dart's] claim is preposterous, the skull is that of a young anthropoid ape . . . and showing so many points of affinity with the two living African anthropoids, the gorilla and chimpanzee, that there cannot be a moment's hesitation in placing the fossil form in this living group."

In addition, there were other unspoken factors: imperialism and racism. The top scholars in Europe did not trust the conclusions of Dart, an obscure anatomist in remote South Africa (even though he had trained in London)—not a recognized expert, but a "country bumpkin." His paper in *Nature* was very short (as they always must be), so the leading paleontologists had only a brief description and a few tiny hand-drawn figures by which to judge the specimen. (Dart later wrote a more detailed description, especially of the brain.) But no one had the time, money, or inclination to take the long sea voyage to South Africa in order to examine the fossil.

So Dart brought it to them. In 1931, he visited Britain and brought the Taung child with him, but to no avail. The racial prejudices of British anthropologists were just too deeply entrenched. In addition, the excitement over the skull of "Peking man" was just reaching Europe as Davidson Black's illustrations and descriptions were published—supporting the earlier evidence for a Eurasian origin offered by "Java man" and "Piltdown man"—so the poor Taung child, the only specimen from Africa, and Dart were overshadowed.

Dart would wait another 20 years before the European anthropological community stopped dismissing him and began to recognize the importance of his find. In 1947, Keith admitted that "Dart was right and I was wrong." But Dart had the last laugh. He lived until 1988, dying at the age of 95, celebrated and honored for his discoveries and for pioneering modern paleoanthropology, while most of his bitter rivals died long before him and are now forgotten.

But in the 1920s and 1930s, other South African scientists were convinced that Dart was right and was being unfairly criticized. Among them was the Scottish-born doctor Robert Broom (chapter 19), who had already made a reputation for himself as an important paleontologist by finding spectacular specimens of Permian reptiles and of the earliest relatives of mammals in the Great Karoo. Some in his network of collectors sent him fossils they had found in the many limestone caves in South Africa. Working in a cave called Kromdraai in 1938, Broom discovered a very robust adult skull he called *Paranthropus robustus* (Greek for "robust near-human"). Later, in the famous cave complex at Swartkrans, he found fossils of more than 130 individuals of *P. robustus*. A recent analysis of their teeth showed that none of these robust, gorilla-like humans lived past 17 years and that they subsisted on a gritty diet of nuts, seeds, and grasses.

Also in 1938, Broom obtained an endocranial cast of a fossil skull with a capacity of 485 cc, far too large to be that of an ape. He called this specimen *Plesianthropus transvaalensis* (Greek for "near ape of the Transvaal"). Then he heard word of fossils coming from a cave called Sterkfontein. On April 18, 1947, Broom and John T. Robinson found the complete skull of what was then considered to be an adult female (now thought to be male) that demonstrated hominin features, yet it was just as primitive as the Taung child (figure 25.6). They nicknamed this specimen "Mrs. Ples," and their discovery showed that South Africa was yielding fossils of hominins that were much more primitive than any that had been found anywhere in Eurasia. Soon other fossils emerged from Sterkfontein, establishing the variability of the

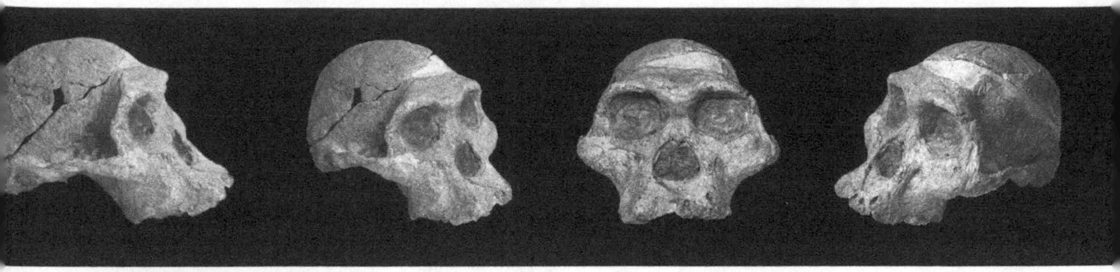

Figure 25.6 ▲
Multiple views of the most complete skull of *Australopithecus africanus*, nicknamed "Mrs. Ples." (Photo courtesy Wikimedia Commons)

population of *Plesianthropus transvaalensis*. Later anthropologists decided that the Sterkfontein adults and the Taung child are the same species, so *Plesianthropus transvaalensis* is now subsumed under Dart's original taxon: *Australopithecus africanus*.

These discoveries in Africa—along with the complete absence of fossils so primitive or ancient in Eurasia—began to shift the momentum of the debate away from the "out-of-Asia" school of thought. The wide spectrum of australopithecines that had been described by 1947 made it appear more and more likely that Darwin was right: humans had originated in Africa. Not only that, but the idea that brains and intelligence drove human evolution, but small teeth and upright posture came later, was also dying (as the older generation of racist anthropologist died off as well). Every fossil found so far demonstrated that upright posture and advanced teeth had evolved first, and the brain began to enlarge much later. So in 1953, when someone decided to examine "Piltdown man" closely, since it no longer fit the emerging picture of human evolution, the hoax finally was exposed—both an embarrassment and a relief.

LEAKEY'S LUCK

Another advocate of the "out-of-Africa" hypothesis was the legendary Louis S. B. Leakey (figure 25.7*A*). By all accounts, he was a charismatic, ebullient, outspoken man who could weave a tremendous tale about his discoveries. Critics also considered him to be somewhat careless and sloppy in his science, and occasionally known for buying into controversial ideas that did not pan out. Nonetheless, he left a permanent legacy in the study of human evolution—not only for discovering many famous fossils, but also for training his wife, Mary (who made most of the discoveries), and his son Richard (who outshone his father) and for mentoring many other important anthropologists. He also inspired primatologists like Jane Goodall, Dian Fossey, and Birute Galdikas to spend years studying wild chimpanzees, gorillas, and orangutans, respectively.

Born into a family of British missionaries in what is now Kenya, Leakey grew up with the wildlife of East Africa and became fluent in the language and culture of the Kikuyu, one of the largest tribes in that region. Although he was partially educated by tutors in Kenya, after World War I he was sent to Cambridge, where he proved to be a brilliant and eager but often eccentric student. He chose a career in anthropology and was already publishing numerous papers on the archeology of Kenya in his twenties. In the early 1930s, his career was nearly derailed when he abandoned his first wife, Frida, after he fell in love with his artist, young Mary Nicol. To escape the censure of the academic community, he and Mary returned to Kenya, where they found a number of sites with primitive apes at Kanam, Kanjera, and Rusinga Island.

While in Kenya before and during World War II, his fluency in the Kikuyu language and his good connections with the native cultures made him an important figure in the politics of the region, not only as a spy during the war, but also as an interpreter and a go-between during tensions between the British and the Kikuyu. He was a key figure in the Mau Mau Revolt and eventually helped resolve the disputes. But he refused to return to Europe, settling instead for a tiny salary at the Coryndon Museum in Nairobi (now the National Museum of Kenya).

Leakey's reputation and finds in Africa were significant, but he was still struggling to discover something spectacular that would not only confirm that humans had indeed arisen and evolved in Africa, but also launch his career and ensure better funding for his work. The specimens in South African caves were important, but they could not be numerically dated. What was needed was a locality where the hominin fossils were buried in the

Figure 25.7 ▶

(*A*) Louis S. B. Leakey, holding an artifact; (*B*) front view of the skull called "*Zinjanthropus*" (now *Paranthropus boisei*), which made the Leakeys world famous and drove research on paleoanthropology back to Africa. ([*A*] courtesy Wikimedia Commons; [*B*] courtesy National Museums of Kenya)

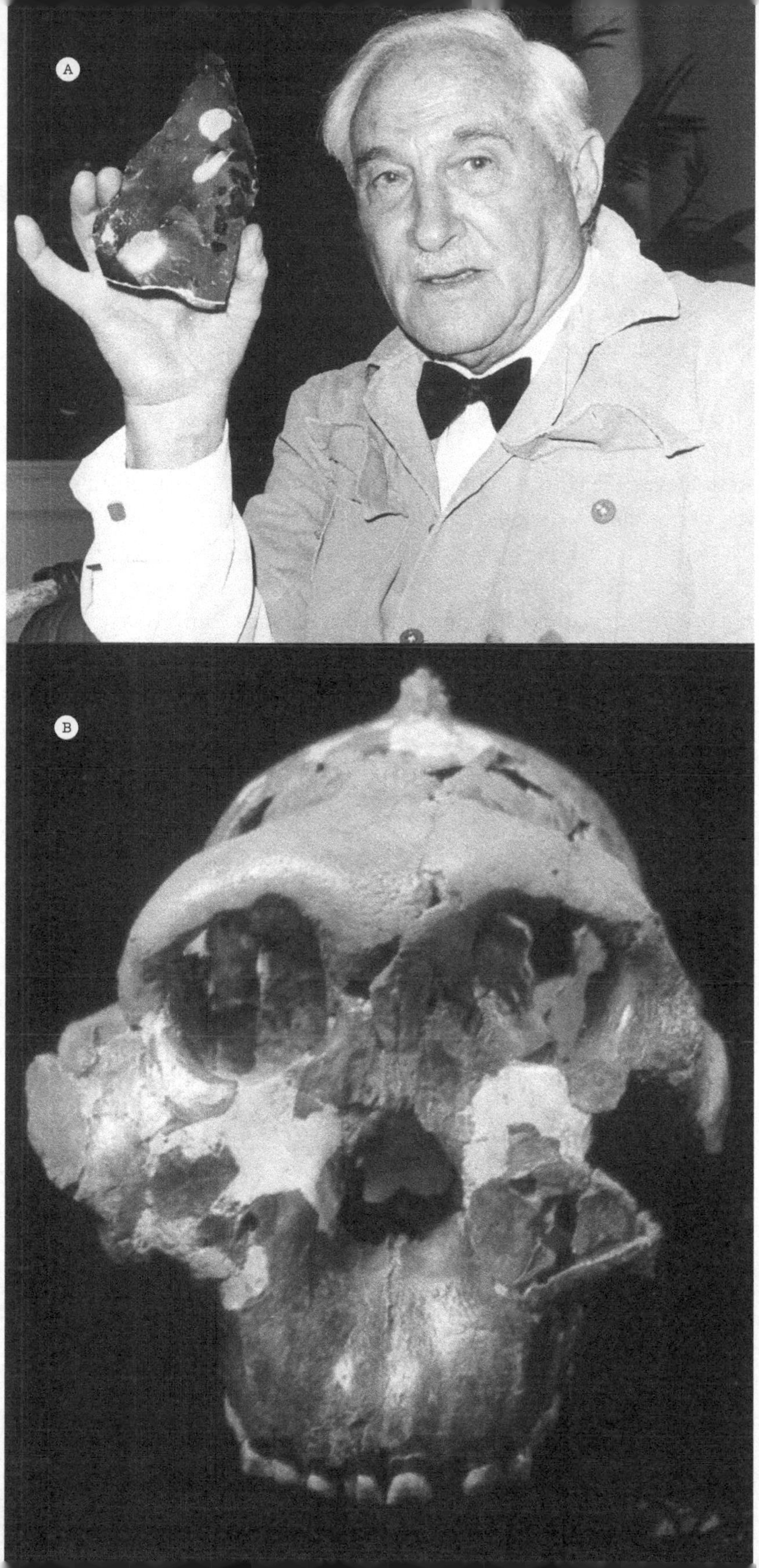
A
B

sediment with age-diagnostic mammal fossils and with volcanic ashes that could provide numerical dates.

In 1913, German archeologist Hans Reck had excavated a fairly modern human skeleton from Bed II of Olduvai Gorge in what is now Tanzania. His find was controversial, since it appeared to be from the middle Pleistocene, much older than European fossils of that level of human evolution. In 1931, Leakey became involved in the debate and managed to convince his colleagues that Reck's specimen was not a modern human buried in ancient strata. In 1951, once World War II and postwar politics were over and he had more time for anthropology, Louis and Mary began full-time work in the lowest levels of Olduvai Gorge. They found many stone tools indicating cultures much more primitive than those in Europe, but no convincing fossils.

Then in 1959, after eight years of hard work (and 30 years after Louis had first worked at Olduvai), Mary found an extraordinary fossil skull (see figure 25.7*B*). It was much more primitive and robust than anything ever unearthed in South Africa or elsewhere, and it came from Bed I, the lowest level in Olduvai Gorge. The spectacular skull was nicknamed "Dear Boy" by the Leakeys and formally named *Zinjanthropus boisei*, or "Zinj" for short. (The genus nickname was the name of a medieval African region, and the species name honors Charles Boise, who funded their research.) Then in 1960, Jack Evernden and Garniss Curtis applied the newly developed technique of potassium-argon dating to an ash layer above Olduvai Bed I and got an age of more than 1.75 million years, far older than anyone believed possible. At that time, most scientists thought that the entire Pleistocene was only a few hundred thousand years old, but the entire timescale was recalibrated in the 1960s with the introduction of more potassium-argon dates. Soon the Leakeys were world famous and championed by the National Geographic Society, which funded their work. More important, anthropologists swarmed to Africa to find more specimens, since it was clear that most of human evolution had indeed occurred in Africa. Only much later did humans migrate to Eurasia and beyond.

LUCY'S LEGACY

The rush to find hominins from the "Dark Continent" soon spread across East Africa, especially in regions with long sedimentary records in fault basins along the Great Rift Valley. Louis Leakey's son Richard, who was initially uninterested in anthropology, eventually adopted his father's mantle. Seeking to escape his father's shadow, he began to excavate in Lake Rudolf (now Lake Turkana) in northern Kenya in the 1970s. There, many more skulls were found, including the best-preserved specimen of *Homo habilis*, the oldest species in our genus, *Homo*. Richard moved on to prominent positions in the Kenyan government (especially fighting the poaching of rhinos and elephants). His wife, Meave, working with local people, carried on the Leakey legacy. His mother, Mary, continued to make significant finds, especially the spectacular trackway of hominins at Laetoli in Tanzania.

Kenya and Tanzania were in the news almost every year with the spectacular finds of the Leakeys. In the late 1960s, Louis Leakey had lunch with President Jomo Kenyatta of Kenya and Emperor Haile Selassie of Ethiopia. The emperor asked Leakey why there had been no discoveries in Ethiopia. Louis quickly persuaded him that fossils would be found if he gave the order to let scientists explore for them. Soon anthropologist F. Clark Howell of Berkeley was working on the northern shore of Lake Turkana, where the Omo River flows out of Ethiopia. Howell and his colleague Glynn Isaac spent many years collecting in the Omo beds, which have abundant volcanic ash dates. Unfortunately, these deposits were formed in flash floods that produced gravelly and sandy streams, which tend to break up and abrade fossils, so no well-preserved hominin specimens were found.

Meanwhile, other rising young anthropologists were eager to make their own discoveries in a region that had been almost exclusively the territory of the Leakeys and their allies. Two of them were Donald Johanson and Tim White. Both were seeking to make their professional fortunes by exploring sites not under the control of the Leakeys. Through French geologists Maurice Taieb and Yves Coppens and anthropologist Jon Kalb, they learned about beds in the Afar Triangle, the rift valley that is opening between the tectonic plates where the Gulf of Aden meets the Red Sea. These beds already had yielded numerous fossils of mammals, suggesting that they were at least 3 million years old, which made them potentially older than any hominin fossil found so far in Kenya or Tanzania. Johanson, White, Taieb, and Coppens received permission to work in these beds and began to excavate at Hadar in 1973.

After months of exploring and prospecting for fossils, and finding a few hominin fragments, on November 24, 1974, Johanson took a break from writing field notes to help his student Tom Gray search an outcrop. He spotted the glint of bone out of the corner of his eye, dug out the fossil, and immediately recognized that it was a hominin bone. They continued to unearth more and more bones, until they found almost 40 percent of a skeleton of a hominin (figure 25.8*A*). It was the first skeleton, rather than isolated bones, found of any hominin older than the Neanderthals of the late Pleistocene. That night as they celebrated over the campfire, they were playing a tape of the Beatles when "Lucy in the Sky with Diamonds" came on. Singing lustily along, a member of the crew named Pamela Alderman suggested that the fossil be nicknamed "Lucy." Later, it was formally named *Australopithecus afarensis*, in reference to the Afar Triangle, where it was found.

A year after the discovery of "Lucy," the crew returned to Hadar, where they found a large assemblage of *A. afarensis* bones. Nicknamed the "First Family," it was the first large sample of fossils of both juvenile and adult hominins from beds dating to 3 million years ago, and it gave anthropologists a look at how much variability was typical in a single population. This can be important when deciding whether a newly discovered fossil that is slightly different from specimens found earlier should be considered a new species or genus or just a member of a variable population.

When the analysis of "Lucy" was conducted, Johanson and White determined that the skeleton was that of an adult female that had stood about 1.1 meters (3.5 feet) tall (see figure 25.8*B*). The most important evidence was the knee joint and the hip bones, which show the critical features that prove that *A. afarensis* walked upright with its legs completely beneath its body, as do modern humans. It had a relatively small brain (380 to 430 cc) and small canines, like those of advanced hominins, yet still had a pronounced snout, rather than a flat face. This was yet another blow to the "big brains first" theory of human evolution, which was still in vogue in the mid-1970s. Its shoulder blade, arms, and hands are quite ape-like, however, so *A. afarensis* still climbed trees, even if it was fully bipedal. Yet the foot shows no signs of a grasping big toe, so its legs and feet were adapted entirely for walking on the ground and its toes could not grasp branches.

Since the discovery of "Lucy" in the mid-1970s, paleoanthropologists have made many more amazing discoveries. "Lucy" was the first ancient hominin (older than 3 million years) known from a skeleton, rather than from a partial skull or a few isolated limb bones. In 1984, Alan Walker and the Leakey team found the "Nariokotome boy" on the shores of West Turkana. About 1.5 million years in age, the skeleton is 90 percent intact and thus is the most complete ancient hominin ever found. It may belong to

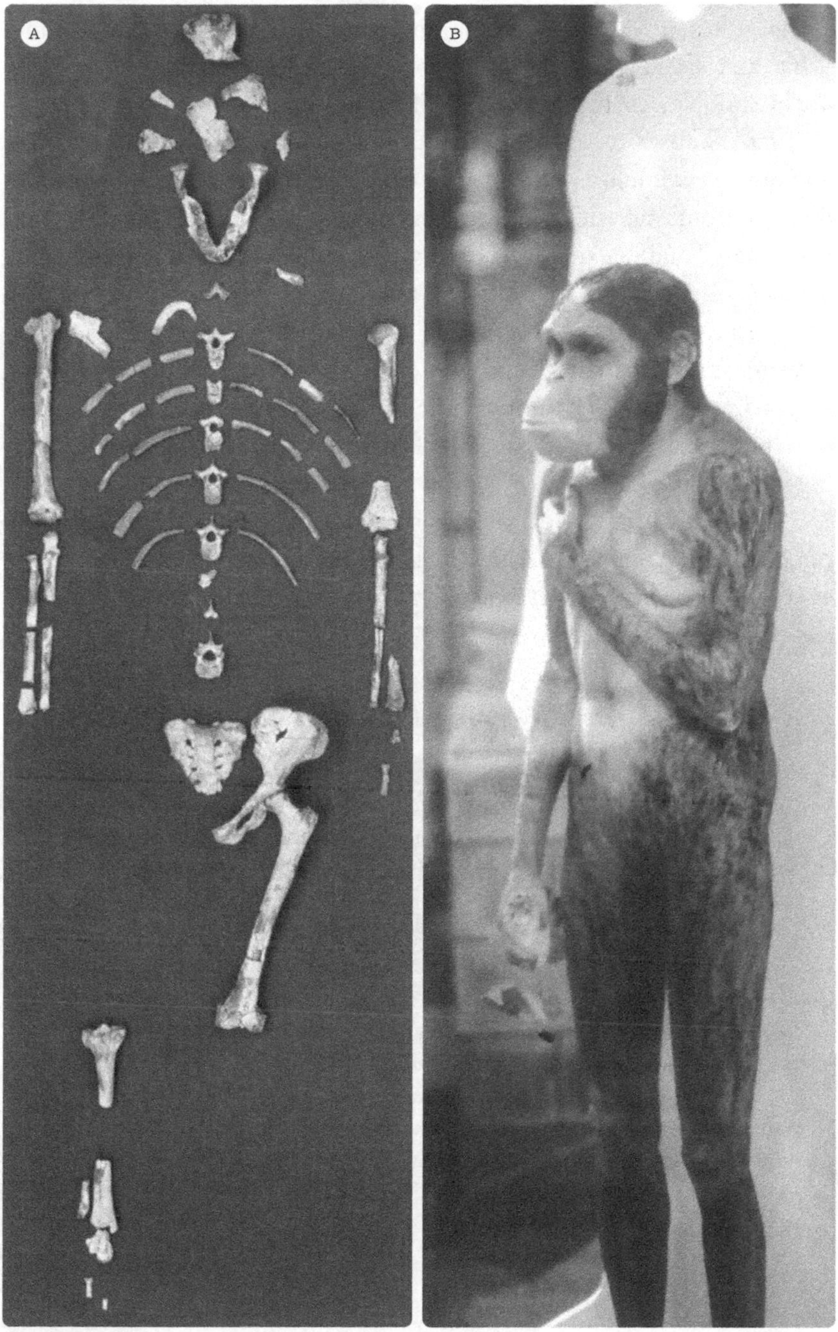

Figure 25.8 ▲

Lucy, an *Australopithecus afarensis*: (*A*) skeleton; (B) reconstruction of her appearance in life. ([*A*] courtesy D. Johanson; [B] photograph by the author)

Homo erectus or *H. ergaster* (its identity is still controversial). And in 1994, White and his crew found a nearly complete skeleton of *Ardipithecus ramidus* in Ethiopia, which dates to 4.4 million years.

Thus the fossil record of hominins gets better year after year, as more and more specimens are found. In a century, we have come an enormous distance from when the only ancient hominins known were Neanderthals, "Java man," and "Peking man" and when "Piltdown man" was still taken seriously. Today, there are six genera of hominins besides *Homo* (*Ardipithecus*, *Australopithecus*, *Kenyanthropus*, *Orrorin*, *Paranthropus*, and *Sahelanthropus*), and more than 12 valid species. From the simplistic idea of a single human lineage evolving through time, the fossil record has revealed a complex, bushy branching pattern of evolution, with multiple lineages coexisting in time and place.

For only the past 30,000 years has there been a single species of hominin dominating the planet. Now *Homo sapiens* threatens to wipe out nearly every other species, as well as itself, making them just as extinct as the fossils described in this book.

SEE IT FOR YOURSELF!

The original fossils of *Ardipithecus*, *Australopithecus*, *Homo habilis*, *H. erectus*, and other earliest hominins are kept in special protected storage in the museums of the countries from which they came (mainly, Ethiopia, Kenya, Tanzania, and Chad). Only qualified researchers are allowed to view these collections or to touch these rarest of treasures.

Many museums have exhibition halls devoted to human evolution, featuring high-quality replicas of the most important fossils. In the United States, they include the American Museum of Natural History, New York; Field Museum of Natural History, Chicago; National Museum of Natural History, Smithsonian Institution, Washington, D.C.; Natural History Museum of Los Angeles County, Los Angeles; San Diego Museum of Man; and Yale Peabody Museum of Natural History, Yale University, New Haven, Connecticut. In Europe, they include the Natural History Museum, London; and Museum of Human Evolution, Burgos, Spain. Farther afield is the Australian Museum, Sydney.

FOR FURTHER READING

Boaz, Noel T., and Russell T. Ciochon. *Dragon Bone Hill: An Ice-Age Saga of* Homo erectus. Oxford: Oxford University Press, 2008.

Dart, Raymond A., and Dennis Craig. *Adventures with the Missing Link*. New York: Harper, 1959.

Johanson, Donald, and Maitland Edey. *Lucy: The Beginnings of Humankind*. New York: Simon & Schuster, 1981.

Johanson, Donald, and Blake Edgar. *From Lucy to Language*. New York: Simon & Schuster, 2006.

Kalb, Jon. *Adventures in the Bone Trade: The Race to Discover Human Ancestors in Ethiopia's Afar Depression*. New York Copernicus, 2000.

Klein, Richard G. *The Human Career: Human Cultural and Biological Origins*. 3rd ed. Chicago: University of Chicago Press, 2009.

Leakey, Richard E., and Roger Lewin. *Origins: What New Discoveries Reveal About the Emergence of Our Species and Its Possible Future*. New York: Dutton, 1977.

Lewin, Roger. *Bones of Contention: Controversies in the Search for Human Origins*. Chicago: University of Chicago Press, 1997.

——. *Human Evolution: An Illustrated Introduction*. 5th ed. New York: Wiley-Blackwell, 2004.

Morell, Virginia. *Ancestral Passions: The Leakey Family and the Quest for Humankind's Beginning*. New York: Touchstone, 1996.

Reader, John. *Missing Links: In Search of Human Origins*. Oxford: Oxford University Press, 2011.

Sponheimer, Matt, Julia A. Lee-Thorp, Kaye E. Reed, and Peter S. Ungar, eds. *Early Hominin Paleoecology*. Boulder: University of Colorado Press, 2013.

Swisher, Carl C., III, Garniss H. Curtis, and Roger Lewin. *Java Man: How Two Geologists Changed Our Understanding of Human Evolution*. Chicago: University of Chicago Press, 2001.

Tattersall, Ian. *The Fossil Trail: How We Know What We Think We Know About Human Evolution*. New York: Oxford University Press, 2008.

——. *Masters of the Planet: The Search for Our Human Origins*. New York: Palgrave Macmillan, 2013.

Wade, Nicholas. *Before the Dawn: Recovering the Lost History of Our Ancestors*. New York: Penguin, 2007.

Walker, Alan, and Pat Shipman. *The Wisdom of the Bones: In Search of Human Origins*. New York: Vintage, 1997.

THE BEST NATURAL HISTORY MUSEUMS

APPENDIX

The museums listed here feature some of the fossils described in this book.

UNITED STATES

The following are the top-10 natural history museums in the United States.

◉ AMERICAN MUSEUM OF NATURAL HISTORY (NEW YORK, NEW YORK)
Widely considered to be the world's greatest natural history museum, the American Museum of Natural History was founded in 1869 and has been the pioneer in paleontological research in the United States since 1895. It has four giant floors of exhibits, millions of specimens, and thousands of fossils that are not on display—including those in the Frick Wing, whose seven floors store fossil mammals that are available for study by researchers. The fourth floor of the museum has housed legendary fossils for more than a century. Renovated in 1996, the galleries are arranged so visitors can follow the branching family tree of life from fossil fish through amphibians, reptiles, and some of the world's best dinosaurs, to primitive and advanced mammals. The first floor features the state-of-the-art Hall of Human Origins, and a huge skeleton of *Barosaurus* rearing up on its hind legs greets visitors in the second-floor Theodore Roosevelt Rotunda.

◉ NATIONAL MUSEUM OF NATURAL HISTORY, SMITHSONIAN INSTITUTION (WASHINGTON, D.C.)
Opened in 1910 as a separate museum of the Smithsonian Institution, the National Museum of Natural History is the most visited natural history museum in the world. The Mammal Hall (with skeletons of most of the famous mammals of the Cenozoic) and the National Fossil Hall (closed for renovation until

2019, with American dinosaurs exhibited in a special show) display some of the best and most important specimens. In addition, the museum has excellent exhibits of invertebrates, including a large collection of Burgess Shale fauna.

◉ FIELD MUSEUM OF NATURAL HISTORY (CHICAGO, ILLINOIS)

The first thing to greet visitors to the Field Museum of Natural History, in the Stanley Field Hall, is the famous *Tyrannosaurus rex* named "Sue." The museum's large modern halls feature many kinds of dinosaurs, spectacular fossil mammals, and the famous paintings of pioneering paleoartist Charles R. Knight. In the Griffin Halls of Evolving Planet, journey through the 4 billion years of life on Earth, including the evolution of humans, and watch museum staff work on fossils in the Fossil Prep Lab. On the ground floor is a century-old exhibit of taxidermied animals, showing many creatures that are not displayed anywhere else in the world.

◉ CARNEGIE MUSEUM OF NATURAL HISTORY (PITTSBURGH, PENNSYLVANIA)

The scientists and collectors at the Carnegie Museum of Natural History, one of the nation's oldest natural history museums, were active in the Rocky Mountain region beginning in the 1890s, so the museum is home to fossils from the beds of what is now Dinosaur National Monument (replicas of the Carnegie's skeleton of *Diplodocus* are in many other museums), from Agate Bone Bed in Nebraska, from the Ice Age caves nearby, and from many other legendary sites (including the type *Tyrannosaurus rex* specimen). The museum also has an excellent exhibit of Paleozoic invertebrate life of the Appalachian region.

◉ DENVER MUSEUM OF NATURE AND SCIENCE (DENVER, COLORADO)

The spectacular fossils of the Rocky Mountains housed at the Denver Museum of Nature and Science are arranged as a trip through time called Prehistoric Journey. As visitors walk through time, they see a three-dimensional diorama of the life in each period, the specimens on which the reconstructions are based, and exhibits showing the localities today, and they learn how paleontologists reconstruct the ancient past and watch scientists as they prepare fossils. The gallery features spectacular sauropods and the most complete stegosaur ever found (showing how its plates and tail spikes were actually arranged). The Cenozoic stretch of the journey has an amazing display of fossils from the Green River Shale, which dates to the Eocene, plus mammals from the Big Badlands, the local Ice Age deposits, and many other places in the Great Plains and Rockies.

◉ NATURAL HISTORY MUSEUM OF LOS ANGELES COUNTY (LOS ANGELES, CALIFORNIA)

Recently renovated, the Dinosaur Hall at the Natural History Museum of Los Angeles County features three specimens of *Tyrannosaurus rex* of different ages, *Triceratops*, *Mamenchisaurus*, *Carnotaurus*, *Stegosaurus*, *Allosaurus*, and a pregnant plesiosaur. The theme of the hall is dinosaur biology, and how paleontologists know about the lives of dinosaurs. The Rotunda features a battling *Tyrannosaurus rex* and *Triceratops*. The Age of Mammals gallery features many spectacular fossils on two different levels, with skeletons of marine mammals hanging from the ceiling.

◉ YALE PEABODY MUSEUM OF NATURAL HISTORY, YALE UNIVERSITY (NEW HAVEN, CONNECTICUT)

One of the first museums to display dinosaurs, the Yale Peabody Museum of Natural History was built on the enormous collections begun in the early 1870s by pioneering paleontologist Othniel Charles Marsh and by generations of Yale paleontologists who followed. The original "Brontosaurus" is here, along with the *Deinonychus* that inspired the *Velociraptor* in *Jurassic Park*, the most complete *Archelon* sea turtle, the first *Stegosaurus* and *Triceratops* ever found, and many other classic dinosaur, bird, and mammal fossils.

◉ MUSEUM OF THE ROCKIES, MONTANA STATE UNIVERSITY (BOZEMAN, MONTANA)

The relatively new Museum of the Rockies was built from the ground up by paleontologist Jack Horner, and its Siebel Dinosaur Complex features many of his discoveries, including dinosaurs eggs, nests, and babies, as well as many specimens of *Tyrannosaurus rex* and *Triceratops* from the nearby Hell Creek Formation.

◉ ACADEMY OF NATURAL SCIENCES OF DREXEL UNIVERSITY (PHILADELPHIA, PENNSYLVANIA)

Home of the first dinosaur and other vertebrate fossils collected by America's first vertebrate paleontologist, Joseph Leidy, in the 1840s and 1850s, the Academy of Natural Sciences of Drexel University features the first dinosaur to be identified in North America (*Hadrosaurus* from New Jersey), plus hundreds of other specimens, including a replica of *Giganotosaurus*, the giant theropod from Argentina.

◉ WYOMING DINOSAUR CENTER (THERMOPOLIS, WYOMING)

A relative newcomer to the scene in an out-of-the-way place, the Wyoming Dinosaur Center has 28 mounted dinosaur skeletons on display, including a 32-meter (106-foot) *Supersaurus*, *Stegosaurus*, *Triceratops*, and *Velociraptor*; fish from the Devonian; and the latest specimen of *Archaeopteryx* to be discovered and described. The museum maintains its own excavation site nearby.

The following are other important natural history museums in the United States.

- **MUSEUM OF COMPARATIVE ZOOLOGY, HARVARD UNIVERSITY (CAMBRIDGE, MASSACHUSETTS)**

  With collections going back to the 1850s, the Museum of Comparative Zoology features the giant *Kronosaurus* along one wall, plus fossils of terrestrial animals from the Permian and of mammals from the Cenozoic.

- **NEW MEXICO MUSEUM OF NATURAL HISTORY AND SCIENCE (ALBUQUERQUE, NEW MEXICO)**

  The core exhibition at the New Mexico Museum of Natural History and Science is Timetracks, which takes visitors from the origin of the universe and the beginning of life; through the Triassic "dawn" of the dinosaurs and the dinosaurs of the Jurassic and Cretaceous, the marine reptiles from the Western Interior Seaway of the Cretaceous, and the birds and mammals of the grasslands of the Paleocene; to the Ice Age and the present day. FossilWorks allows visitors to see fossils being prepared for display.

- **UNIVERSITY OF NEBRASKA STATE MUSEUM (LINCOLN, NEBRASKA)**

  Another classic institution, the University of Nebraska State Museum has a few dinosaurs on display, but is one of the best museums in the United States to see Cenozoic mammals—especially horses, rhinos, and camels—and the Elephant Hall features nothing but mounted skeletons of mastodonts and mammoths.

- **SAM NOBLE OKLAHOMA MUSEUM OF NATURAL HISTORY, UNIVERSITY OF OKLAHOMA (NORMAN, OKLAHOMA)**

  The Hall of Ancient Life at the Sam Noble Oklahoma Museum of Natural History features *Apatosaurus* battling the predator *Saurophaganax*, plus *Tenontosaurus* protecting her young from *Deinonychus*, the full skeleton and massive skull of *Pentaceratops*, many Permian vertebrates (*Dimetrodon*, *Edaphosaurus*, *Cotylorhynchus*, and other archaic amphibians, reptiles, and synapsids) from the red beds of Oklahoma, and spectacular Ice Age mammals. The gallery also has an exhibition on the fauna of the Burgess Shale, and the Paleozoic Gallery showcases dioramas of marine life that look eerily real.

- **MUSEUM OF GEOLOGY, SOUTH DAKOTA SCHOOL OF MINES AND TECHNOLOGY (RAPID CITY, SOUTH DAKOTA)**

  Recently relocated to a new building, the Museum of Geology exhibits fossils of dinosaurs from the Black Hills during the Jurassic and Cretaceous, marine reptiles from the Western Interior Seaway (elasmosaurs, mosasaurs) of the

Cretaceous, and mammals from the Big Badlands during the Eocene and Oligocene.

◉ FLORIDA MUSEUM OF NATURAL HISTORY, UNIVERSITY OF FLORIDA (GAINESVILLE, FLORIDA)

Florida Fossils: Evolution of Life and Land, an exhibition gallery at the Florida Museum of Natural History, covers the past 65 million years of Earth history using Florida as the backdrop. It features different size jaws and teeth of *Carcharocles megalodon* and spectacular mounts of the Cenozoic mammals of Florida.

CANADA

◉ ROYAL TYRRELL MUSEUM (DRUMHELLER, ALBERTA)

The spectacular Royal Tyrrell Museum, built in the heart of the dinosaur-rich Cretaceous badlands of Alberta, features a huge number of Cretaceous dinosaurs, including *Tyrannosaurus rex* and *Triceratops*; many duck-billed dinosaurs; and ankylosaurs, among others. This museum was the first, in the late 1980s, to arrange its galleries as a "trip through time," so visitors can see spectacular displays of prehistoric life from many periods, all in a linear sequence from oldest to youngest. One gallery is devoted entirely to the fauna of the Burgess Shale.

◉ CANADIAN MUSEUM OF NATURE (OTTAWA, ONTARIO)

The Canadian Museum of Nature was the first museum to house the many spectacular dinosaurs from the Cretaceous badlands of Alberta; 30 are on exhibit in the Fossil Gallery, which covers the rise and extinction of the dinosaurs and the rise of the mammals. The hall features many predators, including *Albertosaurus*, *Daspletosaurus*, and *Tyrannosaurus rex*; different ceratopsians, including *Triceratops*, *Monoclonius*, and *Styracosaurus*; as well as duck-bills, ankylosaurs, and dromaeosaurs. In addition, the gallery showcases marine reptiles (*Archelon*, mosasaurs) from the Western Interior Seaway of the Cretaceous, mammals from the Eocene and Oligocene of Canada, and an exhibition on the evolution of whales (*Pakicetus*, *Ambulocetus*, and *Basilosaurus*).

EUROPE, ASIA, AND AFRICA

◉ NATURAL HISTORY MUSEUM (LONDON, ENGLAND)

One of the oldest natural history museums in the world, and home to Richard Owen, Thomas Henry Huxley, and Charles Darwin in the mid-nineteenth century, the cathedral-like Natural History Museum houses most of the specimens of marine reptiles discovered by Mary Anning in Lyme Regis, exhibits fossils of dinosaurs from all over the world, and features a spectacular gallery of living and fossil mammals. In the gallery Our Place in Evolution, visitors follow the story of the evolution of humans.

◉ BEIJING MUSEUM OF NATURAL HISTORY (BEIJING, CHINA)

Home to some of the most important and impressive fossils in the world, the Beijing Museum of Natural History has 11 galleries full of Chinese dinosaurs and fossil mammals as well as displays of the extraordinarily preserved feathered dinosaur and bird fossils from Liaoning and elsewhere.

◉ MUSEUM FÜR NATURKUNDE (BERLIN, GERMANY)

The treasures of German paleontology from the past 200 years are on display at the Museum für Naturkunde, including the largest dinosaur skeleton ever mounted (*Giraffatitan*, formerly *Brachiosaurus*) and other fossils from the Tendaguru beds in Africa, the best specimen of *Archaeopteryx*, and many fossils of marine reptiles (especially ichthyosaurs with body outlines) from Holzmaden.

◉ MUSÉUM DES SCIENCES NATURELLES DE BELGIQUE/KONINKLIJK BELGISCH INSTITUUT VOOR NATUURWETENSCHAPPEN (BRUSSELS, BELGIUM)

The Dinosaur Gallery in the Royal Belgian Institute of Natural Sciences is the largest hall in the world devoted to dinosaurs, featuring fossils from many parts of the planet, but it is most famous for the amazing collection of 30 complete *Iguanodon* skeletons that were found in the 1870s in the Bernissart coal mines and described by Louis Dollo.

◉ MUSÉUM NATIONAL D'HISTOIRE NATURELLE (PARIS, FRANCE)

The Muséum national d'histoire naturelle, whose origin dates to the days before the French Revolution and consists of 14 sites throughout France, was built by the founder of vertebrate paleontology and comparative anatomy: Baron Georges Cuvier. The Gallery of Paleontology features some of the first extinct animals described by Cuvier (the first mosasaur, the Eocene mammal *Palaeotherium*, the first mastodont found, as well as *Megatherium* from South America). The main hall of the museum is still a classic "collector's cabinet" exhibit of comparative anatomy, with hundreds of skeletons of both extinct and living creatures spanning the length of the building.

◉ SOUTH AFRICAN MUSEUM (CAPE TOWN, SOUTH AFRICA)

The world's best collection of Permian reptiles and synapsids are on display at the South African Museum, along with Triassic, Jurassic, and Cretaceous dinosaurs from Africa—from the primitive *Euparkeria* to the huge predator *Carcharodontosaurus* and the sauropod *Jobaria*.

INDEX

Numbers in italics refer to pages on which figures appear.

Printed in the USA
CPSIA information can be obtained
at www.ICGtesting.com
JSHW081759160724
66513JS00002B/8

9 780231 171915